Metal Additive Manufacturing – State of the Art 2020

Metal Additive Manufacturing – State of the Art 2020

Editor

Nader Asnafi

MDPI • Basel • Beijing • Wuhan • Barcelona • Belgrade • Manchester • Tokyo • Cluj • Tianjin

Editor
Nader Asnafi
School of Science and Technolog
Örebro University
Örebro
Sweden

Editorial Office
MDPI
St. Alban-Anlage 66
4052 Basel, Switzerland

This is a reprint of articles from the Special Issue published online in the open access journal *Metals* (ISSN 2075-4701) (available at: www.mdpi.com/journal/metals/special_issues/metal_additive_manufacturing_2020).

For citation purposes, cite each article independently as indicated on the article page online and as indicated below:

LastName, A.A.; LastName, B.B.; LastName, C.C. Article Title. *Journal Name* **Year**, *Volume Number*, Page Range.

ISBN 978-3-0365-1324-9 (Hbk)
ISBN 978-3-0365-1323-2 (PDF)

Contents

About the Editor

Nader Asnafi

Nader Asnafi is a Professor of Mechanical Engineering at Örebro University. Nader has extensive academic and industrial experience as the Director of Studies at Luleå University of Technology, Researcher and Research Group Leader at the Swedish Institute for Metals Research, Head of Mechanical Properties at Sapa Technology, Manager of Product Design, Advanced Engineering and Technology Development at Volvo Cars Body Components, Senior Manager of Research and Advanced Engineering at Volvo Cars, Dean of the School of Engineering at Blekinge Institute of Technology, Vice President—R&D at Uddeholms AB, and Vice President and Head of Development at VA Automotive.

Preface to "Metal Additive Manufacturing – State of the Art 2020"

Additive manufacturing (AM), more popularly known as 3D printing, comprises a group of technologies used to produce objects through the addition (rather than removal) of material. AM is used in many industries—aerospace, defense, automotive, consumer products, industrial products, medical devices, and architecture. AM is transforming the industry, and this industrial transformation is expected to become more comprehensive and quicker during the coming years.

Additive manufacturing of metal components with virtually no geometric limitations has enabled new product design options and opportunities, increased product performance, shorter cycle time in part production, total cost reduction, shortened lead time, improved material efficiency, more sustainable products and processes, full circularity in the economy, and new revenue streams.

This Special Issue of *Metals* focuses on metal additive manufacturing with respect to:

- business models and engineering;
- product/component design (including generative design, topology optimization, and lattice and surface optimization);
- industrial applications (aerospace, defense, automotive, consumer, medical, and industrial products);
- material and process design and engineering;
- new materials;
- powder production and characterization;
- systems and equipment engineering;
- post-processing;
- process control and optimization and quality assurance.

The papers presented in this Special Issue give an up-to-date account of the scientific, technological, and industrial state of the art for metal additive manufacturing from the perspectives mentioned above.

I would like to express my sincere gratitude to the authors, reviewers, Section Managing Editor Ms Lyra Xing, and Editorial Office of *Metals* who have made this 2020 account possible through high-quality research and development, highly valuable perspectives and comments, and highly efficient collaboration in a pleasant and collegial environment.

Nader Asnafi
Editor

metals

Review

Application of Laser-Based Powder Bed Fusion for Direct Metal Tooling

Nader Asnafi

School of Science and Technology, Örebro University, SE-701 82 Örebro, Sweden; nader.asnafi@oru.se; Tel.: +46-19-301-493

Abstract: The journey of production tools in cold working, hot working, and injection molding from rapid tooling to additive manufacturing (AM) by laser-based powder bed fusion (L-PBF) is described. The current machines and their configurations, tool steel powder materials and their properties, and the L-PBF process parameters for these materials are specified. Examples of production tools designed for and made by L-PBF are described. Efficient design, i.e., high tooling efficiency and performance in operation, should be the primary target in tool design. Topology and lattice structure optimization provide additional benefits. Using efficient design, L-PBF exhibits the greatest potential for tooling in hot working and injection molding. L-PBF yields high tooling costs, but competitive total costs in hot working and injection molding. Larger object sizes that can be made by L-PBF, a larger number of powder metals that are designed for different tooling applications, lower feedstock and L-PBF processing costs, further L-PBF productivity improvement, improved surface roughness through L-PBF, and secured quality are some of the targets for the research and development in the future. A system view, e.g., plants with a high degree of automation and eventually with cyber-physically controlled smart L-PBF inclusive manufacturing systems, is also of great significance.

Keywords: rapid tooling; additive manufacturing (AM); laser-based powder bed fusion (L-PBF); production tools; cold working; hot working; injection molding

Citation: Asnafi, N. Application of Laser-Based Powder Bed Fusion for Direct Metal Tooling. *Metals* **2021**, *11*, 458. https://doi.org/10.3390/met11030458

Academic Editor: Massimo Pellizzari

Received: 30 January 2021
Accepted: 6 March 2021
Published: 10 March 2021

1. Introduction

To build objects by adding many very thin layers of material, layer on top of layer, has historically attracted some attention. The progress made during the past 40 years has, however, been the greatest and most profound.

Technologically, it was held that such a layer-by-layer manufacturing was the common denominator of (a) computer aided design (CAD), i.e., solid modelling, (b) enabling technologies, i.e., laser, ink-jet printers and motion control, and (c) traditional technologies, i.e., powder metallurgy, welding, extrusion, computerized or computer numerically controlled (CNC) machining, and lithography.

Industrially, rapid prototyping would enable a fast visualization and evaluation of the product design and a rapid tooling would shorten the time-critical toolmaking in the industrialization of new products (and thereby reduce the time to market).

Rapid tooling, particularly direct rapid tooling, through layer-by-layer toolmaking evolved in the middle of 1990s. Direct rapid tooling was defined as an industrial concept aimed at the realization of production tooling through layered manufacturing directly from CAD data files.

As far as layered manufacturing is concerned, several methods have been developed. Selective laser sintering (SLS), one of the methods that were developed initially, was launched in 1995. This method evolved to selective laser melting (SLM). In these methods, a thin layer of metal powder is sintered or melt. A steel powder based SLM machine was launched in 1999.

At the same, a similar technology, in which an electron beam is used instead of a laser beam, was developed for metallic materials. This paper is focused on metallic materials and SLM.

Now after two decades, the so-called layered manufacturing and SLM are standardized to additive manufacturing (AM) and laser-based powder bed fusion (L-PBF). The remarkable advances in AM and L-PBF in the past two decades have resulted in production tooling that is highly advanced both in design and performance.

This paper is an account of how production tools can be made today thanks to the scientific and technological advances during the past 20 years. To be brief, this account focuses on the "newness". For instance, machining and heat treatment are not mentioned in this paper. (Machining is in many cases used to post-process a production tool made by L-PBF to improve its surface roughness).

This account comprises:

- A brief history of direct rapid tooling by SLS/SLM.
- A brief description of L-PBF.
- A brief account of applications and materials for production tooling.
- The L-PBF process parameters for these materials.
- The new tooling design ingredients, i.e., generative design and topology and lattice structure optimization, primarily as an amplifier of and important complement to L-PBF.
- Examples of tooling applications designed for and made by L-PBF. These examples, which are from cold and hot working and injection molding, illustrate the results of the advances made during the past 20 years and the yet greater future potential in toolmaking.
- Discussions including a future outlook and
- Conclusions.

The methodology used to create this review is described in Appendix A.

2. Direct Rapid Tooling by Selective Laser Sintering/Melting—Brief History as a Baseline

The process of building objects by adding many very thin layers of material, layer on top of layer, has historically had several names. Solid freeform fabrication (SFF) through layered manufacturing in order to accomplish rapid prototyping (RP) can trace its roots back to the 19th century, particularly the fields of topography and photosculpture [1,2].

Early SFF examples were proposed by Swainson in 1968, Ciraud in 1971, Kodama in 1981, and Herbert in 1982 [1]. After a few years, Hull invented the stereolithography machine (SLA) in 1986. This machine is considered to be the first 3D printer [2].

It was held that SFF was the common denominator of the following [1]:

- Computer aided design (CAD): Solid modelling.
- Enabling component technologies: Laser, ink-jet printers and motion control.
- Traditional technologies: Powdered metallurgy, welding, extrusion, CNC machining and lithography.

The various SFF building strategies and deposition/fusion processes included photolithography, laser fusion, lamination, extrusion, and ink-jet printing.

Product creation comprises chiefly product design/development and industrialization. Once the design is accepted, the realization of the production line, in particular the preparation of complex production tooling (tools, dies, and molds), is time-critical in the industrialization phase and has therefore a direct and strong influence on time-to-market. RP through SFF (or layered manufacturing), used to visualize and evaluate the results of product design (or development phase of a product), evolved in the middle of 1990s toward rapid tooling (RT). This extension was caused by the need to further reduce the time-to-market by shortening, not only the development phase, but also the industrialization phase of product creation [3–6].

RT was divided into non-direct RT and Direct RT (DRT):

- Non-direct RT: SFF techniques shortening the classical/traditional toolmaking methods, e.g., using SLA to make a pattern for investment casting, resin tooling, etc., [7]. SLA, FDM (fused material deposition), LOM (laminated object manufacturing), and SLS-P (P = polymers) were used to accomplish non-direct RT.
- DRT: techniques requiring no intermediate steps in the manufacture of tools. In other words, DRT was defined as an industrial concept aimed at the realization of production tooling directly from CAD data files, with the smallest possible process chain (number of operations). Its purpose was the manufacture of tools that can be used under normal production conditions, in terms of durability, accuracy and surface quality. LG (laser generation), SLS-M (M = metals), SLS-PM (selective laser sintering of polymer-coated metal powder), and 3DP (three-dimensional printing) were used to accomplish DRT [8–10].

The present paper is focused on DRT by SLS-M. As far as SLS-M is concerned, several parallel tracks were followed by different researchers. The research and development by Deckard and Beaman at the University of Texas in Austin, [11], resulted in Sinterstation 2000—the first proper commercial system for laser-sintering launched by DTM Corporation of Austin in December 1992 [12,13]. DTM used the Sintersation to laser-sinter a polymer-coated metal powder to form a green part, followed by a furnace process to remove the polymer, bond the metal matrix and infiltrate it with a secondary metal to remove the porosity [14]. This process was called "RapidTool" and commercially released at the end of 1995.

EOS GmbH of Munich collaborated with Electrolux Rapid Development (ERD) of Rusko, Finland, and Nyrhilä that were involved in the development of Direct Metal Laser Sintering (DMLS). ERD was focused on powder metallurgy development and Nyrhilä had invented a novel powder concept for pressureless metal sintering with very low shrinkage in 1989 [15]. This collaboration resulted in development and installation of the first commercial DMLS system, EOSINT M 250, in the summer of 1995 [12,13]. This opened up the first real commercial use of DMLS for rapid tooling [12].

This equipment was used in 1996 to make injecting molding tools. One of these tools (which comprised 11 parts) was laser-sintered in the nickel-bronze mixture (Ni-Bz) developed by ERD and infiltrated with high temperature epoxy resin within 45 h. After sanding, polishing, integrating the gating system and the ejectors, Electrolux assembled the mold and produced approximately 100 mobile phone housings in PC/ABS using standard parameters. An overall accuracy of ±0.1 mm was reached [10]. The same mold was traditionally made in aluminium.

The time from completed CAD data to producing the injected parts in series material took Electrolux four weeks and cost DM 30,000 (DM = Deutsche Mark). This was considered as a remarkable reduction in time and costs for manufacturing injection mold prototypes and small batches of components [10].

SLS-M attracted great attention and many studies were conducted to explore the possibilities, optimize the process and develop DRT further [16–19].

At the same time, the researchers and device/equipment manufacturers were aware of the shortcoming of the DRT techniques that were available in 1995. These techniques allowed only the fabrication of pre-series or short-run tools, could not yield sufficiently high properties for demanding tools, e.g., forging dies, and required time consuming post-sintering through infiltration.

Porosity, part shrinkage, and thermal and mechanical instability were some additional problems. The nickel-bronze mixture developed by Nyrhilä and ERD, [15], exhibited a near-zero shrinkage and minimal curl and warpage [10] when laser-sintered without any protecting gas. However, it was known that oxidation before, during and after sintering of other metal powders would result in lower yield strength, cracks etc., and must therefore be prevented. Parameters influencing the accuracy, such as laser scan speed, laser power, powder delivery, powder properties, powder compaction, post-processing, etc. should be tested, evaluated, and optimized [3,10].

Many research projects were carried out to remedy these shortcomings [12,13,20–22].

The successful efforts at Fraunhofer ILT to accomplish fully dense metal objects by complete melting (SLM) showed how the SLS process needed to be altered. For SLM, the chamber needed to be closed/shielded and flooded with an inert gas like argon so that the wetting properties of the melt were not reduced by formation of an oxide skin layer. The metal powder was, as wished, processed without any binder material and without any kind of pre-treatment. However, the particle shape must be spherical with a size in the range of 20–50 μm. To melt the powder particles completely, as shown in Figure 1, fundamental investigations of the physical interrelations between the process parameters and the behavior of the metal melt were carried out. Laser power, scan speed, hatch distance and layer thickness appeared to be interrelated and crucial for the processing results. The influence of scanning strategies and protection gas flow was determined. In addition, a special scanning strategy with a limited length of each scan vector had to be used to achieve a high density of the parts [20,22].

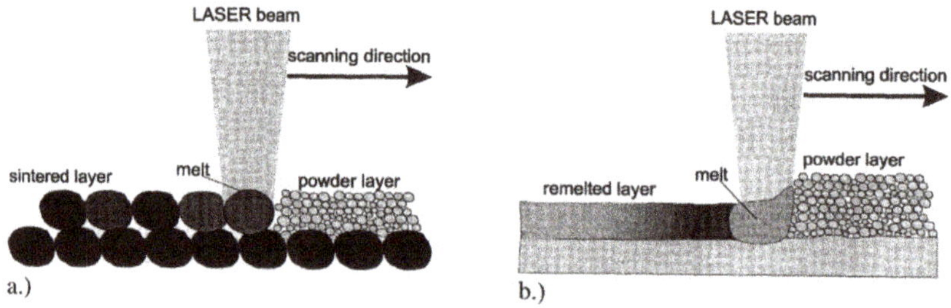

Figure 1. Melting process of single component powder with a laser beam: (**a**) porous structure with SLS and (**b**) dense structure with SLM [22].

With this developed procedure, objects in stainless steel 316L, TiA16V4 and DIN 1.2343 (H11) were made with a density of 99%. Infiltration to improve density and strength was no longer necessary. Figure 2 shows a couple of tools made by the then newly developed SLM. The slider for a die casting tool made by SLM in DIN 1.2343, Figure 2a), was fitted into the conventionally made tool. 4800 housings in AlSi12 were produced in this tool at a casting temperature of 700 °C and a pressure of approx. 800 bar without any visible tool wear. The injection molding tool in Figure 2b) was made by SLM in stainless steel 316L. It took ca 4 h to make the core and ca 7 h to make the cavity. After SLM, the mold release surfaces had to be manually smoothed and the core and cavity had to be fitted into the mother tool. This tool was used to produce thermoplastic parts as well as rubber parts [22]. See also [23].

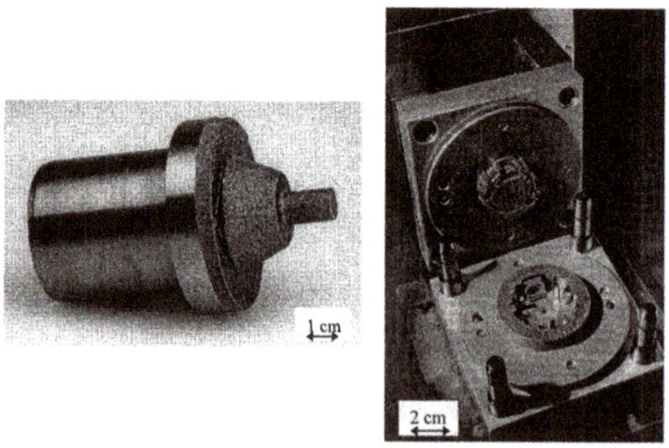

(**a**) (**b**)

Figure 2. Tools made by the then (1999) newly developed selective laser melting (SLM): (**a**) slider for a die casting tool made in 1.2343 and (**b**) injection molding tool made in 316L [22].

3. Laser-Based Powder Bed Fusion (L-PBF)

During the past 15 years, layered manufacturing has been subject to further research studies, standardization and industrialization. According to the developed standards, the technology is named Additive Manufacturing (AM). Laser-based powder Bed Fusion (L-PBF), formerly SLM (Figure 1b), is one of the seven AM basic process categories in the new standards [24–27]. For a detailed description of L-PBF, see [28–32].

Table 1 displays the features of the current (2020) machines for AM by L-PBF. This table, which is based on the machine makers specifications, show that the maximum object size that can be 3D-printed today is 600 × 600 × 600 mm^3 (although the largest height is 850 mm in one of the other machines). This size has more than doubled compared to that in the 1995 machine for selective laser sintering. In addition to higher laser power, some of the current machines have more than one laser and therefore a much higher productivity.

Due to the complexity of L-PBF, the manufacturers of metal AM systems have developed sets of optimized processing conditions for some existing powder metals. The machine manufacturer sets the process parameters for one or some specific powder metals as the default values for additive manufacturing based on the customer preferences before machine delivery and installation, Table 2. The number of these powder metals is much larger than those in 1995. However, this number is still very small, compared to the number of existing materials made and used conventionally.

Table 1. The features of the current L-PBF machines [33–39].

Manufacturer	Model	Number and Type of Lasers	Laser Power (per Laser) (W)	Build Volume (l × w × h) (mm^3)	Build Rate (cm^3/hr)	Layer Thickness (μm)	Scan Speed (m/s)
3D Systems	DMP Factory 500 Printer Module	3 fiber	500	500 × 500 × 500 [1]	-	2–200, Typical: 30, 60 & 90	-
	DMP Factory/Flex 350	1 fiber	500	275 × 275 × 420	-	5–200, Typical: 30, 60 & 90	-
	DMP Flex 100	1 fiber	100	100 × 100 × 90	-	10–100	-
	ProX DMP 300	1 fiber	500	250 × 250 × 330	-	10–100, preset: 40	-
	ProX DMP 200	1 fiber	300	140 × 140 × 115	-	10–100, preset: 30	-
Additive Industries	MetalFAB1	1 to 4 Yb fiber	500	420 × 420 × 400	-	20–100	-
Concept Laser	X Line 2000R	2 (cw) fibre	1000	800 × 400 × 500	-	-	-
	M Line Factory	4 fiber	1000	500 × 500 × 400	-	-	-
	M2 Multilaser	2 (cw) fiber	400	250 × 250 × 350	-	-	-
EOS	EOS M 400-4	4 Yb-fiber	400	400 × 400 × 400 [1]	100	80	7
	EOS M 400	1 Yb-fiber	400	400 × 400 × 400 [1]	50	-	7
	EOS M 300-4	4 Yb-fiber	400	300 × 300 × 400	10	-	7
	EOS M 290	1 Yb-fiber	400	250 × 250 × 325 [1]	-	-	7
	EOS M 100	1 Yb-fiber	200	f100 × 95 [1 & 2]	-	-	7
SLM Solutions	SLM®NXG XII 600	12	1000	600 × 600 × 600	1000	-	-
	SLM®800	4	400 or 700	500 × 280 × 850 [1]	Up to 171	20–90	10
	SLM®500	4 IPG fiber	700	500 × 280 × 365 [1]	Up to 171	20–75	10
	SLM®280	1, 2, 3 or dual	400, 700, or 700 & 1000	280 × 280 × 365 [1]	Up to 113	20–90	10
	SLM®125	1 IPF fiber	400	125 × 125 × 125 [1]	Up to 25	20–75	10
Renishaw	RenAM 500Q/S	1 or 4 Yb-fiber	500	250 × 250 × 350 [1]	Up to 150	20–100	10
	RenAM 500E	1 Yb-fiber	500	250 × 250 × 350 [1]	Up to 150	20–100	10
	RenAM 500M	1 Yb-fiber	500	250 × 250 × 350 [1]	Up to 150	20–100	10
	AM 400	1	400	250 × 250 × 300 [1]	-	-	-
	AM 250	is now replaced by AM 400		250 × 250 × 300 [1]	-	-	-
Trumpf	TruPrint 1000	1 fiber	200	f100 × 100 [1 & 2]	2–18	10–50	-
	TruPrint 2000	1 fiber	300	f200 × 200 [2]	-	20–100	-
	TruPrint 3000	1 fiber	500	f300 × 400 [2]	5–60	20–150	-
	TruPrint 5000	3 fiber	500	f300 × 400 [2]	5–180	30–150	-

[1] Height includes the thickness of the substrate/building plate. [2] Diameter × height.

Table 2. Materials the current machines L-PBF machines can be set in for upon installation [33–39].

Manu-Facturer	Model	Powder Metals Based on . . .						
		Al	Co	Cu	Fe	Ni	Ti	W
3D Systems	DMP Factory 500 Printer Module		By request			Nickel alloys	By request	
	DMP Factory/Flex 350	AlSi7Mg0.6, AlSi10Mg	CoCrF75	-	Maraging Steel, 17-4PH, 316L	Ni625, Ni718	Ti Gr1, Ti Gr5, Ti Gr23	-
	DMP Flex 100	-	CoCr	-	17-4PH, 316L	-	-	-
	ProX DMP 300	AlSi12	CoCr	-	Maraging steel, 17-4PH	-	-	-
	ProX DMP 200	AlSi12	CoCr	-	Maraging steel, 17-4PH, 316 L	-	-	-
Additive Industries	MetalFAB1	AlSi10Mg, ScalmAlloy©	-	-	Tool steel 1.2709, 316L	IN718	Ti6Al4V	-
Concept Laser	X Line 2000R	AlSi10Mg—Balanced & Productivity	-	-	-	Nickel 718	Ti6AL4V Grade 23	-
	M Line Factory	A205	CoCrMo	-	-	Nickel 718 CL	-	-
	M2 Multilaser	AlSi10Mg, AlSi7Mg	CoCrMo	-	Maraging M300, 316L, 17-4PH	Nickel 625, Nickel 718	Ti6AL4V Grade 23	-
EOS	EOS M 400-4	AlSi10Mg	-	-	MS1, 316L	HX, IN718	Ti64, TiCP Grade 2	-
	EOS M 400	AlSi10Mg	-	-	MS1	IN718	Ti64, Ti64ELI	-
	EOS M 300-4	AlSi10Mg	-	-	MS1	IN718	Ti64	-
	EOS M 290	AlSi10Mg	MP1	-	MS1, CX, PH1, 17-4PH, 316L	HX, IN625, IN718	Ti64, Ti64ELI, TiCP Grade 2	-
	EOS M 100	-	SP2	-	316L	-	Ti64	W1
SLM Solutions	SLM®NXG XII 600	ALSi10Mg (No limitations)		No limitations		IN718 (No limit.)	No limitations	
	SLM®800	AlSi10Mg, AlSi7Mg0.6, AlSi9Cu3	CoCr28Mo6, SLM® MediDent	CuSn10, CuNi2SiCr	Maraging 1.2709, 316L (1.4404), 15-5PH (1.4545), 17-4PH (1.4542), H13 (1.2344), Invar 36®	HX, IN625, IN718, IN939	Ti6Al4V ELI (Grade 23), TA15, and Ti (Grade 2)	-
	SLM®500							
	SLM®280							
	SLM®125							
Renishaw	RenAM 500Q/S	AlSi10Mg	CoCr	-	Maraging M300, 316L	IN625, IN718	Ti6Al4ELI	-
	RenAM 500E							
	RenAM 500M							
	AM 400							
	AM 250							
Trumpf	TruPrint 1000	Yes to all except W + precious metal alloys + amorphous metals						
	TruPrint 2000	Yes to all except Cu and W + amorphous metals						
	TruPrint 3000	Yes to all except Co, Cu and W						
	TruPrint 5000	Yes to all except Co, Cu and W						

4. Production Tooling—Applications and Materials

Production tooling is required in many different industrial applications. Cutting (material removal), cold working, hot working, and injection molding are some of these applications. However, cutting tools (material removal) are not covered in this review.

As displayed in Figure 3, the tool material has a large influence on the tool life. Tool material selection is based on (a) the required tool performance during the intended application and (b) the manufacturing of the tool. As far as the tool production is concerned, the tool material machinability, polishability, and heat treatment response are of great significance. Toughness, wear resistance, hot hardness, and resistance to softening are some of the important performance factors.

The failure mechanism encountered in cold working tools, e.g., stamping tools and dies, comprise abrasive and adhesive wear or mixed wear (caused by sliding contact), chipping at cutting edges and corners (fatigue), plastic deformation (exceeding the yield strength locally), cracking (fatigue), and galling (the same mechanism as in adhesive wear). The tool concept (tool material, hardness, surface roughness and treatment) is highly related to the workpiece material (sheet material grade, surface and thickness). The tool concept for 1-mm thick hot-dip galvanized DP600 steel sheet differs therefore from that for 1-mm thick uncoated DP1000 steel sheet [40–42].

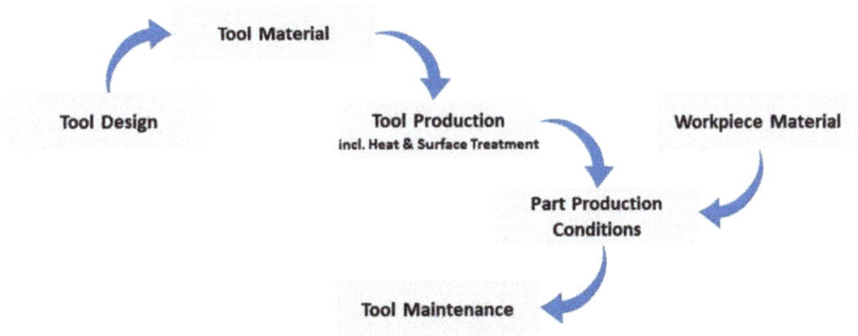

Figure 3. Factors that influence the tool life.

For hot-working tools, i.e., tools and dies in high pressure die casting, hot forging, hot stamping, or extrusion, thermal fatigue (heat checking), corrosion/erosion, cracking (total failure), and indentation are some of the failures that need to be avoided. Thermal fatigue is dependent upon thermal expansion coefficient (should be low), thermal conductivity (should be high), hot yield strength (should be high), temper resistance (a good resistance to softening at high temperature exposure), creep strength (should be high), and ductility. In other words, the tool should display resistance to deformation, softening, wear, impact loading and corrosion/erosion at the working temperature, [43–45].

Some injection molds are likely to be exposed to corrosion, since the plastic materials can produce corrosive by-products, e.g., PVC, and/or due to condensation caused by prolonged production stoppages, humid operating or storage conditions. In such cases, a stainless tool steel is required. Through-hardened molds are used, if the production runs are long, abrasion from certain molding materials needs to be avoided, and/or the closure or injection pressures are high. However, large molds and molds with low demands on wear resistance can be made in pre-hardened steel. Good polishability and excellent surface finish are key requirements for many injection molds [46,47].

Tool steels are classified according to composition, hardness, properties, or application. According to the classification system developed by the American Iron and Steel Institute (AISI), tool steels can be subdivided in seven different groups, based on prominent characteristics such as alloying elements, heat treatment and application. These groups are displayed in Table 3. The properties of some of these conventionally common tool steels are displayed in Table 4.

Table 3. AISI classification of (conventional) tool steels [48,49].

AISI Classification	Prefix (AISI Standards)	Important Group Characteristics	Principal Alloying Elements	Examples of Applications
High speed steels	M (Mo type)	High hardness for high speed cutting of materials, red hardness.	C, Mo, W, V, Cr	Cutting (material removal) tools.
	T (W type)		C, W, V, Cr, Co	
Cold-work steels	A (medium alloy air hardening)	Deep hardening in air (up to 100 mm).	C, Mn, Cr, Mo	Tools/dies/punches for blanking, coining and forming.
	D (high C, high Cr)	High wear resistance at normal temperatures.	C, Cr, (Mo)	
	O (oil hardening)	High wear resistance at normal temperatures.	Varies	
	W (water hardening)	Group W exhibits higher toughness than group L.	C	Woodworking tools, coining tools.

Table 3. *Cont.*

AISI Classification	Prefix (AISI Standards)	Important Group Characteristics	Principal Alloying Elements	Examples of Applications
Hot-work steels	H (Cr type)	Deep hardening, high toughness.	C, Cr, W, (V)	Dies for extrusion of Al and Mg. Hot rolling, hot forging.
	H (W type)	Greater thermal stability than the Cr type. Properties similar to high speed steels.	C, W, V, Cr, Co	Hot rolling, hot forging. Extrusion dies for brass, Ni alloys and steels.
	H, (Mo type)		C, Mo, W, V, Cr, Co	
Shock-resisting steels	S	High shock loading resistance.	Mn, Si, Cr, W, Mo	Chisels, hammers, and punches.
Mold steels	P	Low softening resistance at elevated temperature.	Cr, Ni	Molds, Critical finish molds.
Special purpose steels	L	Low alloyed steels.	Cr, V, Ni, Mo	Various.

Table 5 shows the most common powder steels used to make tools by L-PBF. The number of these tool steels for L-PBF is, as displayed in this table, still very limited. M789 AMPO, L-40, W360 AMPO and AM Corrax were launched within the past 2–3 years, while the maraging steel grade 300, the stainless steels 316L and 17-4PH, and the tool steel H13 have been studied more thoroughly in tooling applications much longer.

Table 4. Some of the properties of some common conventional tool steels. All values labelled with ① are from [50]. All other values are extracted from [48].

AISI Classification	AISI (USA)	W.-Nr. (Germany)	Yield Strength (MPa)	Ultimate Tensile Strength (MPa)	Usual Working Hardness (HRC)	Thermal Conductivity (W/(mK))①	Machinability	Wear Resistance	Tougness	Hot Hardness
High speed steels	M3:2	1.3344	-	2210 ① (hardened to 68 HRC)	63–66	26	4	8	3	8
	T15	1.3202	-	2240 ① (hardened to 69 HRC)	64–68	21	1	9	1	9
Cold-work steels	A2	1.2363	-	1858 ②	57–62	38	8	6	4	5
	D2	1.2379	1510 ① (hardened to 62 HRC)	2000 ① (hardened to 62 HRC)	58–64	31	3	8	2	6
	O1	1.2510	1538 ③	1710 ③	57–62	43	8	4	3	3
	W1	1.1545	-	2320 ① (hardened)	72 ①	48				
Hot-work steels	H11	1.2343	1482	1806	38–55	42	8	3	9	6
	H13	1.2344	-	1820 ① (hardened to 56 HRC)	40–53	29	8	3	9	6
	H21	1.2581	1193	1379	50–55	27	6	4	8	8
	H41	-	-	2210 ① (hardened)	68 ①	28				
Shock-resisting steels	S1	1.2550	-	1840 ① (hardened to 57 HRC)	50–58	41	8	4	8	5
Mold steels	P20	1.2311	1172 ④	1310 ④	30–50	45	8	1a	8	2a
Special purpose steels	L2	1.2210	1546 ⑤ with V	1605 ⑤ with V	45–60	44	8	1	7	2

①The values are from [50]. (All other values are extracted from [48].).②Hardened from 945 °C/Tempered at 540 °C.③Oil quenched/tempered at 425 °C.④After oil quenching from 845 °C and tempering 2 h at 205 °C.⑤Oil quenched from 840 °C/tempered at 425 °C.

Table 5. L-PBF: the major tool powder steels so far and the associated tooling applications.

Class	Name/AISI Designation	DIN Designation	Tooling Applications
Maraging steel	18Ni-Grade 300 (See also Böhler M722 AMPO)	1.2709	Cold and hot forming, injection molding etc.
	Böhler M789 AMPO	-	Injection molding, plastic extrusion, tool holders etc.
	Formetrix L-40	-	Injection molding etc.

Table 5. *Cont.*

Class	Name/AISI Designation	DIN Designation	Tooling Applications
Stainless steel	316L	1.4404	Injection molding etc.
	17-4PH	1.4542	Extrusion (metal powder and additives (polymers and waxes)), Injection molding etc.
	420	1.4021	Injection molding etc.
Tool steel	M2	1.3343	Cutting etc.
	H13	1.2344	Hot forging, die casting, extrusion and injection molding
	Böhler W360 AMPO	-	Extrusion, hot forging, hot stamping, injection molding, high-pressure die casting
	Uddeholm AM Corrax	-	Extrusion of plastic, injection molding

The chemical compositions of these powder steels are shown in Table 6. The properties of the tools made by L-PBF in these powder steels are displayed in Table 7. These tables are based on [33,51–62].

Maraging steel has been used widely in additive manufacturing for applications, where high strength, toughness and ductility at high hardness and dimensional stability are required. Maraging steel's low-carbon soft martensitic matrix is mostly free from interstitial alloying elements, which rank it as an excellent candidate for synthesis by AM and welding [63].

316L is used in a large variety of applications where the tool is in contact with corrosive media at moderate temperatures during service. In, for instance, injection molding, where an overlap of corrosion resistance and increased mechanical strength and hardness (compared to austenitic stainless steels) is required, the so-called PH steel grades such as 17-4PH (i.e., the maraging-type precipitation-hardened martensitic stainless steels) are used [51]. See also Tables 5–7.

420 is the only martensitic stainless steel grade in use so far. The phase content of the microstructure depends strongly on the chosen process parameters. In L-PBF (where the laser power is high), the process can result in retained austenite as well. This will in turn yield a hardness that is comparable with the wrought material [57]. See Tables 6 and 7.

As displayed in Table 6, low carbon maraging steels and carbon-bearing tool steels (e.g., M2 and H13) are used in toolmaking by L-PBF. Martensite with precipitates is the microstructure in both. In the case of carbon-bearing tool steels, the martensitic matrix without precipitates obtained upon quenching is hard and brittle and tempering serves to regain some ductility by precipitating carbon from the martensite matrix in the form of carbides. In the case of the high-Ni maraging steels, quenching leads to a comparably soft and ductile martensitic microstructure which gains its strength by precipitation of intermetallic-phase particles upon ageing [51,63]. See also Table 7.

Since the cooling rates in L-PBF are high, martensite is formed in both maraging and tool steels during AM. Therefore, maraging steels with a ductile martensitic microstructure are well processable in AM while carbon-bearing tool steels (M2 and H13) with a brittle martensitic microstructure tend to crack due to the thermal stresses during the AM. A considerable fraction of the literature on carbon-bearing tool steels regard the optimum process parameters to make crack free, dense samples. In contrast, for maraging steels most studies are aimed at optimizing microstructures, mechanical properties and post process heat treatments [51,63]. See also Tables 6 and 7.

The conducted investigations show that crake free and fully dense tools can be made by L-PBF in the materials displayed in Tables 5–7 [51,54–60,63]. See also [62].

Table 6. The chemical composition of the major tool powder steels for additive manufacturing.

Class	Name/Designation	Alloying Elements (wt.%), (Fe = Balance)																Ref.
		C	Mn	P	Si	Cr	Mo	Co	Ni	V	Al	Ti	Nb	Cu	N	S	W	
Maraging steel	18Ni-Grade 300 (M722 AMPO, DIN 1.2709)	≤0.03	≤0.15	≤0.01	≤0.10	-	4.5–5.20	8.5–10.0	17.0–19.0	-	-	0.8–1.2	-	-	-	≤0.01	-	[33], see also [51]
	M789 AMPO	<0.02	-	-	-	12.2	1.0	-	10.0	-	0.6	1.0	-	-	-	-	-	[52]]
	Formetrix L-40	<0.1	-	-	-	>10.5	<5.0	-	<5.0	-	-	-	<1.0	<1.0	<1.0	-	-	[53]
Stainless steel	AISI 316L (DIN 1.4404)	≤0.03	≤2.0	≤0.045	≤1.0	16.5–18.5	2.0–2.5	-	10.0–13.0	-	-	-	-	-	≤0.11	≤0.03	-	[33], see also [54]
	17-4PH (DIN 1.4542)	<0.07	<1.0	< 0.04	<1.0	15.0–17.5	-	-	3.0–5.0	-	-	-	+Ta 0.15–0.45	3.0–5.0	-	<0.03	-	[33], see also [51]
	AISI 420 (DIN 1.4021)	0.3	0.72	0.012	0.79	12.8	0.5	-	1.0	-	-	-	-	-	0.09	0.008	-	[56]
Tool steel	AISI M2 (DIN 1.3343)	0.9	0.38	-	0.35	3.97	4.89	-	0.3	1.82	-	-	-	0.25	-	0.03	6.15	[58]
	H13 (DIN 1.2344)	0.32–0.45	0.2–0.6	-	0.8–1.2	4.75–5.5	1.1–1.75	-	-	0.8–1.2	-	-	-	-	-	-	-	[51], see also [59]
	W360 AMPO	0.5	0.25	-	0.2	4.5	3.0	-	-	0.55	-	-	-	-	-	-	-	[52]
	P20 * (DIN 1.2311)	0.4	0.83	-	0.45	1.95	0.33	-	-	-	-	-	-	0.02	-	-	-	[60]
	AM Corrax	0.03	0.3	-	0.3	12.0	1.4	-	9.2	-	1.6	-	-	-	-	-	-	[61]

* P20 is a common conventional material for conventionally designed and manufactured injection moulds. Its chemical composition is given here only for comparison.

Table 7. The properties of the tools made by additive manufacturing through L-PBF in the powder steels shown in Table 6.

Class	Name/Designation	Building Direction	Yield Strength (MPa)		Ultimate Tensile Strength (MPa)		Elongation at Break (%)		Hardness (HRC)		Impact Toughness (J)		Thermal Conductivity (W/(m.K)	Ref.
			As-Built	After Heat Treatment [1]	As-Built	After Heat Treatment [1]	As-Built	After Heat Treatment [1]	As-Built	After Heat Treatment [1]	As-Built	After Heat Treatment [1]		
Maraging steel	18Ni-300 (DIN 1.2709)	Horizontal	1080 ± 90	2180 ± 40	1230 ± 70	2260 ± 30	13 ± 2	5 ± 2	35 ± 3	55 ± 3	64 ± 5	7 ± 2	20.9 at 25 °C	[33], see also [51]
		Vertical	1090 ± 50	2070 ± 80	1220 ± 20	2160 ± 90	13 ± 2	2 ± 1						
	M789 AMPO	-	-	1720 ± 50	-	1850 ± 50	-	6 ± 2	-	52 ± 1	-	6–14	-	[52]
	Formetrix L-40	-	1300	1350	1500	1650	14	10	46–48	50–52	60	18	16.3 at 20 °C	[53]
Stainless steel	AISI 316L (DIN 1.4404)	Horizontal	530 ± 20 [2]	370 ± 30 [3]	660 ± 20 [2]	610 ± 30 [3]	39 ± 5 [2]	51 ± 5 [3]	90 ± 6 HRB	83 ± 4 HRB	215 ± 15	220 ± 15	15 at 20 °C	[33], see also [54]
		Vertical	440 ± 20 [2]	320 ± 20 [3]	570 ± 30 [2]	540 ± 30 [3]	49 ± 5 [2]	66 ± 5 [3]						
	17-4PH (DIN 1.4542)	Horizontal	-	1280 ± 30 [4]	-	1450 ± 10 [4]	-	11 ± 1 [4]	32 ± 4	40 ± 2 [4]	71 ± 20	7 ± 2 [4]	18 at 100 °C	[33], see also [55]
		Vertical	830 ± 110	1260 ± 100 [4]	1100 ± 90	1380 ± 20 [4]	19 ± 4	12 ± 2 [4]						
	AISI 420 (DIN 1.4021)	-	700 ± 15	950 ± 20	1050 ± 25	1520 ± 30	2.5 ± 0.2	6.3 ± 0.2	55 ± 1	53 ± 1	-	-	-	[56], see also [57]
Tool steel	AISI M2 (DIN 1.3343)	-	-	-	1280	-	0.8	-	57	64 [5]	-	-	-	[58]
	H13 (DIN 1.2344)	-	1003 ± 8.5	1580 ± 14.7 [2]	1370 ± 175.1	1860 ± 55.8 [2]	1.7 ± 0.6	2.2 ± 0.8 [2]	59 ± 4.6	51 ± 3.7 [2]	-	-	27 at 500 °C	[59], see also [51]
	W360 AMPO	-	-	1500–1670	-	1970–2010	-	6.6–8.1	-	55–57	-	8–14	-	[52]
	AM Corrax	-	760	1600	1150	1700	16	10	34	50	-	18.7	-	[61], see also [62]

[1] Ageing. [2] After stress relief. [3] Full anneal. [4] H900. [5] Every layer is remelted during L-PBF.

5. L-PBF Process Parameters

The quality of the feedstock (metal powder) and the process parameters used in L-PBF are of key significance for the process performance and determine the microstructure and thereby the quality of the end part (in this case, the tool) [64–92].

Table 8 displays the L-PBF processing parameters that result in fully dense objects (tool in this case) in the shown materials. Compare Table 8 with Tables 5–7.

Table 8. The L-PBF processing parameters that result in fully dense objects. See also Tables 5–7.

Class	Material	Used 3D Printer	Layer Thickness (μm)	Laser Power (W)	Scan Speed (mm/s)	Hatch Distance (μm)	Scan Stratgey	Ref.
Maraging steel	18Ni-300 (DIN 1.2709)	3D Systems ProX DMP 300	40	185	1200	70	Line	[62], see also [91]
	M789 AMPO	Trumpf TruPrint 1000	20	160	763	87	NA	[92]
Stainless steel	316L (DIN 1.4404)	Renishaw AM250	50	200	1000	100	Stripe pattern, 20 mm overlap	[87], See also [88,91]
	17-4PH (DIN 1.4542)	3D SystemsProX DMP 300	40	150	1200	50	NA	[89] See also [90]
	420 (DIN 1.4021)	Concept Laser Mlab R	20	90	600	120	Line, 90° rotation (−45°/+45°)	[56]
Tool steel	M2 (DIN 1.3343)	Concept Laser M3 Linear	30	105	150	128	Island (chess)	[58]
	H13 (DIN 1.2344)	SLM250HL	30	175	607.64	120	NA	[59], see also [91]
	P20 (DIN 1.2311)	DiMetal-100	30	160	350	80	NA	[60], see also [91]
	AM Corrax	EOS M290	30	170	1250	100	Line	[62], See also [61]

Table 9 summarizes the proposed measures to combat build and microstructural problems in L-PBF.

Table 9. Measures to combat build and microstructural problems in L-PBF. Based on [74].

Problem	Primary Approach	Secondary Approach
Build defects	-	-
Keyhole porosity	Reduce laser power	Increase laser velocity Remelting/reheating
Lack-of-fusion porosity	Reduce scan line spacing Reduce layer thickness	Reduce laser velocity Increase laser power Remelting/reheating
Composition change	Reduce laser power Increase laser velocity	Increase scan line spacing Increase layer thickness
Residual stress and distortion	Reduce scan vector length Use heated substrate/chamber	Reduce laser velocity
Microstructural/materials related	-	-
Microsegregation	Increase laser velocity	Strongly reduce laser velocity
Undesired texture	Reduce laser velocity	Reduce laser power
Coarse/columnar grains	Reduce laser velocity	Reduce laser power
Meta-stable phases (e.g., martensite)	Remelting/reheating	Reduce laser velocity

The difference between maraging steel, DIN 1.2709, and tool steel, H11 and H13, can be used to illustrate how the measures to combat build and microstructural problems vary from material to material. Since the cooling rates in L-PBF are high, 10^5–10^6 $K{\cdot}s^{-1}$, martensite formation is induced in both steel types. The so-called carbon-free DIN 1.2709 obtains a ductile martensitic microstructure (and is well processable in L-PBF), while the carbon-bearing H11 and H13 display a brittle martensitic microstructure that tends to crack due to the thermal stresses built up by the high cooling rates [51,85]. The conducted studies show that a pre-heating of the base plate (building platform) eliminate the cracking [30,51].

Martensitic microstructure is found in crack-free H13 for all pre-heating temperatures up to 300 °C. Cracking is, in other words, avoided due to lower thermal gradients and reduced residual stresses. A base plate heated to 400 °C results in a change in the phase transformation and a bainitic microstructure (the part is kept above the martensite start temperature) is formed instead. The bainitic microstructure displays higher hardness and tensile strength than the martensitic microstructure formed at lower heating temperatures [51].

The roughness of the functional tool surfaces is of great significance in all tooling applications. Figure 4 displays (a) the different surface types on a part fabricated by and (b) the staircase effect in L-PBF.

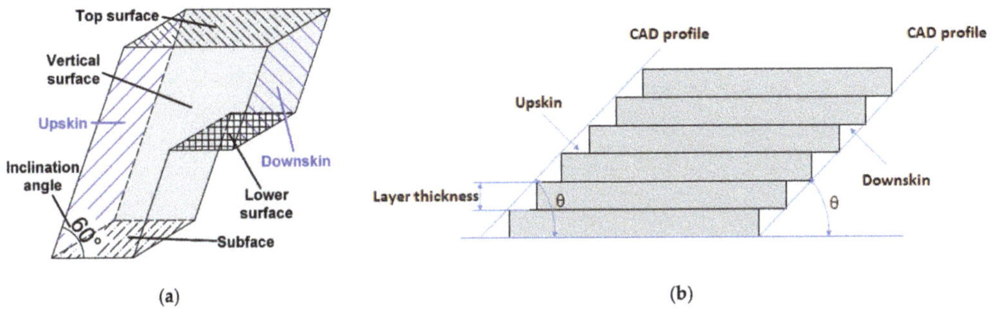

Figure 4. (**a**) schematic presentation of the different surface types on a part fabricated by L-PBF [93]. (**b**) Schematic presentation of the staircase effect.

The staircase effect increases with layer thickness and decreasing θ angle. The conducted studies show that the downskin exhibits the highest roughness in L-PBF, since the material is built upon powder (and not on solid material). The downskin surface exhibits a roughness of 20–65 μm in these investigations [94,95]. For the top surface, a roughness of 5 μm can be achieved in L-PBF [62].

L-PBF is not yet capable to accomplish the surface roughness required in the tooling applications [96–101]. The tool made by L-PBF needs to be machined during post-processing to obtain the required surface roughness. The functional tool surfaces must therefore have an allowance of 0.2–0.5 mm [65].

Benchmarking artifacts have been proposed, designed, fabricated by L-PBF and evaluated during the past years to assess the capabilities and limitations of the process and the performance of different machines. The evaluation of the geometrical performance of L-PBF and the dimensional accuracy of objects made by L-PBF has been one of the purposes for such artifacts [102–105]. Figure 5 displays such artifacts.

Seven (7) state-of-the-art L-PBF machines were used to produce the artifact in Figure 5 in maraging steel 18Ni-300 (DIN 1.2709). Each machine made three complete jobs (artifacts). Table 10 shows the minimum dimensions in the artifact in Figure 5 compared with the minimum dimensions the most capable machine manufacturers claimed to be able to produce. This benchmarking study overcame most of the limitations stated by the machine manufacturers in Table 10. It showed that user experience and expertise were of great significance. The newest machines did not outperform the oldest machine due to this user's experience and expertise. To create an own space in industrial production, L-PBF needs to become more robust, i.e., less dependent upon the user experience and better repeatability [105].

This observation in [105], i.e., the large role played by the user experience and expertise, was made also in [106]. Rings 1-mm thick and 30-mm high were made by L-PBF in the diameters 25, 50 and 85 mm in tool steel H13 (DIN 1.2344). The dimensional accuracy was then measured and analyzed. These rings were made using two different L-PBF scanning strategies—a stripe scan strategy that is automatically generated and provided by the machine manufacturer and a (manually designed) sectional scan strategy, Figure 6. The dimensional accuracy of ring specimens could be improved by up to 81% with the introduced sectional strategy compared to the standard approach [106].

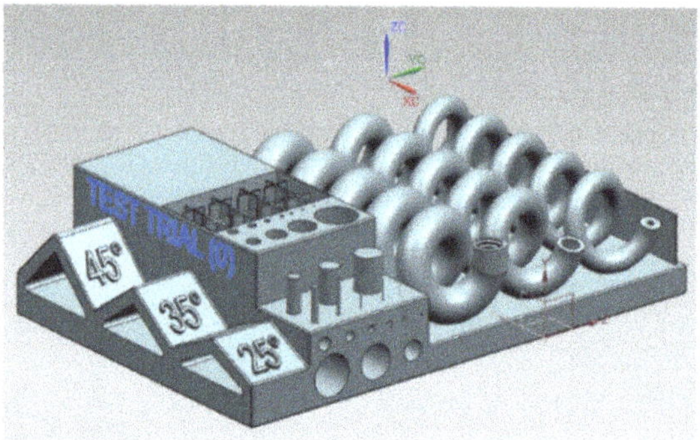

Figure 5. The Artifact with spiral on benchmarking in [105]. The figure is from [104].

Table 10. The minimum dimensions in the artifact in Figure 5 compared with the minimum dimension the most capable machine manufacturers claimed to be able to produce. Material = 18Ni-300 (DIN 1.2709). Seven (7) state-of-the-art L-PBF machines were used to produce 3 complete jobs (artifacts) each [105].

Feature	Minimum Dimension	Min Dimension in Artifact
Wall thickness	0.15 mm	0.10 mm
Overhang structure	45°	25°
Circular holes (diameter)	0.50 mm	0.20 mm
Circular pins (diameter)	0.50 mm	0.10 mm

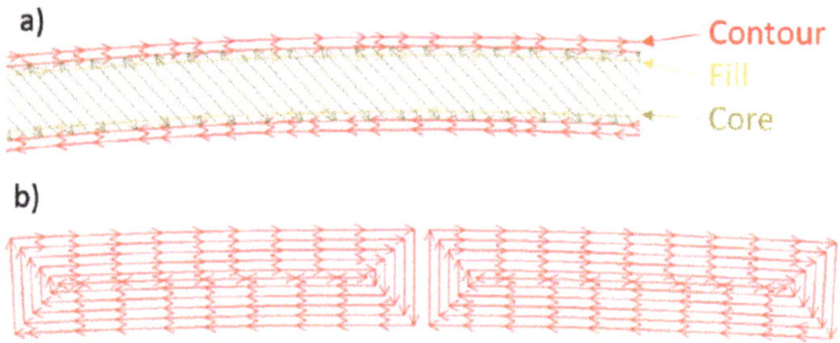

Figure 6. (**a**) The stripe scan strategy automatically generated and provided by the machine manufacturer and (**b**) the manually designed sectional scan strategy [106].

6. Tool Design for Metal Additive Manufacturing by L-PBF

Additive manufacturing provides a significantly larger design freedom (compared to conventional manufacturing), the benefits of which can be maximized by the emerging computer-aided design (CAD) technologies like generative design (GD). It is now possible to:

(1) Generate a wide range of design alternatives by artificial intelligence-based algorithms (GD software) after setting the part/object design space, constraints, criteria and objectives. The designer reviews the different design options and chooses the best-suited for the application [107,108].

(2) Topology optimize the selected design alternative, the purpose of which is to remove unnecessary material while meeting (or exceeding) the performance criteria. The goal is to optimize a part properly (weight, stiffness, frequency . . .) while respecting a certain set of constraints. The topology optimization process uses various mathematical algorithms and methods (each having several versions) [109–112].
(3) Optimize the internal lattice and surface structure of the topology optimized object by creating an internal mesh while meeting (or exceeding) the performance criteria [113–116].
(4) Produce this complex object by L-PBF (and post-processing).

During the past two decades, different new design methodologies have emerged, among them the Design for X (DfX). X represents a particular perspective to improve during the product design as well as the design process [117,118]. DfX applied to the Additive Manufacturing process is named DfAM. It aims at using the full potential of the AM technologies for design, for which there are two methods [108,119,120].

A new detailed DfAM process has been proposed including the available design support (methods, design rules, guidelines and software tools), the tools and methods that are best suited at different stages of the design process are specified, and the possibility to achieve a more automated DfAM is indicated [121].

As far as the L-PBF process design (i.e., the preparation of the L-PBF process) is concerned, the part orientation, support structures, overhangs and part supporting angles, channels and holes, wall thickness, tolerances and offsets are addressed in [122–124]. VDI 3405 Part 2, [122], addresses the qualification, quality assurance and post processing. The procedure described in this standard is applied to DIN 1.2709 in a VDI inter-laboratory test. The obtained properties in this inter-laboratory test are accounted for in [122]. The values for DIN 1.2709 in Tables 6 and 7 are in parity with the values stated for the same material in [122].

The angle between the building platform and the building direction is also of great significance for the properties of the as-built object. In other words, the properties of the as-built object are anisotropic due to the layered manufacturing, the loading direction etc. Table 11 and Figure 7 display 7 different configurations, in which tensile specimens in 316L (1.4404) were built with an oversize of 0.4 mm in width and thickness. After milling to final shape, the specimens were tested. Table 12 displays the tensile properties of these specimens. The maximum strength is obtained at a 45° layer versus loading offset [125].

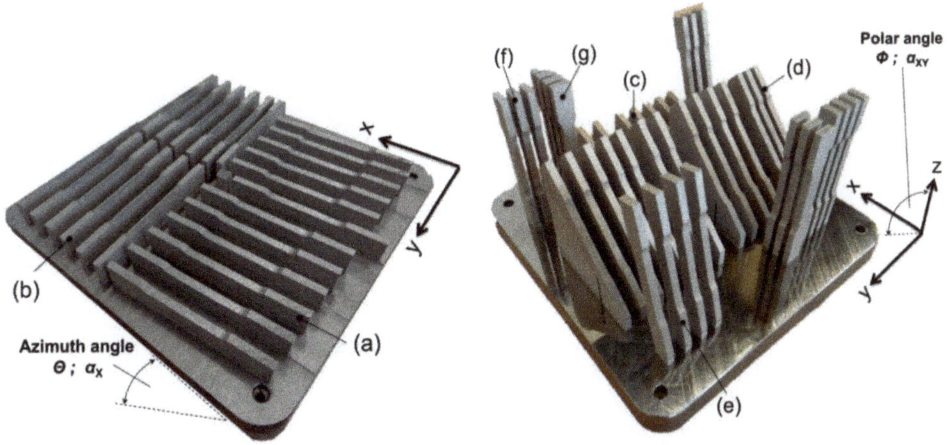

Figure 7. The tensile samples in 1.4404 (316L) on the building platform: overview of the positioning, building angles and arrangement (configurations). See also Tables 11 and 12. The figure is from [125].

Table 11. The tested configurations. See also Figure 7 and Table 12. The table is from [125].

Config.	Polar Angle Φ; α_{XY} (°)	Azimuth Angle Θ; α_X (°)	Total Runtime (h)
(a)	0	0	-
(b)	0	90	39.5
(c)	15	0	-
(d)	45	0	-
(e)	75	0	86.5
(f)	90	0	-
(g)	90	90	-

Table 12. The tensile properties of 1.4404 (316L) in different building configurations. See also Table 11 and Figure 7. The specimens were built with an oversize of 0.4 mm in width and thickness and milled to final shape before tensile testing. The table is from [125].

Config.	Young's Modulus E (GPa)		Yield Strength $R_{p0.2}$ (MPa)		Ultimate Tensile Strength R_m (MPa)		Elongation at Failure At (%)		Poisson's Ratio ν (-)	
	Average	STDEV	Average	STDEV	Average	STDEV	Average	STDEV	Average	STDEV
(a)	151.01	25.56	516.51	7.16	634.43	7.39	33.24	0.57	0.444	0.031
(b)	207.57	24.22	539.47	3.29	643.67	3.25	42.74	0.82	0.155	0.014
(c)	147.87	23.59	501.32	7.70	624.65	4.36	34.09	1.12	0.479	0.058
(d)	227.35	25.12	589.89	11.86	698.98	23.65	32.56	10.17	0.203	0.024
(e)	151.43	18.80	485.65	11.93	571.23	18.63	22.84	7.27	0.558	0.020
(f)	137.78	14.25	438.60	9.69	511.99	17.95	11.76	5.38	0.453	0.005
(g)	137.83	16.25	457.21	17.29	530.22	8.09	17.46	4.42	0.170	0.085

Design demands knowledge of the precise material properties. These properties are anisotropic and the component orientation on the build platform needs therefore to be considered [125]. (Post L-PBF heat treatment affects also these properties. See the properties of 316L (1.4404) before and after heat treatment in Table 7.)

It is possible to achieve ±0.2% in tolerance, with the minimal value being 0.2 mm. It is also possible to achieve a wall thickness of 0.5 mm. However, the wall shape is dependent upon the orientation, its height (in relation to the thickness), and the possibility to have support if needed. As far as gaps (between walls or solid portions) are concerned, the minimum gap width should be larger than at least the melt pool width. It is also important to add an allowance of 0.3–0.5 mm to the locations and surfaces that require post-processing by machining to reach the required tolerance and surface roughness [114,122].

7. Examples of Tooling Applications Designed for and Made by L-PBF

Research and development have been conducted on toolmaking for cold working, hot working and injection molding by an L-PBF inclusive process, i.e., design for L-PBF (metal AM), L-PBF, and post-processing. In the following, some examples of the studied tooling applications are accounted for. These examples display the scientific and technological achievements and the commercial benefits of an L-PBF inclusive toolmaking.

7.1. Cold Working

7.1.1. Piercing Punches

100-mm long cylindrical punches with a piercing diameter of φ10 mm were made by L-PBF in H13, M300, and KP4. These punches were made vertically with a layer thickness of 50 μm. Durability studies were conducted by using these punches to make round holes in 1.2-mm thick CP1180 sheet steel with a tensile strength of 1200 MPa. Conventionally manufactured punches in conventional solid SKD11 and HWS were used in this investigation for comparison. The M300 punch had the best mechanical properties among the additively manufactured punches and exhibited the same performance, i.e., managed

10,000 strokes, as the conventional SKD11 and HWS punches. Process optimization and the application of other powder metals are indicated as topics for further investigations [126].

7.1.2. Stamping: Case 1

In a comparison of different additive manufacturing methods, a small prototyping sheet metal forming tool (bending matrix, holding block and die) was made in ABS, VeroGrey, PEAK, and H13. The variant made by L-PBF in H13 (vertically built with a layer thickness of 30 μm) exhibited the smallest form deviations, took the longest time to print, and had the highest production costs [127].

7.1.3. Stamping: Case 2

To study the potential of additive manufacturing in automotive stamping, 3 cutting inserts (the largest being 200 × 135 × 65 mm^3), a calibration insert (205 × 97 × 56 mm^3) and a blankholder insert (95 × 44 × 44 mm^3) were made by an L-PBF inclusive process in DIN 1.2709. These inserts were printed as solid objects and hade a hardness of 38 HRC after L-PBF. After heat treatment in a vacuum at 490 °C in 6 h, the inserts were machined to a surface roughness of 0.2 μm. The final hardness was 57 HRC and the inserts managed one million strokes/hits per insert. The costs were more than doubled and the lead time was more than halved, compared to conventional toolmaking [128].

7.1.4. Stamping: Case 3

To study the potential of topology optimization of the stamping tools, certify such tools, and study the options in design and manufacturing of such tools, semi-industrial forming (u-bending) and trimming tools were made by an L-PBF inclusive process [129–132].

The left U-bending tool half in Figure 8 was topology optimized and 3D-printed (L-PBF). The right U-bending tool half in the same figure was 3D-printed solidly. Both tool halves were 3D-printed in DIN 1.2709. LS-TaSC (LS-Dyna) was used for the topology optimization and the Z-displacement at the tool half profile radius was the compliance criterion. The volume fraction of 0.45 was selected, since the maximum Z-displacement at this fraction is very close to that of the solid [129–131]. See Figures 8–10.

After L-PBF, the U-bending tool halves were heat treated at 490° C in 6 h and then machined to a surface roughness of 0.2 μm. The hardness was 55 HRC. This tool was used to U-bend 2-mm thick hot-dip galvanized DP600 steel sheet with a minimum tensile strength of 600 MPa. This tool managed 50,000 strokes (U-bends) without any problems and was approved [129]. See Figure 8.

Compared to the 3D-printed solid tool half, the topology optimized and 3D-printed tool half exhibited a weight reduction (or improved material usage) by 19.4% and a lead time reduction by 11.1%. Initially, the profile radius of the left tool half (topology optimized) was 5.05 mm and that of the right tool half (solid) was 5.04 mm. After 50,000 strokes, the maximum wear measured as a change in the profile radius was 0.0186 mm [129]. See Figure 8.

Figure 11 displays the 3D-printed (L-PBF) semi-industrial solid and topology optimized trimming tools. The hardness varies between 54 and 56 HRC. The surface roughness R_a = 0.2 μm. Both the solid and topology optimized tools managed 100,000 strokes in 2-mm thick DP600 sheet steel without any problems and were approved. Figure 12 shows the wear after 100,000 strokes at the edge (profile) radius on the trim dies. Compared to the 3D-printed solid tool, the topology optimized and 3D-printed tool exhibits a weight reduction by 47% and a lead time reduction by 29.6%, Figure 11, [131,132].

Based on the approved semi-industrial results above, a station in an industrial progressive die used to stamp 1-mm thick DP600 sheet steel was chosen and 3D-printed (L-PBF) in DIN 1.2709 with a honeycomb inner structure. This station comprises a punch and a puller, as shown in Figure 13 [62] (see also [129]).

The punch in this industrial stamping tool, Figure 13, was topology optimized using LS-TaSC with a volume fraction of 0.45 as the target, and using NX12 with a weight reduction by 70% as the target. The shape and the resultant displacements prior to unloading (in stamping) in the punch with the original design and the topology optimized versions are displayed in Figure 14 [62,131]. These topology optimized versions, both 3D-printed in DIN 1.2709, are shown in Figure 15.

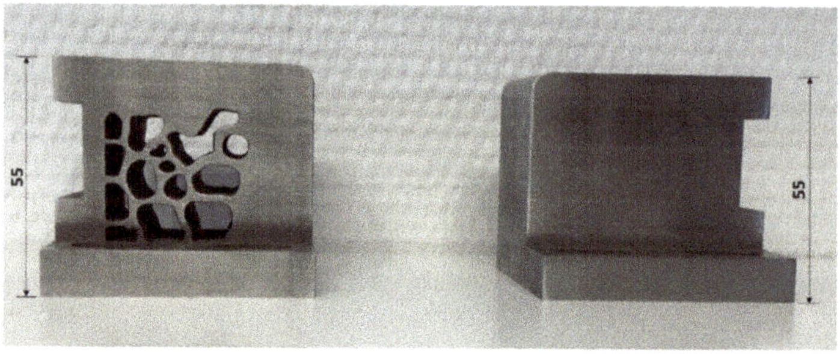

Figure 8. The U-bending tool: the right tool half was 3D-printed (L-PBF) as a solid piece. The left tool half was topology optimized at the volume fraction of 0.45 (Figure 9) and 3D-printed (L-PBF). Both tool halves were 3D-printed in maraging steel DIN 1.2709. Both tool halves managed 50,000 U-bends and were approved [129]. All displayed dimensions are in mm.

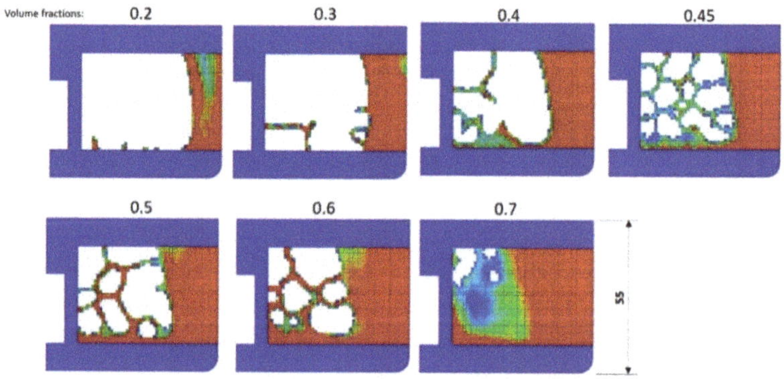

Figure 9. Topology optimization of the U-bend tool (Figure 11) with different volume fractions [130].

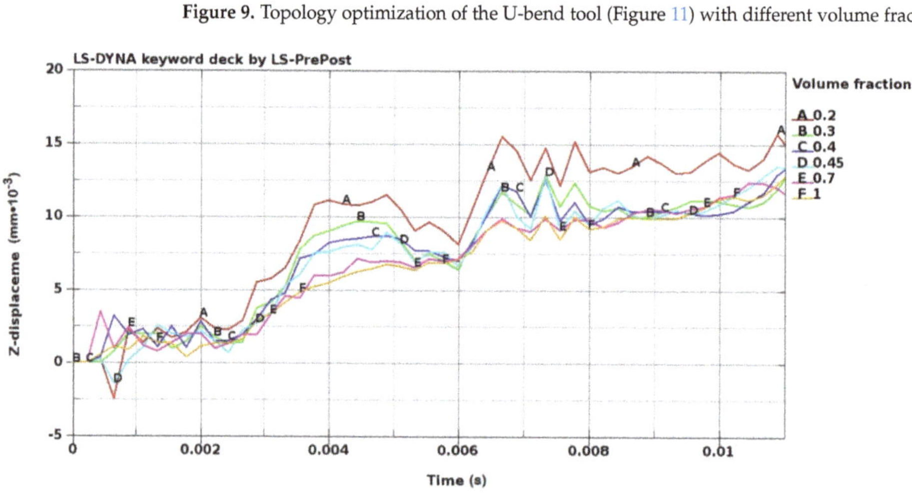

Figure 10. The Z-displacement at the U-bending tool profile radius for different volume fractions (Figure 9) [131].

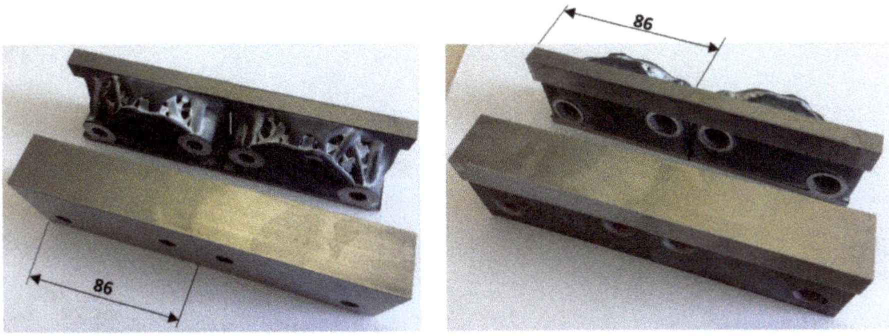

Figure 11. The 3D-printed (L-PBF) solid and topology optimized trimming tools. Both versions were 3D-printed (L-PBF) in DIN 1.2709 [131]. All displayed dimensions are in mm.

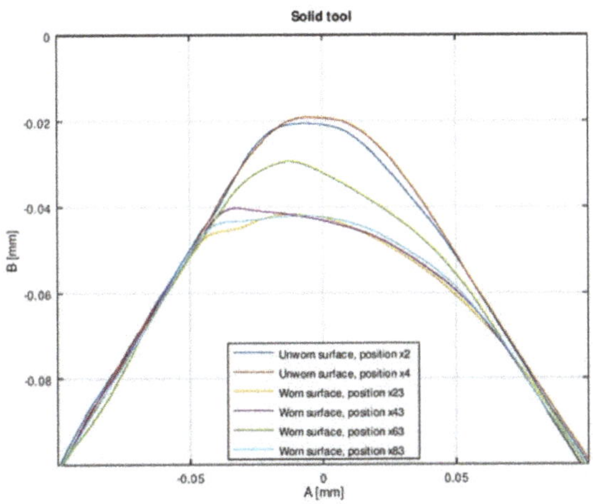

(a) Solid tool.

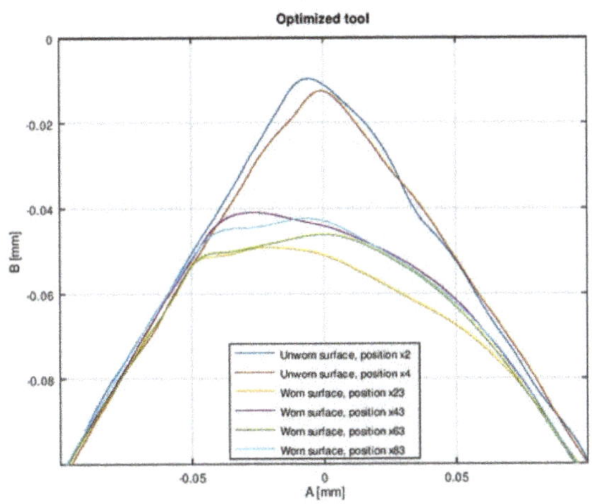

(b) Optimized tool.

Figure 12. The wear after 100,000 strokes at the edge (profile) radius on the trim dies [132].

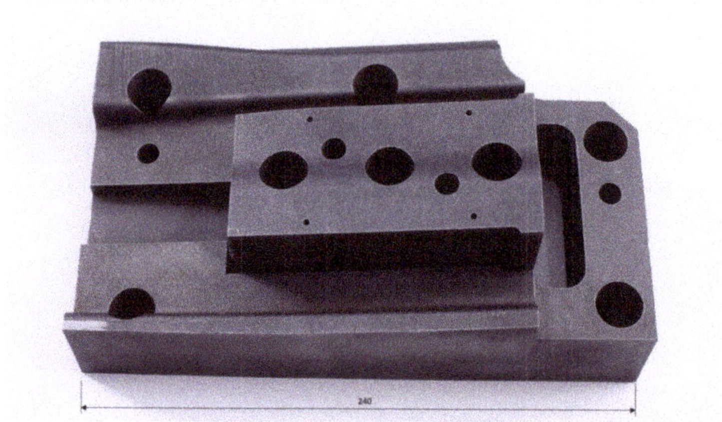

Figure 13. A station in an industrial progressive die, used to stamp 1-mm thick DP600 sheet steel, 3D-printed (L-PBF) in DIN 1.2709 with a honeycomb inner structure [62] (see also [129]). The displayed dimension is in mm.

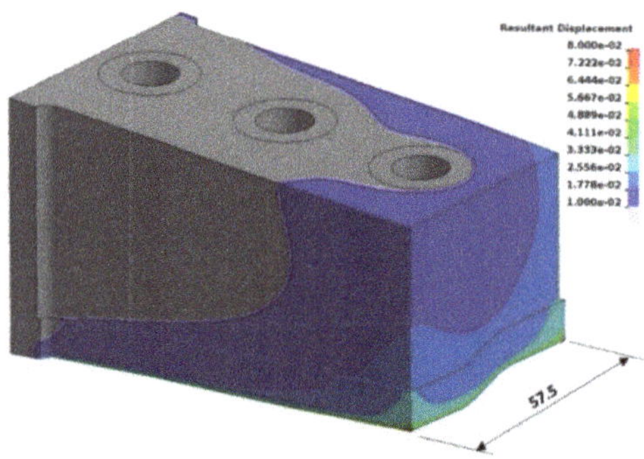

(a)

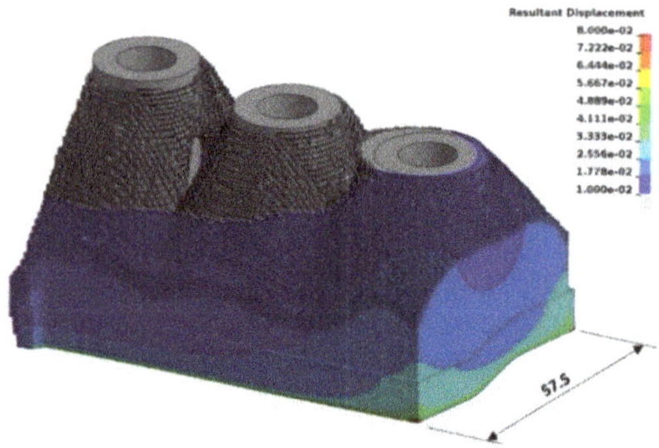

(b)

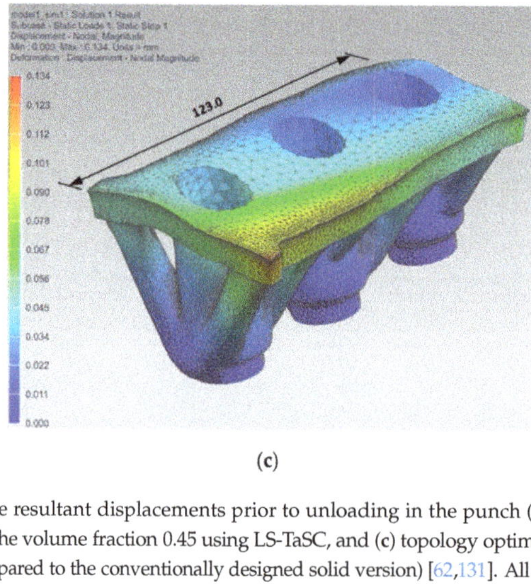

(c)

Figure 14. The shape and the resultant displacements prior to unloading in the punch (**a**) with the original design and (**b**) topology optimized with the volume fraction 0.45 using LS-TaSC, and (**c**) topology optimized using NX12 and targeting a weight reduction by 70% (compared to the conventionally designed solid version) [62,131]. All displayed dimensions are in mm.

Figure 15. Three (3) versions of the industrial punch (Figure 13)—all 3D-printed (L-PBF) in DIN 1.2709: (**left**) with a honeycomb inner structure, (**middle**) after topology optimization using LS-TaSC and (**right**) after topology optimization using NX12 [62]. See also Figure 14.

Compared to a 3D-printed solid punch, both the honeycomb inner structure and the topology optimization with the volume fraction of 0.45 improved the material usage (and thereby reduced the weight) and the printing time by ca 45% & ca 34% respectively. This means that the same printing time reduction and improved material efficiency can be accomplished in at least two different fashions—topology optimization and a honeycomb inner structure [62]. See Figure 15.

The topology optimization targeting a weight reduction by 70% (compared to the solid version made conventionally) led to much larger displacements held to have a large negative impact on the trimming result and the die life length [62]. See Figure 15.

7.2. Hot Working

7.2.1. Hot Stamping

Hot stamping, or press hardening, is a process used to form lightweight high strength (2000 MPa) structural car body parts. The thin steel sheet is heated to the austenitizing temperature (ca 950 °C), formed, and rapidly cooled down to 200 °C in the forming tool. This leads to martensitic high strength structure. The cooling, the time the die must be kept closed after forming and before unloading (the holding time), dominates the cycle time. To accomplish the cooling, channels are drilled (ca one hour per meter borehole) [114].

To accomplish improved cooling and maximize the benefits, a so-called hybrid tooling approach was selected. The tool base was made by conventional milling of a conventionally made solid block. The functional structure was made by L-PBF in DIN 1.2709 over this tool base, as shown in Figures 16 and 17. Before L-PBF, this functional structure was optimized through thermo-fluidic simulations and by placing the cooling channels very close and conformal to the surface. Due to the optimized cooling, the part was cooled down more rapidly and evenly. The holding time could be reduced from initially 10 to 5 s [133]. See Figure 18.

7.2.2. Extrusion: Case 1

During extrusion, the feedstock is heated up and pushed through a forming tool. The formed shape needs to be cooled down and maintained. To do so, a calibration tool is used directly after the extrusion. The cooling is accomplished by a fluid flowing through the cooling channels. The vacuum, accomplished through the vacuum channel, forces the extruded material to adopt the cylindrical inner surface of the calibration tool. Thereby, the extruded geometry is fixed and "calibrated" [114].

The tool described above was specially designed for AM by integrating the complex internal channels in a limited design space. After design, the calibration tool was made by L-PBF in 17-4PH. For the internal channels, tolerances in the range of ± 0.5 mm were met. The channel diameter had to be about 3 mm to be able to get rid of the not bound powder [114]. See Figures 19 and 20.

There is no conventionally manufactured predecessor. Comparable conventionally manufactured tools use compressed air for calibration. Better thermal and calibration performance and the reduced size are the main benefits of the calibration tool in Figures 19 and 20 compared to the conventional less effective tools [114].

Figure 16. Tool insert for hot stamping made by the so-called hybrid tooling. The tool base was made by conventional milling of a conventionally made solid block (**left**). The functional structure including the optimized cooling channel was made by L-PBF in DIN 1.2709 over this tool base (**right**). The figure (**right**) shows the hybrid tool ready for final machining [133].

Figure 17. The assembled hot stamping punch [114].

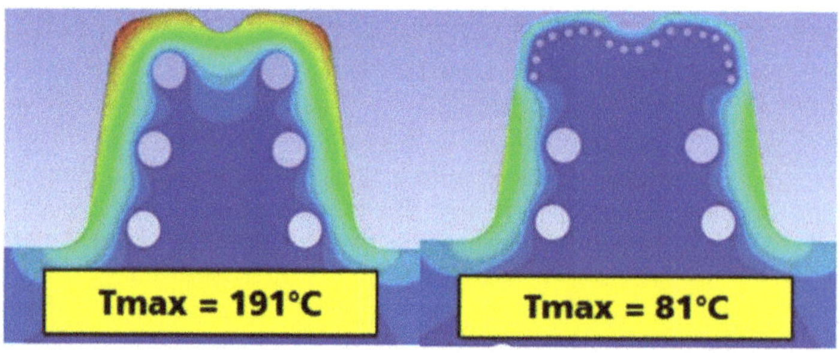

Figure 18. The temperature distribution in the hot stamping punch: (**left**) the conventional tool and (**right**) the tool designed for and made by L-PBF [133].

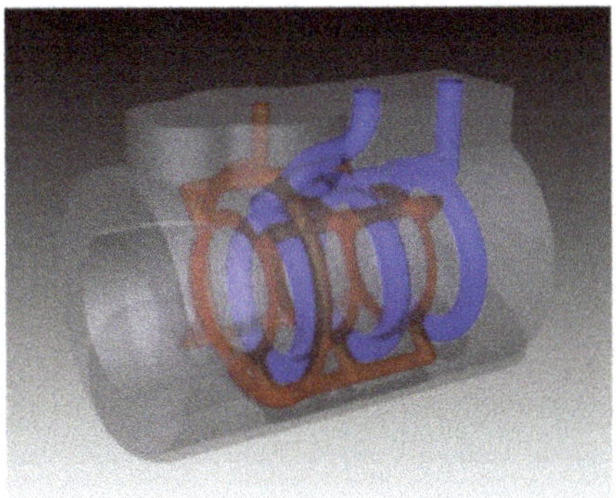

Figure 19. The calibration tool: The internal cooling and vacuum channels were integrated in a very compact design space. The minimum distance between the channels is about 1 mm [114].

Figure 20. The calibration tool was made by L-PBF in 17-4PH [114].

7.2.3. Extrusion: Case 2

In another study, the thermal control and the maintenance of a uniform temperature in the extrusion process of aluminum alloys was at the focus. The purpose was to generate sound profiles with high press productivities. Numerical studies were conducted to check the mechanical and thermal performances of die inserts both in uncooled and cooled conditions. Conformal cooling channels through which liquid nitrogen flowed was the targeted cooling system. A thermally controlled die insert was designed and made in H13 by L-PBF. Initially, cracks due to thermally induced stresses caused problems. This problem was solved by reducing the waiting time for one layer to another layer melting. The crack-free insert made by L-PBF is shown in Figure 21. This insert was then integrated into a conventionally machined steel housing [134]. The height and outer diameter of the insert in Figure 21 are 20 mm and 25 mm respectively [135].

7.2.4. Die Casting: Case 1

The bedplate for a passenger car engine was to be manufactured regularly by aluminium die casting in the alloy EN AC-Al Si9Cu3 (Fe). Gas and shrink porosity were some of the chief problems encountered in the initial testing phases. The conducted investigations resulted in an insert made by L-PBF in DIN 1.2709 mounted as a component in the die [133], as shown in Figure 22.

The optimization of the cooling channels in this insert comprised [133]:

- FE-simulations to obtain the shortest possible distance between the cooling and the cavity on one hand and sufficient insert strength on the other.
- Thermal simulations to optimize the cooling channels in steps, ranging from spiral design to a complex system of small channels, as shown in Figure 23.

The optimized insert reduced the gas and shrinkage porosity to a permissible level, the scrap rate by 10% and the cycle time by 3% [133].

Figure 21. An extrusion die insert with conformal cooling channels through which liquid nitrogen flow. This die insert was made by L-PBF in H13 and integrated into a conventionally machined steel housing [134]. The height = 20 mm and the outer diameter = 25 mm [135].

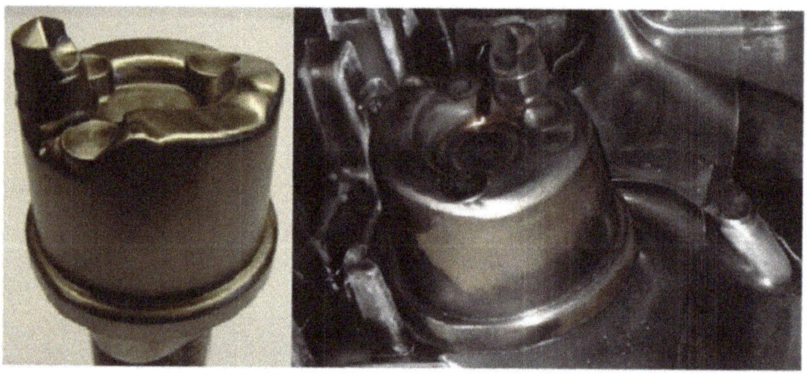

Figure 22. Aluminium die casting: **(left)** The tool insert was made in DIN 1.2709 by L-PBF, after which it was installed as a component in the die [133].

7.2.5. Die Casting: Case 2

In another study, the main objective was to investigate the benefits of conformal cooling in the tool used to manufacture sample batches of zinc alloy castings. A hybrid approach was chosen to reduce the process and material costs. The impression block in Figure 24 was made of two pieces separated at the dashed line [136]:

- The upper piece, which includes the cores and the conformal cooling channels, was built by L-PBF from powders of AISI H11 (DIN 1.2343). An allowance of 0.3 mm was left for CNC finish milling after a thermal treatment.
- The lower piece, which is 25 mm thick and includes the inlet and outlet ports for the coolant, was built by CNC machining from a blank made of the same steel.

The main lines of the cooling system were placed on the parting plane to ease powder removal. The whole process took 180 h, of which 91 h for the additive manufacturing phase [136].

Figure 25 displays the redesigned impression block with the conformal cooling channels. The conformal cooling improved the surface finish of the castings (due to a reduced need of spray cooling) and reduced the time and shrinkage porosity [136].

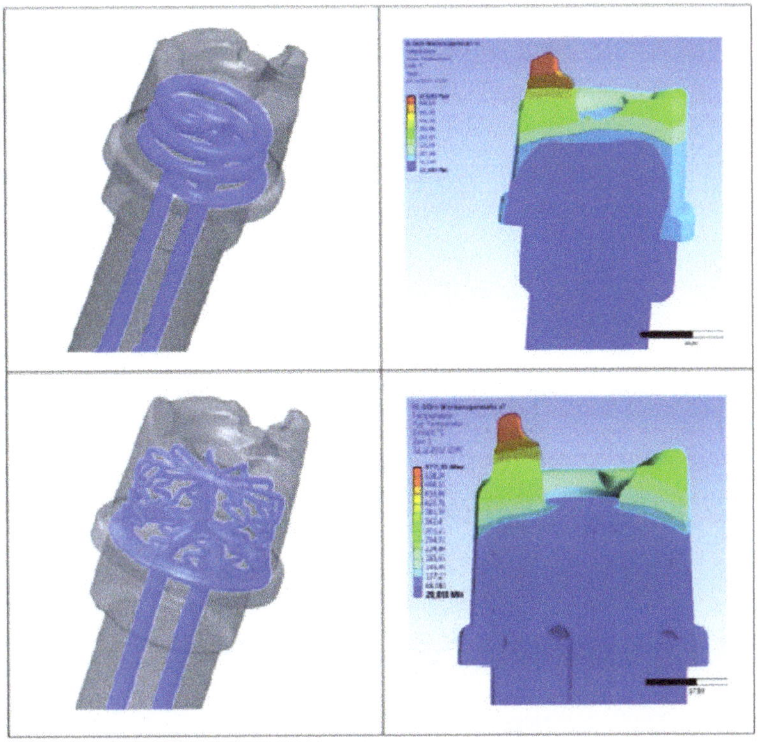

Figure 23. The insert in Figure 22: Stepwise optimization of the cooling channels, ranging from spiral design to a complex system of small channels [133].

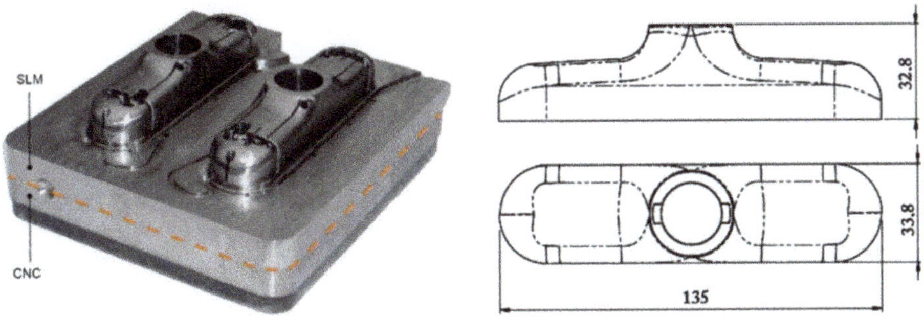

Figure 24. The impression block, a die for zinc alloy castings, was made by applying a hybrid approach. The upper piece was made by selective laser melting (SLM) or L-PBF from powders of AISI H11 (DIN 1.2343). The lower piece was built by CNC machining from a blank made of the same steel [136]. See also Figure 25. All displayed dimensions are in mm.

7.2.6. Sheet Molding Compound (SMC) Technology

Isothermally heated compression mold for an aircraft component was subject to a study. The pre-impregnated chopped long fiber fabrics are cut, stacked, placed into the heated mold and extruded by high pressure and at temperatures between 130 and

145 °C. The modular tooling concept, selected and used in this study, comprised standard tool rack with four guide pins made of hardened steel and a quick-change mold system manufactured by L-PBF [137]. See Figure 26.

The mold system consisted of a die and a punch. For an optimum heat conduction and temperature distribution (on the surfaces of the shape) the heated channels were integrated close to the cavity surface and meander-shaped. The mold was made by L-PBF with island strategy in DIN 1.2709. After heat treatment and milling, the surfaces of the die and punch were chromized with a coating thickness of 10–15 μm. This mold was used to make the carbon fiber SMC cleat in Figure 26 [137].

Improved FRP (fiber reinforced plastics) part quality, reproducibility, and temperature distribution and heat conduction due to surface compliant heating channels were the obtained benefits [137].

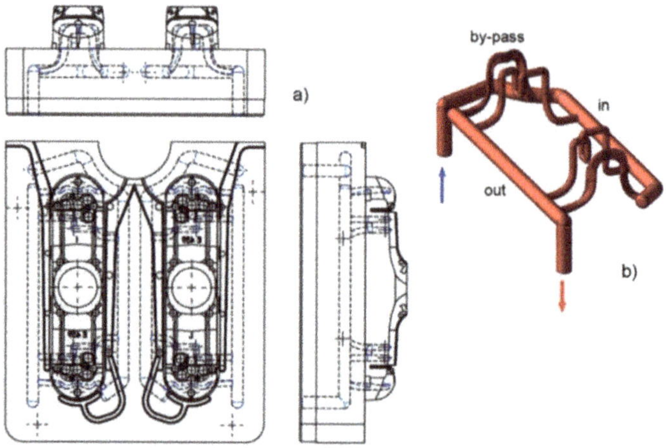

Figure 25. Redesigned impression block, a die for zinc alloy castings, with conformal cooling channels [136]. (**a**) The impression block. (**b**) The conformal cooling channels inside the impression block. See also Figure 24.

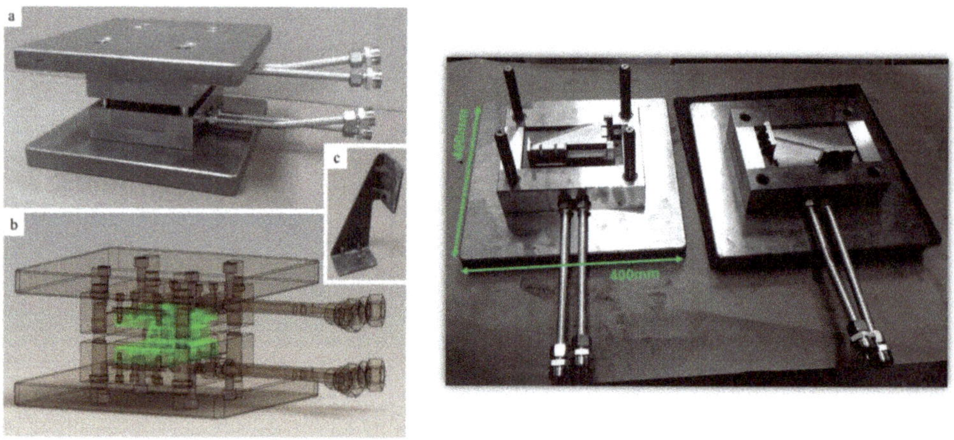

Figure 26. The selected modular tooling concept comprised standard tool rack with four guide pins made of hardened steel and a quick-change mold system manufactured by L-PBF. (**a**) the tool, (**b**) CAD model of the tool, and (**c**) carbon fiber SMC cleat [137,138].

7.2.7. Vulcanisation Mold for Tyres

The last step in production of tyres (for cars, motorcycles, trucks and agricultural vehicles) is the vulcanisation in a closed mould, which gives the tyre its final shape and tread pattern. In this process, the rubber material of the tyre is cured and converted into an elastomer at approximately 200 °C. The tyre mould consists of eight to sixteen segments and each segment is conventionally made in a wear resistant aluminium–silicon alloy by casting and 5 axis CNC milling [139,140], Figure 27.

Tyre profiles comprise an array of broad and narrow grooves (also called sipes). Broad grooves can be integrated in the aluminium mold. However, the hardness and stiffness of aluminium is insufficient for narrow grooves with a width of less than approximately 3 mm. Instead, thin steel inserts called sipes are mounted in the aluminium mold [140], Figure 27.

The manufacture and installation of sipes is extremely costly and time consuming. Sipes are made by stamping and cold bending. Four to five sets of stamping and bending tools are therefore required for a specific sipe geometry. Each tyre mold contains various sipe geometries. Sipes (sipes with undercut) made by L-PBF have therefore attracted a large interest, particularly for the running performance of snow tyres, during the past years [140].

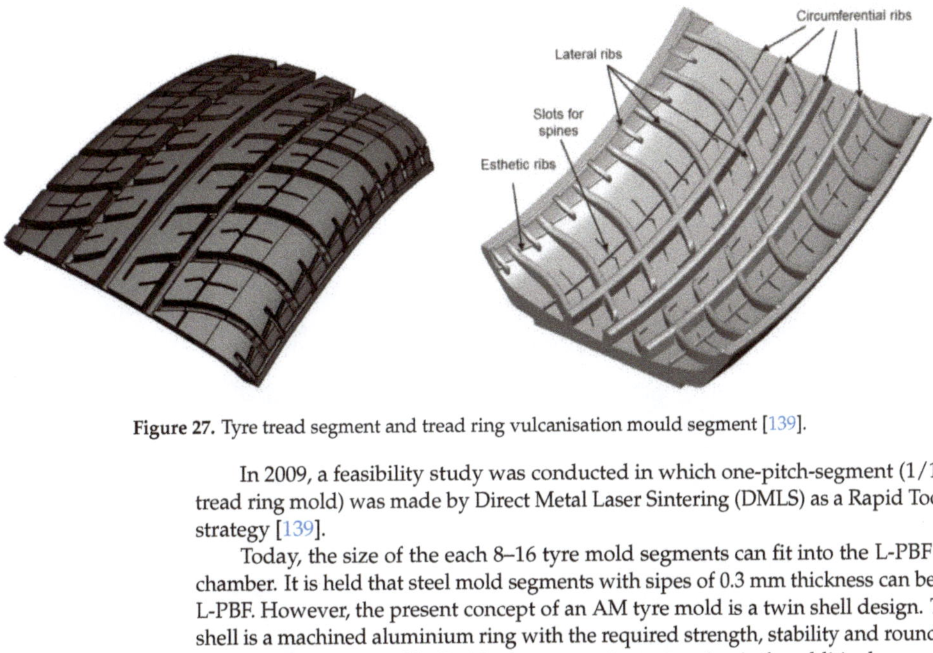

Figure 27. Tyre tread segment and tread ring vulcanisation mould segment [139].

In 2009, a feasibility study was conducted in which one-pitch-segment (1/128 of the tread ring mold) was made by Direct Metal Laser Sintering (DMLS) as a Rapid Tooling (RT) strategy [139].

Today, the size of the each 8–16 tyre mold segments can fit into the L-PBF building chamber. It is held that steel mold segments with sipes of 0.3 mm thickness can be made by L-PBF. However, the present concept of an AM tyre mold is a twin shell design. The outer shell is a machined aluminium ring with the required strength, stability and roundness that supports the inner profile. Inside the supporting outer ring is the additively manufactured shell forming the tyre profile. This is as thin as possible for cost reduction [140].

7.3. Injection Moulding

An L-PBF inclusive manufacturing of the tooling for injection molding has been addressed in many studies. The possibility to design, manufacture and use conformal cooling channels has been subject to investigation from different perspectives in the many of these studies [59,141–145].

7.3.1. Injection Molding: Case 1

Figures 28 and 29 display a case in which an existing injection molding core was redesigned and made by L-PBF in a new powder material. Simulations were conducted to optimize the injection molding core (inserts) with respect to cooling and solidification of the molded part. Conformal cooling channels were designed based on solidification, cooling, heat flux, average mold temperature per cycle, average mold temperature at the end of the cycle, and warpage of the molded part [62], Figure 28. The inserts optimized by these simulations were 3D-printed (L-PBF) in AM Corrax, hardened to 48 HRC and post-machined to the surface roughness R_a = 0.2 μm [62], Figure 29.

The 3D-printed inserts were tested in real production and compared with the existing conventionally designed and manufactured core. This comparison showed that the water flow was reduced by 86.4% and the cycle time somewhat with the core shown in Figure 29 [62].

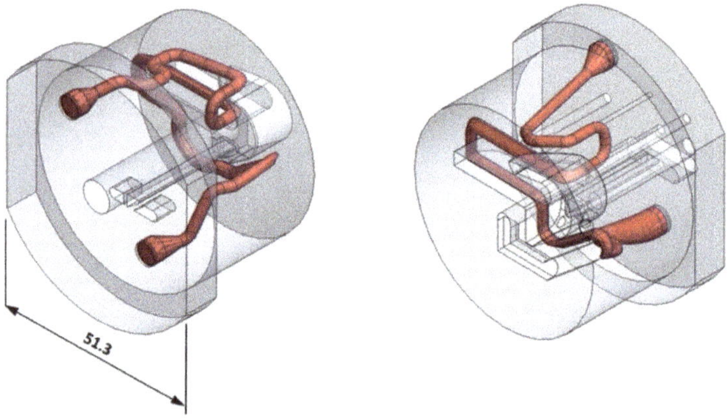

Figure 28. The core/inserts for injection molding optimized by the simulations. Red color = the cooling channels after optimization [62]. The displayed dimension is in mm.

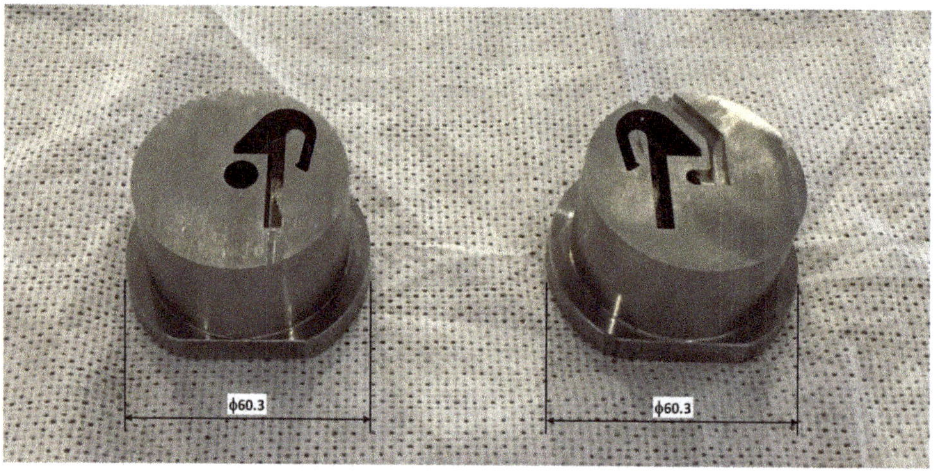

Figure 29. The core/inserts for injection molding optimized by simulations (Figure 28) 3D-printed (L-PBF) in AM Corrax [62]. The displayed dimensions are in mm.

7.3.2. Injection Molding: Case 2

High performance thermal management of molds for injection molding and tools for die casting with conformal cooling cavities has been studied in several investigations and noted as one of the great benefits of L-PBF. In addition, the use of flexible lattice structure inside the mould provides better thermal fatigue resistance, leak protection, defined thermal management and weight reduction. The use of internal lattice structure leads to a reduction of the L-PBF production time and material usage, which in turn reduce the costs [143], Figure 30. The height = the outer diameter = 200 mm, the outer skin thickness = 6 mm, and the lattice diameter = 0.5 mm in Figure 30 [144].

(a) (b)

Figure 30. A split mould tool with internal lattice structure in green (**a**) and a close view of the same mold (**b**) [143]. (**a**) Height = outer diameter = 200 mm. (**b**) Outer skin thickness = 6 mm and lattice structure = 0.5 mm [144].

7.3.3. Injection Molding: Case 3

As mentioned above, the implementation of conformal cooling channels to enhance the cooling efficiency has been the most common example of L-PBF design freedom in the injection molding sector and corresponding academic research. To benefit further from the L-PBF potential, the removal of excessive non-contributing material, topology optimization, should also be considered [145].

In this study, two topology optimization methods were considered and used [145]:

- A commercially available software, which adopts the geometry-based constraints and achieves a mass reduction of ≈50%.
- A topology optimization method developed in-house, which integrates a simplified AM process model within the standard topology optimization algorithm. The underlying thermal aspect of L-PBF is therefore included with a sequential localized steady-state thermal analysis and, thus, local overheating issues during manufacturing are avoided.

Figure 31 displays the original design of the studied injection molding insert. Figure 32 shows the "same" injection molding insert topology optimized with the standard density-based topology optimization (TO) algorithm and with the in-house novel TO algorithm with the AM thermal constraint (hotspot detection) [145]. The topology optimized inserts in Figure 32 have the same dimensions as the original, i.e., 220 mm × 110 mm × 25 mm. For simplicity, the authors display only half of each insert (110 mm × 110 mm × 25 mm) in Figure 32 [146].

The use of the novel in-house algorithm with the thermal AM constraint limits the occurrence of heat accumulation, local overheating and accumulated residual stresses, and results in a greater geometrical accuracy and surface quality [145].

The proposed algorithm that includes AM related thermal aspects indicates that there is room for further improvements [145].

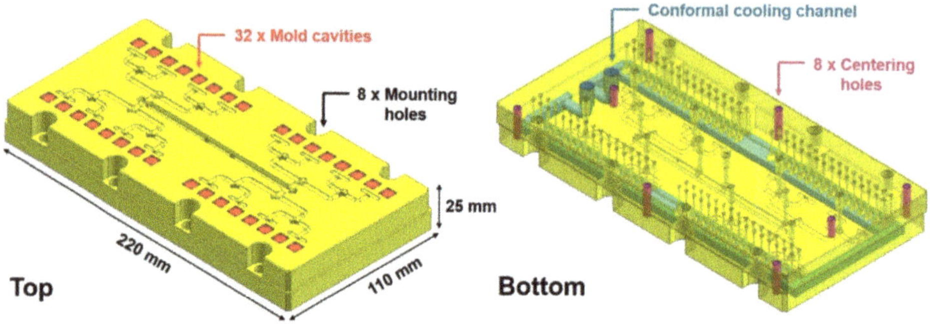

Figure 31. Original design of the injection molding insert [145].

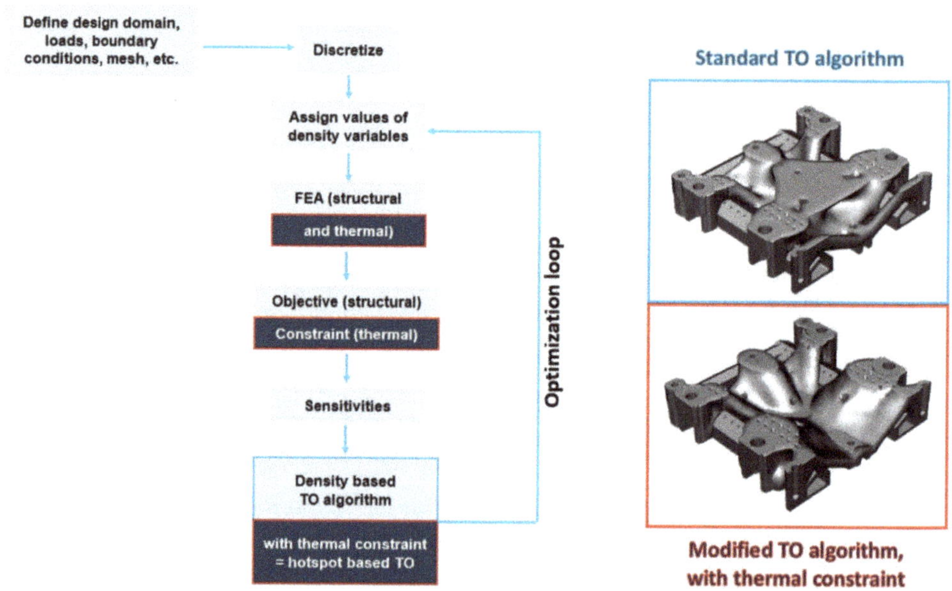

Figure 32. The injection molding insert, topology optimized with the standard density-based topology optimization (TO) algorithm and the in-house novel TO algorithm with AM thermal constraint (hotspot detection) [145]. The topology optimized inserts have the same dimensions as the original, Figure 31. For simplicity, the authors display only half of each insert (110 mm × 110 mm × 25 mm) in this figure [146].

8. Discussion

Three categories have been proposed for dividing the research in DfAM (design for additive manufacturing)—system design, part design, and process design [121]. The following selection criteria can be applied to identify whether a redesign for additive manufacturing would be beneficial [147]:

- Integrated design, the objective of which is to identify assemblies or groups of parts that can be re-designed into one single part.
- Individualization, which is driven by the wish to meet different customers' needs.
- Lightweight design, where the reduced weight improves the performance of the product.
- Efficient design, the objective of which is to improve the efficiency and performance of the product (the tool in this paper) in operation.

The tooling examples in the previous section illustrate, emphasize the correctness of, and confirm the importance of using the system, part and process design approach and having efficient design as the selection criteria. Efficient design is of particular significance for the production tools in hot working and injection molding. The importance of process design and its close relationship to part and system design is illustrated in Figure 32 and [145].

While the primary target is high efficiency and performance in operation, using generative design and topology and lattice structure optimization will also lead to lightweight design.

Using the efficient design, i.e., high efficiency and performance in operation, as the criterion, the tooling examples in the previous section show that L-PBF (combined with conformal cooling and topology and lattice structure optimization) has its greatest potential in production tools for hot working and injection molding.

In injection molding, cooling stands for 60–90% of the total cycle time [143]. The related tooling examples in the previous section display the importance of efficient design. Conformal cooling channels in topology and lattice structure optimized inserts/core provide high efficiency and performance in injection molding.

In a 2003 review, design for layer manufacturing (LM) was identified as a topic that "required attention in the future". To take full advantage of the design freedom offered by LM, the need for "research and development in the area of dedicated CAD design software incorporating design for LM modules" was highlighted [148]. Today, it is interesting to note the accuracy of this prediction.

In hot working, it is important to avoid thermal fatigue. The linear tensile or compressive stress that occurs in the tool can be calculated by [143]

$$\sigma = \alpha E (T_i - T_f) = \alpha E \Delta T \quad (1)$$

in which σ = thermal stress developed due to the thermal change ΔT, T_i = initial temperature, T_f = final temperature, α = expansion coefficient and E = modulus of elasticity.

Thermal fatigue is related to the parameter [143]

$$\sigma_f k / \alpha E \quad (2)$$

where σ_f = the mean fatigue strength and k = the thermal conductivity. An increase in this parameter indicates better resistance to thermal fatigue.

Fast cooling by a large temperature difference ΔT will increase the thermal stress. Choosing a material with low α and E will decrease the thermal stresses, while an increased tensile strength will directly increase the effect the thermal stress has on the material.

AISI H13 has a thermal stress of $\sigma = 2.604\Delta T$ N/mm^2 °C and a thermal conductivity of k = 27 W/mK at 500 °C. DIN 1.2709 has the properties $\sigma = 2.124\Delta T$ N/mm^2 °C and k = 24 W/mK at 500 °C. Both steels have an almost equal ability to withstand thermal fatigue [143].

Samples tested with as-produced surface conditions usually show a poor fatigue life, since fatigue strength depends mostly on surface and internal defects. Therefore and for instance, surface machining improves fatigue properties considerably (in some cases to a comparable fatigue life as conventionally-produced specimens) [51]. See also [149].

H11 and H13 have superior wear resistance compared to 1.2709 [150]. As discussed in Section 5, the carbon-bearing H11 and H13 tend, however, to crack during L-PBF. Preheating the base plate, which lowers the thermal gradients and reduces the thermal stresses, has been identified as a measure to obtain crack-free H11 and H13 in L-PBF.

High tool efficiency and performance in operation is also affected by the corrosion properties in injection molding (of for instance reinforced polyamide). The corrosion properties of AM-produced steels, mainly stainless steels, is typically superior to those of conventionally produced material, mainly due to the fine microstructure and sometimes also due to the special texture and the two-phase nature of AM-produced stainless steels. The research into the reasons for the improved corrosion resistance is still in its early stages [51].

DIN 1.2709 (18Ni-300), the most established maraging grade, combines high values of mechanical properties (strength, hardness, toughness and ductility) with very good processability in L-PBF. However, due to the lack of chromium this steel grade does not exhibit corrosion resistance. The stainless steels 316L and 17-4PH are also well established as powder materials for additive manufacturing but have relatively low mechanical strength and hardness [151]. See also Table 7.

The recently developed M798 AMPO displays mechanical properties that are close to those of DIN 1.2709 but has better corrosion resistance than 17-4PH [151]. See also Table 7.

The recently developed AM Corrax displays also high strength and hardness, Table 7. Both untreated and surface treated AM Corrax samples were subject to corrosion tests, the salt spray testing and the cyclic polarization. The samples were surface treated with physical vapor deposition (PVD) coatings—TiN, CrN, and diamond-like carbon (DLC)—at temperatures below 450 °C, as well as with plasma nitriding treatments for conventional steel grades (PN1) and suitable for stainless steel grades (PN2). These tests showed that the corrosion resistance is high and can be summarized as untreated > PN2 > CrN > TiN > DLC > PN1 [152].

The number of powder steels for L-PBF of production tools is however very limited today, both in general and particularly for hot working (and injection molding). See Tables 5–7 and [51].

It was noted in the 2003 review that "a lot remains to be done to develop better base materials (powders, liquid photopolymers, etc.) … " for LM [148]. Metals, plastics, ceramics and composites are mentioned [148]. Many materials have been developed since then. However, the number of powder steels for L-PBF of production tools is, as noted above, still very limited.

As far as the feedstock (powder) is concerned, the metal powder should, according to some investigations, exhibit high sphericity, a clean and smooth surface, and low amount of satellites—features that are associated with gas-atomized powder [66]. The particle size distribution of the powder used in L-PBF is nominally 10–45 μm [65]. Some other investigations show that fine particles (D_{90} = 24.17 μm in the used powder) can easily be melted and are therefore beneficial for high part density and mechanical strength, while bigger particles (D_{90} = 55.54 μm in the used powder) are beneficial for higher breaking elongations. The selection of the appropriate particle size distribution must therefore take care of these contrary effects [153]. 316L is the powder material used in [66] and [153].

Water-atomized powder results in irregular shapes and low flowability but lower atomization costs. Water-atomized Fe-4800 powder has been successfully used to make objects by L-PBF. The low powder flowability was overcome through in-situ-alloying of carbon black nanoparticles in the range of 20–100 nm [154]. Compared to gas atomization, water atomization reduces the costs significantly. The costs will be discussed more later on in this section.

Spherical, ovoidal, and irregular particle shapes using powders atomized with argon, nitrogen and air (non-spherical N/air atomized powder were not possible to use before due to severe wiper blocking) have been successfully applied in L-PBF using nanoparticles as spacers. This process has been used for aluminium-silicon powder. However, since the principle of using nanoparticles is based on geometry, the same principle can be transferred to other material systems [155].

Research and development in the area of feedstock (powder) production and quality is therefore highly important from process, property, part quality, cost, and lead time perspectives.

Figure 33 displays a comparison of different metal AM methods with respect to part performance, cost and lot size. Production tools are normally made in single or few units and required to perform well in operation to avoid stops, minimize or eliminate rejections and maximize the production efficiency. L-PBF (PBF-L in Figure 33) is, other words, in a good position from these perspectives. The high L-PBF costs indicated in Figure 33 are observed in many AM tooling related investigations. See, for instance, [62,128].

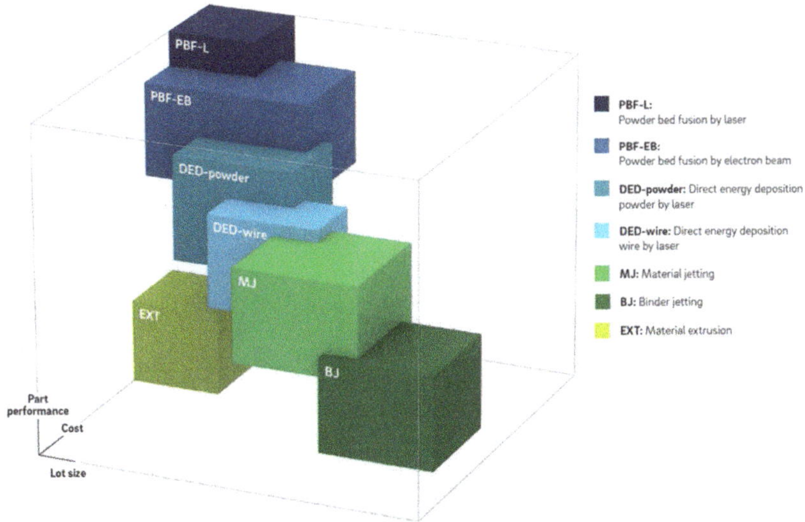

Figure 33. A 2018 comparison of different metal AM methods with respect to part performance, cost and lot size [32].

For the parts that require production tooling, the costs of this tooling constitute an initial investment which in majority of the cases are larger for an AM inclusive process. However, the total cost per produced part can be reduced due to e.g., improved cooling and shorter cycle time using production tools made by an AM inclusive process [62], Figure 34.

For hot working and injection molding, the tools, dies, molds, cores and inserts made by L-PBF cost more but lead to lower total costs (the part costs) since such tools enable shorter cycle time, improved quality etc.

For cold working, both the tooling and part costs are, however, higher (than the conventionally made tool and the part made in it), in case the tools are made by L-PBF. A lead time reduction in toolmaking is reported particularly for the (cold) stamping tools made by an L-PBF inclusive process [62,128]. For (cold) stamping tools, the cost increase can in some cases be motivated by the significant lead time reduction. A so-called late change in the industrialization of a car body component can be considered as such a case.

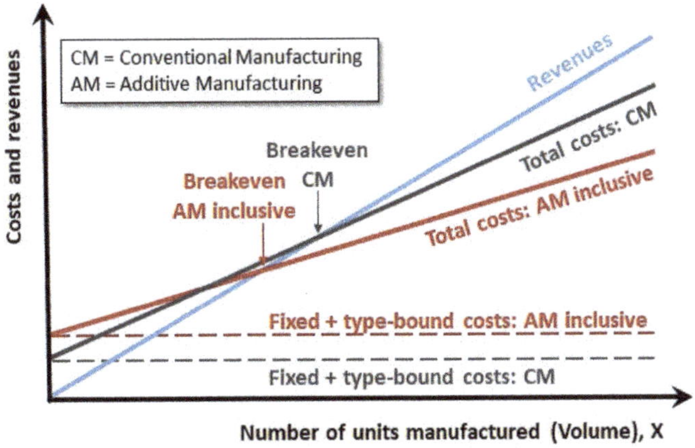

Figure 34. The part costs and revenues: conventional versus AM inclusive fabrication. The lower running AM costs is valid only for the cases in which the tooling enables a shorter cycle time, e.g., conformally cooled injection molding cores [62].

The cost of feedstock (powder) production was mentioned above. The cost and productivity of L-PBF is closely related to the number of lasers that are active at the same time, Table 1. As displayed in this table, up to 12 lasers can be active at the same time. The costs for L-PBF have been reduced significantly during the past six years and will continue to fall in the coming years. However, these costs are still much higher than the costs for conventional toolmaking [32].

To accomplish a significant improvement of the process efficiency (productivity and cost), different technologies and methods need to be studied. As an example, the multi-spot array system based on multi-diode lasers can be mentioned. The machine design uses no scanner systems and instead relies on a printer head featuring several individually controllable diode lasers that is moved using linear axes [156]. See also [157].

It was noted in the 2003 review that productivity had been improved partially and that "a lot of effort is still required to further boost production rate" [148]. In addition, "exaggerated equipment and material cost" was observed due to the "great divergence of machines offered and their costs" [148]. The 2003 review concerned all LM methods (not only SLM/L-PBF). A great progress has been made since then. The build rate has improved significantly. In one of the current L-PBF machines, the build rate is 71 times higher than that in 1995. However, the productivity and cost need, as indicated above, to be improved further.

The 2003 review indicated "From near to net-shape" as another topic that "required attention in the future" [148]. Improved accuracy, repeatability and secured quality were mentioned in this context. "The combination with traditional processes in multiprocessing equipment might be explored" [148]. This topic has been addressed and a large progress has been made since 2003. However, post-processing is and will be an important topic also during the coming years.

In tooling applications, the roughness of the functional tool surfaces is of great significance. As mentioned in Section 5, L-PBF is not yet capable of providing the surface roughness required in tooling applications. The tool made by L-PBF needs therefore to be machined during post-processing to obtain the required surface roughness. Further improvement of L-PBF is, in other words, required in the future from this perspective. Laser surface re-melting might be a track to explore further to accomplish improved surface roughness.

Heat treatment is frequently needed in different tooling applications.

Multiprocessing equipment has been subject to exploration and will be important to explore further also in the future. Plants with a L-PBF inclusive production process, high degree of automation, digitization and digitalization is also a track that needs to be explored further in the future.

The size of the object than can be made by L-PBF is of great significance for many tooling applications. As mentioned in Section 3, the maximum object size that can be made by L-PBF today is $600 \times 600 \times 600$ mm^3 (although the largest height is 850 mm in one of the other current machines), Table 1. This size has more than doubled compared to that in the 1995 machine for SLS.

However, many production tools, particularly (cold) stamping, press-hardening, die casting tools are larger than $600 \times 600 \times 600$ mm^3. L-PBF can be used to make tool/die/mold inserts, which then are mounted in a core or shoe that is made by e.g., casting. Another option is to design the tool modularly and make each module by L-PBF. The object size that can be made by L-PBF will still be an important topic in the future. The challenge will then be to realize a homogenous shielding gas flow over the entire powder bed in combination with high productivity (due to the large object size).

According to a non-peer-reviewed assessment in 2018, metal AM is on the brink of industrialization [32]. The manufacturing readiness level (MRL) is the highest for L-PBF compared to the other metal AM methods [32]. According to another non-peer-reviewed assessment in 2020, L-PBF has reached the highest technology maturity index and the highest industrialization index (compared to other metal AM methods) and thereby a widespread industrial use [158]. L-PBF is almost fully industrialized [158]. Two other non-peer-reviewed assessments from 2013 and 2015 show that [159,160]

- dental/medical applications have reached the highest MRL (full- rate production),
- tooling applications the second highest MRL (low-rate production),
- aerospace industry the third highest MRL (capability in operational environment demonstrated—low rate production), and
- automotive industry the lowest MRL (technology validated in laboratory environment—basic capabilities shown).

The assessment in [160] is stated to be based on expert interviews.

In spite of this high MRL for L-PBF of tooling (low-rate production), further research and development is required to improve toolmaking by a L-PBF inclusive process: larger maximum objects sizes than be made by L-PBF, a larger number of powder metals that are designed for different tooling applications, lower feedstock (powder) costs, lower L-PBF processing costs, further L-PBF productivity and lead time improvement, minimized post-processing, improved surface roughness (both through L-PBF and in combination with efficient post-processing), and secured quality (including the L-PBF portion of the process). As indicated above, these improvements require research and development both within "single" topics, e.g., L-PBF in a multi-spot array system based on multi-diode lasers or laser re-melting, and from a toolmaking system view, e.g., plants with a high degree of automation and eventually cyber-physically controlled smart L-PBF inclusive manufacturing systems.

9. Summary and Conclusions

For laser-based powder bed fusion (L-PBF) of production tooling, the following summary and conclusions apply:

- Generative design and topology and lattice structure optimization enable, amplify and are a complement to the design freedom provided by L-PBF.
- A focus on efficient design, the objective of which is to improve the efficiency and performance of the tooling in operation, should be the primary target in tool design. Lightweight solutions can be obtained through topology and lattice structure optimization as a complement.
- Using high tooling efficiency and performance as the primary target, L-PBF exhibits the greatest potential for hot working and injection molding tools, dies and molds.
- The tools made by L-PBF cost more than those made conventionally. However, the total costs (the part costs) can be lower than those of the parts made in conventional tools, since the cycle time is reduced significantly in the tools made by L-PBF. Hot working and injection molding are the processes in which lower total costs can be obtained.
- For cold working, e.g., (cold) stamping, both the tooling and part costs are higher with tools made by L-PBF. However, the reported lead time reduction for toolmaking through L-PBF can be used to motivate higher tooling costs for e.g., the so-called late changes in the industrialization of car body components.
- The largest objects that can be made by L-PBF are currently $600 \times 600 \times 600$ mm^3 and $800 \times 400 \times 500$ mm^3.
- The number of metal powder alloys for tooling applications is limited. Currently, maraging steels 18Ni-300 (or DIN 1.2709) and M789 AMPO, stainless steels 316L, 17-4PH and 420, and the tool steels M2, H13, P20, and AM Corrax can be selected as tool material.
- According to some investigations, the metal powder should exhibit high sphericity, a clean and smooth surface, and low amount of satellites—features that are associated with inert gas atomization. The nominal particle size distribution used for L-PBF is 10–45 µm. Other investigations show that it is possible to L-PBF metal powder with irregular particle shapes—metal powder made by water or N/air atomization—successfully. This improves the resource efficiency, since larger percentages of atomized powder can be used in L-PBF.

- Laser power (W), scan velocity (mm/s), and hatch distance (mm) between neighboring melt tracks, and the layer thickness (μm) determine the process window in L-PBF.
- The path of the scan vector and thereby the scan strategy are essential for the final quality, i.e., defect types and numbers, microstructure, residual stresses, and surface roughness.
- A chess strategy (or island scanning) is effective in reducing the residual stresses.
- User experience and expertise play a large roll in some cases, e.g., selection of a scan strategy other than the one pre-set by the machine builder might lead to better dimensional accuracy and higher quality.
- The values of the process parameters and the size of the process window which provide fully dense and high quality objects in the above-mentioned metal powder tool materials can be found in this paper (Table 8).
- The downskin surface exhibits a roughness of 20–65 μm after L-PBF. For the upskin and the top surface, a roughness of 5 μm can be achieved in L-PBF. In other words, L-PBF is not yet capable of providing the surface roughness required in tooling applications. Smaller roughness requires post-processing by machining.
- The properties of the built object are anisotropic and dependent upon the build directions and angles.
- Faces with the downskin angle <45° require support structure.
- Internal channels should be built vertically.
- Several examples from different tooling applications in cold working, hot working and injection molding made by L-PBF are showcased in this paper.
- The corrosion properties of AM-produced steels, mainly stainless steels, are superior to those of the conventionally produced materials. The research into the reasons for the improved corrosion resistance is still in its early stages.
- The so-called carbon-free DIN 1.2709 obtains a ductile martensitic microstructure and is well processable in L-PBF, while the carbon-bearing H11 and H13 display a brittle martensitic microstructure that tends to crack due to the thermal stresses built up by the high cooling rates. A pre-heating of the base plate (building platform) eliminates, however, this cracking. H11 and H13 have superior wear resistance compared to DIN 1.2709. L-PBF yields excellent performance and is suitable for small lot sizes (in tooling, only a single or a few objects are made).
- According to non-peer-reviewed assessments, metal AM is on the brink of industrialization. The manufacturing readiness level is the highest for L-PBF compared to the other metal AM methods and the second highest for L-PBF of tooling (low-rate production) compared to L-PBF in other industrial/application segments.
- Rapid tooling has made a remarkable journey to AM of production tools by L-PBF thanks to the enormous efforts in research, development, entrepreneurial endeavors and company building.
- Further research and development is required to improve toolmaking by a L-PBF inclusive process: larger object sizes that can be made by L-PBF, a larger number of powder metals that are designed for different tooling applications, lower feedstock (powder) costs, lower L-PBF processing costs, further L-PBF productivity and lead time improvement, minimized post-processing, improved surface roughness (both through L-PBF and in combination with efficient post-processing) and secured quality (including the L-PBF portion of the process). These improvements require research and development both on "single" topics, e.g., L-PBF in a multi-spot array system based on multi-diode lasers, and from a toolmaking system view, e.g., plants with a high degree of automation and eventually cyber-physically controlled smart L-PBF inclusive manufacturing systems.

Funding: This research received no external funding.

Institutional Review Board Statement: Not applicable.

Informed Consent Statement: Not applicable.

Data Availability Statement: The data presented in this study are openly available in the referred publications. See the reference list.

Acknowledgments: The author expresses his sincere gratitude to Vegard Brøtan, SINTEF Manufacturing AS, Norway, Marc Fette, Airbus/CTC GmbH, Germany, Gerhard Hackl, ASMET, Austria, Professor Em. Jean-Pierre Kruth, KU Leuven, Belgium, Bernhard Langefeld, Roland Berger GmbH, Germany, Alexander Leicht, RISE Research Institutes of Sweden, Bernhard Mueller, Fraunhofer IWU, Germany, Professor Barbara Reggiani, UNIMORE, Italy, and Mirko Sinico, KU Leuven, Belgium, for providing valuable information and in some cases permissions.

Conflicts of Interest: The author declares no conflict of interest.

Appendix A

In this appendix, the methodology and process used to create and structure this review manuscript is described.

The purpose of this study was to conduct a review of the journey of the production tools for cold working, hot working and injection molding from rapid tooling to Additive Manufacturing (AM) by Laser-based Powder Bed Fusion (L-PBF).

To conduct this review, the keywords and the search profile were discussed with the university library staff resulting in the following search profile:

> *metals ("direct rapid tooling" OR "metal additive manufacturing" OR "laser powder bed fusion" OR "selective laser melting" OR "selective laser sintering") (tools OR dies OR molds)*

This search profile was then used in the available engineering databases at the university library: Materials Science and Engineering Database/Collection, Science Direct, Springer Link and Wiley Online.

Table A1 displays the number of search results in this first step. See also Figure A1.

Table A1. The number of search results in the first step.

The Database	Number of Results
Materials Science and Engineering Collection	401
Science Direct	5366
Springer Link	929
Wiley Online	972
Total	**7668**

To reduce the number of results, the obtained results (Table A1) were limited by using a number of selection criteria. Table A2 displays the used selection criteria per database and the number of the obtained results. See also Figure A1.

In step 3, the abstracts were reviewed and the duplications (or papers with "similar" contents), papers that were not related to direct rapid tooling, papers that were not related to production tools for cold working, hot working and injection molding, and the papers that were related to selective laser sintering (SLS) published after 2005 were removed. See also Figure A1.

In Step 4, the full papers were reviewed and

- similar papers were "neglected", review papers were given a higher priority and industrial tooling applications were focused, and
- the remaining papers were classified based on Metallic materials for tooling applications, Design, simulations and modelling, Equipment for SLS/SLM/L-PBF, Process design and parameters, Properties and performance, and Examples of tooling applications.

Table A2. The selection criteria per database and the number of results. The indicated selection criteria were applied to the results obtained in the first step, i.e., Table A1.

The Database	Number of Results
Materials Science and Engineering Collection *Peer-reviewed, English*	239
Science Direct *Open access,* *Procedia CIRP, Additive Manufacturing,* *Journal of Materials Processing Technology*	325
Springer Link *Subdiscipline Manufacturing,* *Machines, Tools, Processes,* *English*	212
Wiley Online *Journals,* *Advanced Engineering Materials,* *English*	112
Total	**888**

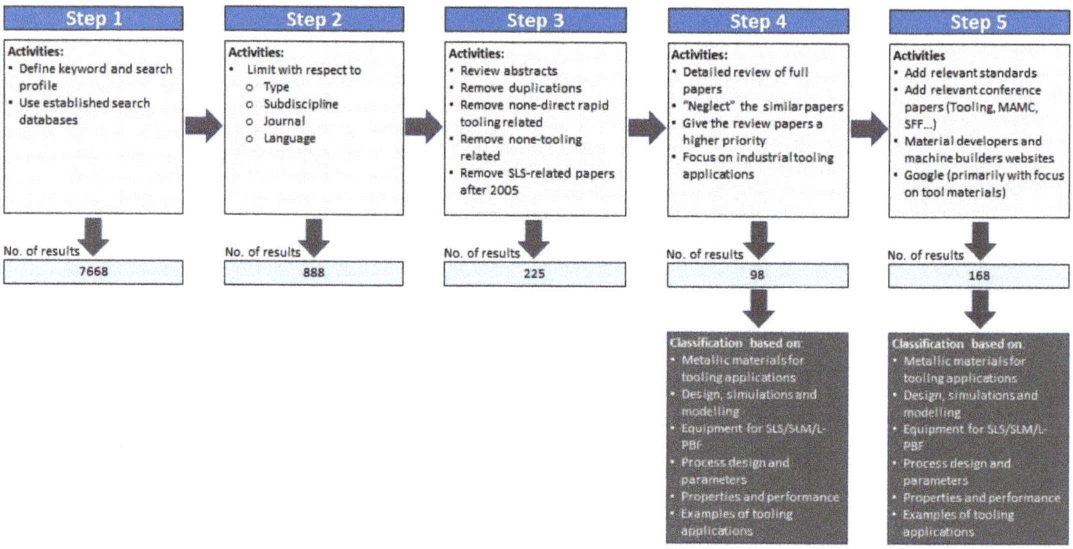

Figure A1. The method applied for creating this review.

The selected review targets at this stage were to

1. set a baseline for the review by providing a brief historical background until 2005,
2. provide the capabilities of the 2020 L-PBF equipment/machines,
3. provide the chemical compositions and properties of the current materials for L-PBF of production tools in cold working, hot working and injection molding,
4. describe the impact of feedstock and process parameters on the properties and performance,
5. provide the L-PBF process parameters for the above-mentioned tool materials,
6. provide a brief description of system, part and process design,
7. review some examples of tooling applications designed for and made by L-PBF,
8. discuss the advances made during the past 20 years and
9. draw some conclusions.

Based on these selected targets, the remaining papers were kept in accordance with the above-mentioned classification.

In step 5, relevant standards were reviewed. The papers presented at some relevant conferences, such as the Tooling Conference, Metal Additive Manufacturing Conference (MAMC), and Solid Freeform Fabrication symposium (SFF) were selected based on the above-mentioned criteria and targets and reviewed. The websites of the material developers/producers and L-PBF machine builders were visited and the relevant material and machine information were gathered from these websites. A google search was conducted, the purpose of which was to find papers about the tooling materials mentioned above. These papers were then reviewed. The same classification was used for this information. See also Figure A1.

The review (manuscript) was then outlined based on the above-mentioned procedure and submitted. As this manuscript was peer-reviewed, the reviewers suggested:

- that items 4 and 6 above should be focused on tooling applications, and
- some improvements. To facilitate these improvements, the reviewers provided additional 10 publication titles.

These publications were reviewed and the manuscript was modified in accordance the reviewer suggestions.

As far as the machine builders are concerned, it is worth noting that Table 1 in the manuscript is limited to seven machine builders.

The present review manuscript does not comprise tools made by "conventional" powder metallurgy-based processes.

There is no bias in this review.

References

1. Prinz, F.B.; Atwood, C.L.; Aubin, R.F.; Beaman, J.J.; Brown, R.L.; Fussel, P.S.; Lightman, A.J.; Sachs, E.; Weiss, L.E.; Wozny, M.J. *Japanese and World Technology Evaluation Centers Panel Report on Rapid Prototyping in Europe and Japan*; Analytical Chapters; Rapid Prototyping Association of the Society of Manufacturing Engineers: Baltimore, MD, USA, 1997; Volume 1, ISBN 1-883712-44-0.
2. Matias, E.; Rao, B. 3D Printing: On Its Historical Evolution and the Implications for Business. In *Management of the Technology Age, Proceedings of the PICMET '15 Conference, Portland International Center for Management of Engineering and Technology (PICMET), Portland, OR, USA, 2–6 August 2015*; IEEE: Portland, OR, USA, 2015; pp. 551–558.
3. Karapatis, N.P.; van Griethuysen, J.-P.S.; Glardon, R. Direct rapid tooling: A review of current research. *Rapid Prototyp. J.* **1998**, *4*, 77–89. [CrossRef]
4. Kruth, J.-P. Rapid prototyping, a new application of physical and chemical processes for material accretion manufacturing. In Proceedings of the 11th International Symposium for Electromachining, Lausanne, Switzerland, 17–21 April 1995; pp. 3–28.
5. Gebhardt, A. Rapid prototyping– laser-gestützte Revolution der Produktentwicklung. *Laser Magazin* **1996**, *1*, 6–9.
6. Wohlers, T. *Rapid Prototyping State of the Industry: 1995–96 Worldwide Progress Report*; Society of Manufacturing Engineers Symposium: Dearborn, MI, USA, April 1996.
7. Styger, L. Rapid Prototyping and Tooling: The Enabling Technology of the '90s. IEE Manufacturing Division Colloquium on Rapid Prototyping in the UK: Coventry, UK, 1994.
8. *Rapid Prototyping: Rewriting the Rule Book*; IEE Manufacturing Division Colloquium on Rapid Prototyping in the UK: Coventry, UK, 1994.
9. *CRIF-Liège Information Leaflet*; CRIF: Liège, Belgium, 1996.
10. Lohner, A. Laser sintering ushers in new route to PM parts. *Metal Powder Rep.* **1997**, *52*, 24–30.
11. Girouard, D. The SLSTM selective laser sintering process: Technology, applications, and materials for advanced rapid prototyping. In Proceedings of the IBEC '93, Detroit, MI, USA, 21–23 September 1993; pp. 10–14.
12. Shellabear, M.; Nyrhilä, O. Dmls—Development history and state of the art. In Proceeding of the 4th LANE (Laser Assisted Net Shape Engineering), Erlangen, Germany, 21–24 September 2004.
13. Wohler, T.; Gornet, T. History of additive manufacturing. *Wohler Rep.* **2016**, *24*, 118.
14. Hejmadi, U.; McAlea, K. Selective laser sintering of metal molds: The RapidToolTM process. In Proceedings of the Annual International Solid Freeform Fabrication Symposium, Austin, TX, USA, 12–14 August 1996; pp. 97–104.
15. Nyrhilä, O.; Syrjälä, S. Manufacture of Dimensionally Precise Pieces by Sintering. U.S. Patent 5,061,439, 5 December 1990.
16. Holger, D.; Pilz, R.; Eleser, N.S. Rapid tooling of EDM electrodes by means of selective laser sintering. *Comput. Ind.* **1999**, *39*, 35–45.

17. Majewski, C.; Hopkinson, N. The Effect of Layer Thickness on Ejection Forces for Injection Moulded Parts Made Using DMLS Tools. In Proceedings of the Annual International Solid Freeform Fabrication Symposium, Austain, TX, USA, 5–7 August 2002; pp. 73–82.
18. King, D.; Tansey, T. Rapid tooling: Selective laser sintering injection tooling. *J. Mater. Process. Technol.* **2003**, *132*, 42–48. [CrossRef]
19. Kruth, J.-P.; Mercelis, P.; Van Vaerenbergh, J.; Froyen, L.; Rombouts, M. Advances in selective laser sintering. In Proceedings of the 1st International Conference on Advanced Research in Virtual and Rapid Prototyping (VR@P), Leiria, Portugal, 1–4 October 2003; pp. 59–69.
20. Meiners, W.; Wissenbach, K.; Gasser, A. Selective Laser Sintering at Melting Temperature. U.S. Patent 6,215,093, 10 April 2001.
21. Neikov, O.D. Powders for Metal Additive Processing. In *Handbook for Non-Ferrous Metal Powders—Technologies and Applications*, 2nd ed.; Neikov, O.D., Naboychenko, S.S., Yefimov, N.A., Eds.; Elsevier: Amsterdam, The Netherlands, 2019; pp. 373–399, ISBN 978-0-08-100543-9.
22. Meiners, W.; Over, C.; Wissenbach, K.; Poprawe, R. Direct Generation of Metal Parts and Tools by Selective Laser Powder Remelting (SLPR). In Proceedings of the Annual International Solid Freeform Fabrication Symposium, Austain, TX, USA, 9–11 August 1999; pp. 655–661.
23. Kruth, J.-P.; Vandenbroucke, B.; Van Vaerenbergh, J.; Mercelis, P. Benchmarking of different sls/slm processes as rapid manufacturing techniques. In Proceeding of International Conference Polymers & Moulds Innovations (PMI), Gent, Belgium, 20–23 April 2005; pp. 1–7.
24. Sprinkle, T. The 5 Most Important Standards in Additive Manufacturing. *ASTM Standardization News*. 11 May 2020. Available online: https://www.astm.org/standardization-news/?q=features/5-most-important-standards-additive-manufacturing-.html (accessed on 25 November 2020).
25. ISO. *International Standard ISO 52900, 2016—Additive manufacturing—General principles—Terminology (ISO/ASTM 52900, 2015, IDT)*; ISO: Geneva, Switzerland, 2016.
26. ISO. *International Standard ISO 17296-2, 2016—Additive manufacturing—General principles—Part 2, Overview of process categories and feedstock (ISO 17296-2, 2015)*; ISO: Geneva, Switzerland, 2016.
27. Ålgårdh, J.; Strondl, A.; Karlsson, S.; Farre, S.; Joshi, S.; Andersson, J.; Ågren, J. *State-of-the-Art for Additive Manufacturing of Metals*; Report 2016-03898—State-of-the-Art—Version 2.1; Swedish Arena for Additive Manufacturing of Metals: Stockholm, Sweden, 2017.
28. Criales, L.E.; Arısoy, Y.M.; Lane, B.; Moylan, S.; Donmez, A.; Özel, T. Laser powder bed fusion of nickel alloy 625, Experimental investigations of effects of process parameters on melt pool size and shape with spatter analysis. *Int. J. Mach. Tool Manuf.* **2017**, *121*, 22–36. [CrossRef]
29. Schwarze, D. *Auswirkungen der Bauraumvorheizung auf die thermomechanischen Bauteileigenschaften*; Anwender Zentrum Augsburg, IWB: München, Germany, 2010.
30. Over, C. *Generative Fertigung von Bauteilen aus Werkzeugstahl X38CrMoV5-1 und Titan TiAl6V4 mit "Selective Laser Melting"*; Shaker: Aachen, Germany, 2003.
31. Niendorf, T.; Leuders, S.; Riemer, A.; Richard, H.A.; Tröster, T.; Schwarze, D. Highly Anisotropic Steel Processed by Selective Laser Melting. *Metall. Mater. Trans. B* **2013**, *44B*, 794–796. [CrossRef]
32. Langefeld, B.; Moehrle, M.; Balzer, C.; Schildbach, P. *Advancements in Metal 3D Printing—Beyond Powder Bed—Additive Manufacturing on the Brink of Industrialization*; Roland Berger: Munich, Germany, 2018.
33. 3D Systems, Printers and Materials Specifications. Available online: https://www.3dsystems.com (accessed on 27 November 2020).
34. Additive Industries, Printers and Materials Specifications. Available online: https://www.additiveindustries.com (accessed on 27 November 2020).
35. Concept Laser, Printers and Materials Specifications. Available online: https://www.ge.com/additive/who-we-are/concept-laser (accessed on 27 November 2020).
36. EOS (Electro Optical Systems), Printers and Materials Specifications. Available online: https://www.eos.info/en (accessed on 27 November 2020).
37. SLM Solutions, Printers and Materials Specifications. Available online: https://www.slm-solutions.com/ (accessed on 27 November 2020).
38. Renishaw, Printers and Materials Specifications. Available online: https://www.renishaw.com/en/metal-3d-printing--32084 (accessed on 27 November 2020).
39. Trumpf, Printers and Materials Specifications. Available online: https://www.trumpf.com/en_INT/products/machines-systems/additive-production-systems (accessed on 27 November 2020).
40. Liljengren, M.; Kjellsson, K.; Johansson, T.; Asnafi, N. Die Materials, Hardening Methods and Surface Coatings for Forming of High, Extra High & Ultra High Strength Steel Sheets (HSS/EHSS/UHSS). In Proceedings of the Annual Conference of the International Deep Drawing Research Group's (IDDRG), Porto, Portugal, 19–21 June 2006; pp. 507–603.
41. Asnafi, N.; Johansson, T.; Miralles, M.; Ullman, A. Laser Surface-Hardening of Dies for Cutting, Blanking or Trimming of Uncoated DP600. Proceeding of the Recent Advances in Manufacture & Use of Tools & Dies and Stamping of Steel Sheets, Olofström, Sweden, 5–6 October 2004; pp. 169–190.
42. Asnafi, N. Tooling and Technologies for Processing of Ultra High Strength Sheet Steels. In Proceedings of the Tools and Technologies for Processing Ultra High Strength Materials, Graz University of Technology, Graz, Austria, 19–21 September 2011.

43. Sjöström, J. Chromium Martensitic Hot-Work Tool Steels—Damage, Performance and Microstructure. Ph.D. Thesis, Karlstad University Studies, Karlstad, Sweden, 2004; p. 52, ISBN 91-85335-21-5.
44. Fuchs, K.-D. Hot-Work Tool Steels with Improved Properties for Die casting Applications. In Proceedings of the 6th International Conference on Tooling, Karlstad, Sweden, 10–13 September 2002; pp. 17–26.
45. Klobčar, D.; Tušek, J.; Taljat, B. Thermal fatigue of materials for die-casting tooling. *Mater. Sci. Eng. A* **2008**, *472*, 198–207. [CrossRef]
46. Vasco, J.; Capela, C.; Bártolo, P.; Granja, D. Material Selection for High Performance Moulds. In Proceedings of the SPE Meeting Improving Quality and Tool Efficiency within Injection Moulding K2007, Düsseldorf, Germany, 25 October 2007; pp. 1–7.
47. Rebeggiani, S. *On Polishability of Tool Steels*; Chalmers University of Technology: Göteborg, Sweden, 2013; ISBN 978-91-7385-828-1.
48. Roberts, G.; Krauss, G.; Kennedy, R. *Tool Steels*, 5th ed.; ASM International: Novelty, OH, USA, 1998; ISBN 978-0-87170-599-0.
49. Højerslev, C. *Tool steels. Risoe-R, No. 1244(EN)*; Risø National Laboratory, Forskningscenter Risoe: Roskilde, Denmark, 2001.
50. The online Material Properties Database MakeItFrom.com. Available online: https://www.makeitfrom.com (accessed on 12 September 2020).
51. Bajaj, P.; Hariharan, A.; Kini, A.; Kürnsteiner, P.; Raabe, D.; Jägle, E.A. Steels in additive manufacturing: A review of their microstructure and properties. *Mater. Sci. Eng. A* **2020**, *772*, 138633. [CrossRef]
52. Böhler Edelstahl, Material Specifications. Available online: https://www.bohler-edelstahl.com/en/applications/3d-printing-selective-laser-melting/ (accessed on 27 November 2020).
53. FORMETRIX, L-40 Specification. Available online: https://www.formetrixmetals.com/products/formetrix-l-40/ (accessed on 27 November 2020).
54. Eriksson, P. *Evaluation of Mechanical and Microstructural Properties for Laser Powder-Bed Fusion 316L*; Uppsala University: Uppsala, Sweden, 2018.
55. Zai, L.; Zhang, C.; Wang, Y.; Guo, W.; Wellmann, D.; Tong, X.; Tian, Y. Laser Powder Bed Fusion of Precipitation-Hardened Martensitic Stainless Steels: A Review. *Metals* **2020**, *10*, 255. [CrossRef]
56. Nath, S.D.; Irrinki, H.; Gupta, G.; Kearns, M.; Gulsoy, O.; Are, S. Micostructure-property relationships of 420 stainless steel fabricated by laser-power bed fusion. *Powder Technol.* **2019**, *343*, 738–746. [CrossRef]
57. Zhao, X.; Wei, Q.; Song, B.; Liu, Y.; Luo, X.; Wen, S.; Shi, Y. Fabrication and Characterization of AISI 420 Stainless Steel Using Selective Laser Melting. *Mater. Manuf. Process.* **2015**, *30*, 1283–1289. [CrossRef]
58. Kempen, K.; Vrancken, B.; Buls, S.; Thijs, L.; Van Humbeek, J.; Kruth, J.-P. Selective Laser Melting of Crack-Free High Density M2 High Speed Steel Parts by Baseplate Preheating. *J. Manuf. Sci. Eng.* **2014**, *136*, 061026-1. [CrossRef]
59. Mazur, M.; Brincat, P.; Leary, M.; Brandt, M. Numerical and experimental evaluation of conformally cooled H13 steel injection mould manufactured with selective laser melting. *Int. J. Adv. Manuf. Technol.* **2017**, *93*, 881–900. [CrossRef]
60. Li, H.X.; Qi, H.L.; Song, C.H.; Li, Y.L.; Yan, M. Selective laser melting of P20 mould steel: Investigation on the resultant microstructure, high-temperature hardness and corrosion resistance. *Powder Metall.* **2018**, *61*, 21–27. [CrossRef]
61. Uddeholm, AM Corrax Specification. Available online: https://www.uddeholm.com/sweden/sv/products/uddeholm-am-corrax (accessed on 27 November 2020).
62. Asnafi, N.; Rajalampi, J.; Aspenberg, D.; Alveflo, A. Production Tools Made by Additive Manufacturing Through Laser-based Powder Bead Fusion. *Berg Huettenmaenn Mon.* **2020**, *165*, 125–136. [CrossRef]
63. Mooney, B.; Kourousis, K.I. A Review if Factors Affecting the Mechanical Properties of Maraging Steel 300 Fabricated via Laser Powder Bed Fusion. *Metals* **2020**, *10*, 1273. [CrossRef]
64. Vock, S.; Klöden, B.; Kirchner, A.; Weißgärber, T.; Kieback, B. Powders for powder bed fusion: A review. *Prog. Addit. Manuf.* **2019**, *4*, 383–397. [CrossRef]
65. Sames, W.J.; List, F.A.; Pannala, S.; Dehoff, R.R.; Babu, S.S. The metallurgy and processing science of metal additive manufacturing. *Int. Mater. Rev.* **2016**, *61*, 315–360. [CrossRef]
66. Leicht, A. *Aspects of Building Geometry and Powder Characteristics in Powder Bed Fusion*; Tech Technical report no. IMS-2018-1; Ph.D. Thesis; Chalmers University of Technology: Göteborg, Sweden, 2018.
67. Li, R.; Shi, Y.; Wang, Z.; Wang, L.; Liu, J.; Jiang, W. Densification behavior of gas and water atomized 316 L stainless steel powder during selective laser melting. *Appl. Surface Sci.* **2010**, *256*, 4350–4356. [CrossRef]
68. Hoeges, S.; Zwiren, A.; Schade, C. Additive manufacturing using water atomized steel powders. *Metal Powder Rep.* **2017**, *72*, 111–117. [CrossRef]
69. Abdelwahed, M.; Casati, R.; Bengtsson, S.; Larsson, A.; Riccio, M.; Vedani, M. Effects of Powder Atomization on Microstructural and Mechanical Behaviour of L-PBF Processed Steels. *Metals* **2020**, *10*, 1474. [CrossRef]
70. ASTM. *F3049-14 Standard Guide for Characterizing Properties of Metal Powders Used for Additive Manufacturing Processes*; ASTM International: West Conshohocken, PA, USA, 2014.
71. Spierings, A.B.; Voegtlin, M.; Bauer, T.; Wegener, K. Powder flowability characterization methodology for powder-bed-based metal additive manufacturing. *Prog. Addit. Manuf.* **2016**, *1*, 9–20. [CrossRef]
72. A Guidebook to Particle Size Analysis. Horiba Scientific. Available online: https://www.horiba.com (accessed on 27 November 2020).

73. Karg, M.C.H.; Hentschel, O.; Ahuja, B.; Junker, D.; Hassler, U.; Schäperkötter, C.S.; Haimerl, A.; Arnet, H.; Merklein, M.; Schmidt, M. Comparison of process characteristics and resulting microstructure of maraging steel 1.2709 in Additive Manufacturing via Laser Metal Deposition and Laser Beam Melting in Powder Bed. In Proceedings of the 6th International Conference on Additive Technologies, ICAT2016, Nürnberg, Germany, 29–30 November 2016; pp. 39–50.
74. Oliveira, J.P.; LaLonde, A.D.; Ma, J. Processing parameters in laser powder bed fusion metal additive manufacturing. *Mater. Design* **2020**, *193*, 108762. [CrossRef]
75. Metens, R.; Dadbakhsh, S.; Van Humbeeck, J.; Kruth, J.-P. Application of base plate preheating during selective laser melting. *Procedia CIRP* **2018**, *74*, 5–11. [CrossRef]
76. Leicht, A. Laser Powder Bed Fusion of 316L Stainless Steel—Microstructure and Mechanical Properties as a Function of Process Parameters, Design and Productivity. Ph.D. Thesis, Chalmers University of Technology, Göteborg, Sweden, 2020; ISBN 978-91-7905-289-8.
77. Ali, H.; Ghadbeigi, H.; Mumtaz, K. Effect of scanning strategies on residual stress and mechanical properties of Selective Laser Melted Ti6Al4V. *Mater. Sci. Eng. A* **2018**, *712*, 175–187. [CrossRef]
78. Bartlett, J.L.; Li, X. An overview of residual stresses in metal powder bed fusio. *Addit. Manuf.* **2019**, *27*, 131–149.
79. Saunders, M. How process parameters drive successful metal ALM part production. *Metal AM* **2018**, *4*, 127–135.
80. Kou, S. *Welding Metallurgy*, 2nd ed.; John Wiley & Sons: Hoboken, NJ, USA, 2003.
81. Wang, Z.; Denlinger, E.; Michaleris, P.; Stoica, A.D.; Ma, D.; Beese, A.M. Residual stress mapping in Inconel 625 fabricated through additive manufacturing: Method for neutron diffraction measurements to validate thermomechanical model predictions. *Mater. Design* **2017**, *11*, 3169–3177. [CrossRef]
82. Rodrigues, T.A.; Duarte, V.; Avila, J.A.; Santos, T.G.; Miranda, R.; Oliveira, J. Wire and arc additive manufacturing of HSLA steel: Effect of thermal cycles on microstructure and mechanical properties. *Addit. Manuf.* **2019**, *27*, 440–450. [CrossRef]
83. Bobbio, L.D.; Bocklund, B.; Otis, R.; Borgonia, J.P.; Dillon, R.P.; Shapiro, A.A.; McEnerney, B.; Liu, Z.-K.; Beese, A.M. Characterization of a functionally graded material of Ti-6Al-4V to 304L stainless steel with an intermediate V section. *J. Alloys Compd.* **2018**, *742*, 1031–1036. [CrossRef]
84. Carroll, B.E.; Otis, R.A.; Borgonia, J.P.; Suh, J.O.; Dillon, R.P.; Shapiro, A.A.; Hofmann, D.C.; Liu, Z.K.; Beese, A.M. Functionally graded material of 304L stainless steel and inconel 625 fabricated by directed energy deposition: Characterization and thermodynamic modeling. *Acta Mater.* **2016**, *108*, 46–54. [CrossRef]
85. DebRoy, T.; Wei, H.L.L.; Zuback, J.S.S.; Mukherjee, T.; Elmer, J.W.W.; Milewski, J.O.O.; Beese, A.M.M.; Wilson-Heid, A.; De, A.; Zhang, W. Additive manufacturing of metallic components—Process, structure and properties. *Prog. Mater. Sci.* **2018**, *92*, 112–224. [CrossRef]
86. Dai, D.; Gu, D. Effect of metal vaporization behavior on keyhole-mode surface morphology of selective laser melted composites using different protective atmospheres. *Appl. Surface Sci.* **2015**, *355*, 310–319. [CrossRef]
87. Zhong, Y.; Liu, L.; Wikman, S.; Cui, D.; Shen, Z. Intragranular cellular segregation network structure strengthening 316L stainless steel prepared by selective laser melting. *J. Nucl. Mater.* **2016**, *470*, 170–178. [CrossRef]
88. Kluczyński, J.; Śnieżek, L.; Grzelak, K.; Janiszewski, J.; Płatek, P.; Torzewski, J.; Szachogłuchowicz, I.; Gocman, K. Influence of Selective Laser Melting Technological Parameters on the Mechanical Properties of Additively Manufactured Elements Using 316L Austenitic Steel. *Materials* **2020**, *13*, 1449. [CrossRef] [PubMed]
89. Burns, D.E.; Kudzal, A.; McWilliams, B.; Manjarres, J.; Hedges, D.; Parker, P.A. Investigating Additively Manufactured 17-4 PH for Structural Applications. *J. Mater. Eng. Perform.* **2019**, *28*, 4943–4951. [CrossRef]
90. Irrinki, H.; Dexter, M.; Barmore, B.; Enneti, R.; Pasebani, S.; Badwe, S.; Stitzel, J.; Malhotra, R.; Are, S. Effects of Powder Attributes and Laser Powder Bed Fusion (L-PBF) Process Conditions on the Densification and Mechanical Properties of 17-4 PH Stainless Steel. *JOM* **2016**, *68*, 860–868. [CrossRef]
91. Yan, J.; Zhou, Y.; Gu, R.; Zhang, X.; Quach, W.-M.; Yan, M. A Comparison Study of Steel Powders (316L, H13, P20 and 18Ni300) for Their Selective Melting Additive Manufacturing. *Metals* **2019**, *9*, 86. [CrossRef]
92. Turk, C.; Zunko, H.; Aumayr, C.; Leitner, H.; Kapp, M. Advances in Maraging Steels for Additive Manufacturing. *Berg-und Hüttenmännische Mon.* **2019**, *164*, 112–116. [CrossRef]
93. Jiang, J.; Chen, J.; Ren, Z.; Mao, Z.; Ma, X.; Zhang, D.Z. The influence of Process Parameters and Scanning Strategy on Lower Surface Quality of TA 15 Parts Fabricated by Selective Laser Melting. *Metals* **2020**, *10*, 1228. [CrossRef]
94. Chen, Z.; Wu, X.; Tomus, D.; Davies, C.H.J. Surface roughness of Selective Laser Melted Ti-6Al-4V alloy components. *Addit. Manuf.* **2018**, *21*, 91–103. [CrossRef]
95. Snyder, J.C.; Thole, K.A. Understanding Laser Powder Bed Fusion Surface Roughness. *J. Manuf. Sci. Eng.* **2020**, *142*, 71003. [CrossRef]
96. Wang, D.; Liu, Y.; Yang, Y. Theoretical and experimental study on surface roughness of 316L stainless steel metal parts obtained through selective laser melting. *Rapid Prototyp. J.* **2016**, *22*, 706–716. [CrossRef]
97. Spierings, A.B.; Levy, G. Comparison of density of stainless steel 316L parts produced with selective laser melting using different powder grades. In Proceedings of the Annual International Solid Freeform Fabrication Symposium, Austin, TX, USA, 3–5 August 2009; pp. 342–353.
98. Yang, T.; Liu, T.; Liao, W.; Macdonald, E.; Wei, H.; Chen, X. The Influence of Process Parameters on Vertical Surface Roughness of the AlSi10Mg Parts Fabricated by Selective Laser Melting. *J. Mater. Process. Technol.* **2019**, *266*, 26–36. [CrossRef]

99. Casalino, G.; Campanelli, S.L.; Contuzzi, N.; Ludovico, A.D. Experimental Investigation and Statistical Optimisation of the Selective Laser Melting Process of a Maraging Steel. *Opt. Laser Technol.* **2015**, *65*, 151–158. [CrossRef]
100. Liu, B.; Wildman, R.; Tuck, C.; Ashcroft, I.; Hague, R. Investigation the Effect of Particle Size Distribution on Processing Parameters Optimisation in Selective Laser Melting Process. In Proceedings of the Annual International Solid Freeform Fabrication Symposium, University of Texas at Austin, Austin, TX, USA, 8–10 August 2011; pp. 227–238.
101. Sendino, S.; Martinez, S.; Lartategui, F.; Diaz, J.; Lamikiz, A.; Ukar, E. Surface Roughness Dependant Factors in Metal Bed Fusion Processes. In Proceedings of the Joint Special Interest Group meeting between euspen (the European Society for Precision Engineering and Nanotechnology) and ASPE (the American Society of Precision Engineering) on Advancing Precision in Additive Manufacturing, Ecole Centrale de Nantes, Nantes, France, 16–18 September 2019.
102. Moylan, S.; Slotwinski, J.; Cooke, A.; Jurrens, K.; Donmez, M.A. An Additive Manufacturing Test Artifact. *J. Res. Natl. Inst. Stand. Technol.* **2014**, *119*, 429–459. [CrossRef] [PubMed]
103. Rebaioli, L.; Fassi, I. A review on benchmark artifacts for evaluating the geometrical performance of additive manufacturing processes. *Int. J. Adv. Manuf. Technol.* **2017**, *93*, 2571–2598. [CrossRef]
104. Moshiri, M.; Tosello, G.; Mohanty, S. A new design for an extensive benchmarking of additive manufacturing machines. In Proceedings of the 18th International Conference of the European Society for Precision Engineering and Nanotechnology (Euspen 18), Venice, Italy, 4–8 June 2018.
105. Moshiri, M.; Candeo, S.; Carmignato, S.; Mohanty, S.; Tosello, G. Benchmarking of Laser Powder Bed Fusion Machines. *J. Manuf. Mater. Process.* **2019**, *3*, 85. [CrossRef]
106. Tomas, J.; Hitzler, L.; Köller, M.; von Kobylinski, J.; Sedlmajer, M.; Werner, E.; Merkel, M. The Dimensional Accuracy of Thin-Walled Parts Manufactured by Laser-Powder Bed Fusion Process. *J. Manuf. Mater. Process.* **2020**, *4*, 91. [CrossRef]
107. Krish, S. A practical generative design method. *Comput. Aided Des.* **2011**, *43*, 88–100. [CrossRef]
108. Briard, T.; Segonds, F.; Zamariola, N. G-DfAM: A methodological proposal of generative design for additive manufacturing in the automotive industry. *Int. J. Interact. Des. Manuf.* **2020**, *14*, 875–886. [CrossRef]
109. Bendsoe, M.P.; Sigmund, O. *Topology Optimization: Theory, Methods and Applications*; Springer: Berlin/Heidelberg, Germany, 2004.
110. Liu, X.; Yi, W.J.; Li, Q.S.; Shen, P.S. Genetic evolutionary structural optimization. *J. Constr. Steel Res.* **2008**, *64*, 305–311. [CrossRef]
111. Zhao, F. A nodal variable ESO (BESO) method for structural topology optimization. *Finite Elem. Anal. Des.* **2014**, *86*, 34–40. [CrossRef]
112. Guo, X.; Zhang, W.; Zhang, J.; Yuan, J. Explicit structural topology optimization based on moving morphable components (MMC) with curved skeletons. *Comput. Methods Appl. Mech. Eng.* **2016**, *310*, 711–748. [CrossRef]
113. Niendorf, T.; Brenne, F.; Schaper, M. Lattice Structures Manufactured by SLM: On the Effect of Geometrical Dimensionson Microstructure Evolution During Processing. *Metall. Mater. Trans. B* **2014**, *45B*, 1181–1185. [CrossRef]
114. Schnabel, T.; Oettel, M.; Mueller, B.; Hoschke, K.; Pfaff, A.; Amund-Kopp, C.; Klöden, B.; Gebauer, M.; Töppel, T. *Design for Additive Manufacturing, Guidelines and Case Studies for Metal Applications*; Fraunhofer IWU, EMI and IFAM: Dresden, Germany, 2017.
115. Pan, C.; Han, Y.; Lu, J. Design and Optimization of Lattuce Structures: A Review. *Appl. Sci.* **2020**, *10*, 6374. [CrossRef]
116. Dong, G.; Tang, Y.; Zhao, Y.F. A Survey of Modeling of Lattice Structures Fabricated by Additive Manufacturing. *J. Mech. Des.* **2017**, *139*, 100906. [CrossRef]
117. Kuo, T.C.; Huang, S.H.; Zhang, H.C. Design for manufacture and design for 'X': Concepts, applications, and perspectives. *Comput. Ind. Eng.* **2001**, *41*, 241–260. [CrossRef]
118. Tomiyama, T.; Gu, P.; Jin, Y.; Lutters, D.; Kind, C.; Kimura, F. Design methodologies: Industrial and educational applications. *CIRP Ann. Manuf. Technol.* **2009**, *58*, 543–565. [CrossRef]
119. Laverne, F.; Segonds, F.; Answer, N.; Le Coq, M. Assembly Based Methods to Support Product Innovation in Design for Additive Manufacturing: An Exploratory Case Study. *J. Mech. Des.* **2015**, *137*, 121701. [CrossRef]
120. Reiher, T.; Lindemann, C.; Jahnke, U.; Deppe, G.; Koch, R. Holistic approach for industrializing AM technology: From part selection to test and verification. *Prog. Addit. Manuf.* **2017**, *2*, 43–55. [CrossRef]
121. Wiberg, A.; Persson, J.; Ölvander, J. Design for additive manufacturing—A review of available design methods and software. *Rapid Prototyp. J.* **2019**, *25*, 1080–1094. [CrossRef]
122. VDI. *VDI Guideline 3405 Part 2—Additive Manufacturing Processes, Rapid Manufacturing—Beam Melting of Metallic Parts—Qualification, Quality Assurance and Post Processing*; VDI: Düsseldorf, Germany, 2013.
123. International Standard ISO/ASTM 52911-1. *Additive Manufacturing—Design—Part 1, Laser-Based Powder Bed Fusion of Metals*; ISO: Geneva, Switzerland, 2019.
124. Reinarz, B.; Sehrt, J.T.; Witt, G.; Deiss, O.; van Kampen, J.; Münzer, J.; Ott, M. Optimization of media feed channels in laser beam melting. *Proc. ASPE. Berkeley* **2014**, *57*, 13–18.
125. Hitzler, L.; Hirsch, J.; Heine, B.; Merkel, M.; Hall, W.; Öchsner, A. On the Anisotropic Mechanical Properties of Selective Laser-Melted Stainless Steel. *Materials* **2017**, *10*, 1136. [CrossRef] [PubMed]
126. Li, R.; Kim, Y.S.; Tho, H.V.; Yum, Y.J.; Kim, W.J.; Yang, S.Y. Additive manufacturing (AM) of piercing punches by the PBF method of metal 3D printing using mold steel powder materials. *J. Mech. Sci. Technol.* **2019**, *33*, 809–817. [CrossRef]
127. Zelený, P.; Váňa, T.; Stryal, J. Application of 3D Printing for Specific Tools. *Mater. Sci. Forum* **2016**, *862*, 316–323. [CrossRef]
128. Leal, R.; Barreiros, F.M.; Alves, L.; Romeiro, F.; Vasco, J.C.; Santos, M. Additive manufacturing tooling for the automotive industry. *Int. J. Adv. Manuf. Technol.* **2017**, *92*, 1671–1676. [CrossRef]

129. Asnafi, N.; Shams, T.; Aspenberg, D.; Öberg, C. 3D Metal Printing from an Industrial Perspective—Product Design, Production and Business Models. *BHM Berg-und Hüttenmännische Mon.* **2019**, *164*, 91–100. [CrossRef]
130. Aspenberg, D.; Asnafi, N. Topology optimization of a U-bend tool using LS-TaSC. In Proceedings of the 12th European LS-DYNA Conference, Koblenz, Germany, 14–16 May 2019.
131. Asnafi, N.; Rajalampi, J.; Aspenberg, D. Design and Validation of 3D-Printed Tools for Stamping of DP600. In Proceedings of the IOP Conference Series: Materials Science and Engineering, 3–7 June 2019; Volume 651, p. 12010.
132. Skåre, T.; Asnafi, N. Forming and trimming of 2-mm thick DP600 sheet steel in tools and dies 3D-printed in maraging steel by laser-based powder bed fusion. In Proceedings of the IOP Conference Series: Materials Science and Engineering, 26–30 October 2020; Volume 67, p. 12040.
133. Müller, B.; Gebauer, M.; Hund, R.; Malek, R.; Gerth, N. Metal Additive Manufacturing for tooling applications—Laser Beam Melting technology increases efficiency of dies and molds. In Proceedings of the Metal Additive Manufacturing Conference 2014 (MAMC 2014) Meet the Future of Industrial Manufacturing Now! Vienna, Austria, 20–21 November 2014.
134. Reggiani, B.; Todaro, I. Investigation on the design of a novel selective laser melted insert for extrusion dies with conformal cooling channels. *Int. J. Adv. Manuf. Technol.* **2019**, *104*, 815–830. [CrossRef]
135. Reggiani, B.; (University of Modena and Reggio Emilia (UNIMORE), Reggio Emilia, Italy). Personal communication, 15 February 2021.
136. Armillotta, A.; Baraggi, R.; Fasoli, S. SLM tooling for die casting with conformal cooling channels. *Int. J. Adv. Manuf. Technol.* **2014**, *71*, 573–583. [CrossRef]
137. Fette, M.; Sander, P.; Wulfsberg, J.; Zierk, H.; Herrmann, A.; Stoess, N. Optimized and Cost-Efficient Compression Molds Manufactured by Selective Laser Melting for the Production of Thermoset Fiber Reinforced Plastic Aircraft Components. *Procedia CIRP* **2015**, *35*, 25–30. [CrossRef]
138. Fette, M.; (Airbus/CTC GmbH, Stade, Germany). Personal communication, 15 February 2021.
139. Milovanovic, J.; Stojkovic, M.; Trajanovic, M. Rapid tooling of tyre treat ring mould using direct metal laser sintering. *J. Sci. Ind. Res.* **2009**, *68*, 1038–1042.
140. Schlieper, G. Metal Additive Manufacturing grains ground in the tyre industry. *Metal AM* **2017**, *3*, 81–86.
141. Shellabear, M.; Weilhammer, J. *Tooling Applications with EOSINT M*; EOS Whitepaper: München, Germany, 2007.
142. Mazur, M.; Leary, M.; McMillan, M.; Elambasseril, J.; Brandt, M. ALM additive manufacture of H13 tool steel with conformal cooling and structural lattices. *Rapid Prototyp. J.* **2016**, *22*, 504–518. [CrossRef]
143. Brøtan, V.; Åsebø Berg, O.; Sørby, K. Additive manufacturing for enhanced performance of molds. *Procedia CIRP* **2016**, *54*, 186–190. [CrossRef]
144. Brøtan, V.; (SINTEF Manufacturing AS, Trondheim, Norway). Personal communication, 15 February 2021.
145. Sinico, M.; Ranjan, R.; Moshiri, M.; Ayas, C.; Langelaar, M.; Witvrouw, A.; van Keulen, F.; Dewulf, W. A Mold Case Study on Topology Optimized Design for additive Manufacturing. In Proceedings of the 30th Annual International Solid Freeform Fabrication Symposium, Solid Freeform Fabrication 2019—An Additive Manufacturing Conference 2019, Austin, TX, USA, 12–14 August 2019; pp. 1921–1931.
146. Sinico, M.; (KU Leuven, Leuven, Belgium). Personal communication, 15 February 2021.
147. Klahn, C.; Leutenecker, B.; Meboldt, M. Design for additive manufacturing—Supporting the Substitution of Components in Series Products. *Procedia CIRP* **2014**, *21*, 138–143. [CrossRef]
148. Levy, G.N.; Schindel, R.; Kruth, J.P. Rapid manufacturing and rapid tooling with layer manufacturing (LM) technologies, state of the art and future perspectives. *CIRP Ann.* **2003**, *2*, 589–609. [CrossRef]
149. Spierings, A.B.; Starr, T.L.; Wegener, K. Fatigue performance of additive manufactured metallic parts. *Rapid Prototyp. J.* **2013**, *19*, 88–94. [CrossRef]
150. Florian, H.; Bischof, C.; Hentschel, O.; Heberle, J.; Zettl, J.; Nagulin, K.Y.; Schmidt, M. Laser beam melting and heat treatment of 1.2343 (AISI H11) tool steel—Microstructure and mechanical properties. *Mater. Scienec Eng. A* **2019**, *742*, 109–115.
151. Zunko, H.; Turk, C.; Kapp, M.; Hofer, C. Corrosion resistant maraging steel for AM-produced plastic mold inserts. In Proceedings of the 11th Tooling 2019 Conference & Exhibition—Communication along the Supply Chain in the Tooling Industry, Aachen, Germany, 12–16 May 2019.
152. Oikonomou, C.; Karamchedu, S. A complete solution in metal AM for tooling applications—A case study on plastic injection molding. In Proceedings of the 11th Tooling 2019 Conference & Exhibition—Communication Along the Supply Chain in the Tooling Industry, Aachen, Germany, 12–16 May 2019.
153. Spierings, A.B.; Herres, N.; Levy, G. Influence of the particle size distribution on surface quality and mechanical properties in additive manufactured stainless steel parts. *Rapid Prototyp. J.* **2011**, *17*, 195–202. [CrossRef]
154. Bischof, C.; Nitsch, G.; Scheitler, C.; Dressler, A.; Schmidt, M. Laser beam melting of water atomized iron base alloy FE 4800 with in-situ alloying of carbon nanoparticles. In Proceedings of the 17th Annual international Conference of RAPDASA, Building on the Foundations, Johannesburg, South Africa, 2–4 November 2016.
155. Karg, M.C.H.; Munk, A.; Ahuja, B.; Backer, M.V.; Schmitt, J.P.; Stengel, C.; Kuryntsev, S.V.; Schmidt, M. Expanding particle size distribution and morphology of aluminium-silicon powders for Laser Beam Melting by dry coating with silica nanoparticles. *J. Mater. Process. Technol.* **2019**, *264*, 155–171. [CrossRef]
156. Hinke, C.; Merkt, S.; Eibl, F.; Schrage, J.; Bremen, S. Additive Manufacturing: Perspectives for Diode Lasers. In Proceedings of the IEEE High Power Diode Lasers and Systems Conference 2015, Coventry, UK, 14–15 October 2015; pp. 39–40.

157. Zavala-Arredondo, M.; Boone, N.; Willmott, J.; Childs, D.T.D.; Ivano, P.; Groom, K.M.; Mumtaz, K. Laser diode area melting for high speed additive manufacturing of metallic components. *Mater. Des.* **2017**, *117*, 305–315. [CrossRef]
158. AMPOWER GmbH & Co. KG. *Additive Manufacturing, New Metal Technologies, Ampower Insights*; AMPOWER GmbH & Co. KG: Hamburg, Germany, 2020; Volume 6.
159. Roland Berger. *Additive Manufacturing—A Game Changer for the Manufacturing Industry?* Roland Berger: Munich, Germany, 2013.
160. Roland Berger. *Additive Manufacturing (AM)—Opportunities in a Digitalized Production*; Roland Berger: Munich, Germany, 2015.

metals

Article

Additive Manufactured 316L Stainless-Steel Samples: Microstructure, Residual Stress and Corrosion Characteristics after Post-Processing

Suvi Santa-aho [1,*], Mika Kiviluoma [1], Tuomas Jokiaho [1], Tejas Gundgire [1], Mari Honkanen [2], Mari Lindgren [3] and Minnamari Vippola [1]

1 Materials Science and Environmental Engineering, Tampere University, P.O. Box 589, 33014 Tampere, Finland; mika.kiviluoma@hotmail.com (M.K.); Tuomas.jokiaho@tuni.fi (T.J.); tejas.gundgire@tuni.fi (T.G.); Minnamari.vippola@tuni.fi (M.V.)
2 Tampere Microscope Center, Tampere University, P.O. Box 692, 33014 Tampere, Finland; mari.honkanen@tuni.fi
3 Metso Outotec Research Center, Kuparitie 10, 28330 Pori, Finland; mari.lindgren@mogroup.com
* Correspondence: suvi.santa-aho@tuni.fi; Tel.: +358-408-4901-45

Abstract: Additive manufacturing (AM) is a relatively new manufacturing method that can produce complex geometries and optimized shapes with less process steps. In addition to distinct microstructural features, residual stresses and their formation are also inherent to AM components. AM components require several post-processing steps before they are ready for use. To change the traditional manufacturing method to AM, comprehensive characterization is needed to verify the suitability of AM components. On very demanding corrosion atmospheres, the question is does AM lower or eliminate the risk of stress corrosion cracking (SCC) compared to welded 316L components? This work concentrates on post-processing and its influence on the microstructure and surface and subsurface residual stresses. The shot peening (SP) post-processing levelled out the residual stress differences, producing compressive residual stresses of more than −400 MPa in the AM samples and the effect exceeded an over 100 μm layer below the surface. Post-processing caused grain refinement and low-angle boundary formation on the sample surface layer and silicon carbide (SiC) residue adhesion, which should be taken into account when using the components. Immersion tests with four-point-bending in the heated 80 °C magnesium chloride solution for SCC showed no difference between AM and reference samples even after a 674 h immersion.

Keywords: powder methods; additive manufacturing (AM); post-processing; residual stresses; 316L stainless-steel; electron microscopy

Citation: Santa-aho, S.; Kiviluoma, M.; Jokiaho, T.; Gundgire, T.; Honkanen, M.; Lindgren, M.; Vippola, M. Additive Manufactured 316L Stainless-Steel Samples: Microstructure, Residual Stress and Corrosion Characteristics after Post-Processing. *Metals* **2021**, *11*, 182. https://doi.org/10.3390/met11020182

Received: 22 December 2020
Accepted: 18 January 2021
Published: 20 January 2021

1. Introduction

The additive manufacturing (AM) of metals is becoming more and more popular due to its ability to produce complex geometries and parts on-demand from many different available powder materials with efficient material use [1,2]. Powder bed fusion (PBF) is one of the metal AM methods which uses either laser (L-PBF) or electron beam (EB-PBF) as a heat source to build the component in a layer-by-layer fashion and can produce accurate, sub-millimeter features [3,4]. One widely used powder material in PBF (especially L-PBF) is austenitic stainless-steel 316L. Due to its superior corrosion resistance, 316L is widely used in different components facing demanding corrosion atmosphere in chemical, petrochemical, process and marine industry [5]. The microstructure of PBF 316L is usually composed of columnar [6–9] grains oriented parallel to the thermal gradients, i.e., the build direction. These microstructural features are originated by the anisotropic heat removal and re-melting of previous layers, also leading to anisotropic mechanical properties [8,10]. Compared to the wrought sheet materials, AM-produced 316L components have better mechanical properties [8,9,11,12]. In the PBF manufactured components, the motion of dislocations is hindered by various methods that all strengthen the 316L stainless-steel material. The grain boundary strengthening according to the Hall-Petch effect [8] of the very fine microstructure is formed due to the high cooling rate of the melt pool. In addition, other strengthening mechanisms are the production of high dislocation density in the as-built material [8], as well as the high amount of low-angle grain boundaries [9] and the fine cellular structure with cell sizes from 0.5 to 1 μm [9,10,13,14].

In addition to the distinct microstructural features, the residual stresses and their formation are also inherent to L-PBF manufactured components [15] due to the rapid localized heating–cooling thermal cycling of the material. At first, the laser beam heats the material and causes it to expand against the surrounding material and thus creates compressive stresses. Then, while the laser beam moves on, the material cools rapidly and its shrinking is constrained by the surrounding material, generating tensile residual stresses [16]. However, the produced residual stress field is said to be very complicated due to the complex heating and heat transfer [16]. The generating residual stresses may lead the material to crack, delaminate [17] and deform [18], and thus the result may be broken or warped components. Therefore, the controlling of the generation and end effects of residual stresses both in-situ and after the actual manufacturing process are of great interest.

During the AM process, the controlling of residual stresses is done in-situ with either feedback control, thermal gradient control, scanning strategy control and/or mechanical control [15,19]. However, often the scan strategy is an important parameter in controlling the formulating residual stresses in the printing [16,20,21]. For example, Robinson et al. [16,21] noticed that the greatest magnitudes of residual stresses are parallel to the scan vector and the scan vector direction determines the direction of primary residual stress. Residual stresses can be controlled by the scanning strategy, but it requires a high amount of designing and planning. Many characteristics of the component can be controlled during the printing but usually, the printed components are treated with different post-processing methods. These methods to control and mitigate residual stresses include heat treatments, shot and laser shock peening [22,23] and machining [15]. In addition, other traditional surface treatments could be applied, such as abrasive blasting, electropolishing and plasma spraying. Surface finish optimization can also be done to improve the corrosion resistance of the component [24]. Other reasons for post-processing are to increase the density of component, tailor the mechanical properties [15] or homogenize the microstructure [3,15], for instance.

The post-processing for residual stress controlling of AM 316L often includes a stress relief annealing heat treatment ("a build-on-plate stress relief heat treatment" [15]). Shiomi et al. [25] stated that the post-annealing reduces the residual stresses with 70% of the as-built AM component manufactured from chrome molybdenum steel (JIS SCM440) powder mixed with copper phosphate and nickel powders. The stress relief annealing can be done in different steps: at first, the lower temperatures are used for relieving the residual stresses and then higher temperatures for microstructure manipulation to produce microstructures involving recrystallized equiaxed grains [26]. Kirka et al. [27] reported for EB-PBF built Alloy 718 that the equiaxed grain structure could be beneficial for the applications demanding isotropic mechanical properties. Cruz et al. [28] studied AM 316L specimens subjected to three different stress-relieving heat treatments generating different residual stresses after printing. They concluded that after 4 h at 400 °C, no microstructural variations were found, and the cellular substructure became invisible after 5 min at 1100 °C [28], whereas the magnitude of the residual stress was closely related to the temperature and duration of the heat treatment and the varying temperatures produced 23%, 64% and 92% stress relief at 400 °C for 4 h, 650 °C for 2 h, and 1100 °C for 5 min heat-treated specimens, respectively.

Mechanical treatments such as a shot peening (SP) as a post-treatment are used to improve the as-built surface quality [29]. The shot peening also offers beneficial compressive residual stresses to the component surface and subsurface. In addition, some novel post-processing methods have been studied. Chen et al. [24] used laser polishing for a stainless-steel 316L surface and studied the corrosion behavior. The laser polishing melts the top surface and therefore reduces the surface roughness. Additionally, it was observed to refine the columnar structure in as-built sample into a fine cellular structure. The surface roughness reduction and a fine cellular structure were also noticed to improve the corrosion resistance. Wang et al. [23] utilized an ultrasonic impact peening and a laser shot peening for AM 316L. They noticed the different effects of the two methods on the surface hardness, yield strength and tendency of producing mechanical twins to the surface layer. Therefore, it can be stated that the different post-processing methods would provide lots of tools to manipulate or to tailor the surface layer characteristics [22,23].

Even though traditional stainless-steel 316L has a good corrosion resistance and it is used in demanding applications where superior corrosion resistance is needed, the very aggressive environments may cause pitting corrosion in chloride-bearing environments [30]. The corrosion responses of AM 316L have not been widely studied and more research in this field is definitely needed. For example, the porosity has been noticed to affect the corrosion properties of the AM component. Cruz et al. [28] used AM 316L specimens (three different stress-relieving heat treatments generating different residual stresses after printing) for pitting corrosion behavior tests in acidic $FeCl_3$ solution. The weight loss measurements after exposure showed that the residual stress did not significantly impact the kinetics of pitting to AM 316L samples. New properties of the AM components have merited testing under simulated operation conditions. Lou et al. [31] studied the AM 316L in a high-temperature annealed condition and its stress corrosion cracking (SCC) in high-temperature water (288 °C) with either dissolved oxygen or hydrogen. They did not find any significant differences in crack growth rate between AM and traditionally manufactured 316L. However, the intragranular oxide inclusions on grain boundaries were found to cause oxidation and cracking.

In the present study, basic sheet AM 316L components were characterized by the presence of residual stress, distinct microstructural features, etc., which could be detrimental in an actual working environment. In order to change the traditional manufacturing method of certain 316L components to AM, thorough characterization is needed to verify the suitability of AM components. Therefore, it is important to study these components and the post-processing methods and investigate their effect on the microstructure, residual stress formation and corrosion behavior. On very demanding corrosion atmospheres, the question is does AM lower or eliminate the risk of SCC compared to the welded 316L component? Here, the AM 316L samples were tested in different conditions: as-built, stress relief annealed and shot peened. A systematic investigation of surface and subsurface residual stresses as well as microstructure was also done in as-built and post-treated conditions. The SCC behavior of the selected samples was also investigated. In our future work, we will study more complex AM 316L shapes after different shot peening post-processing variations to consider the effects of the shape. The complex shapes might need more adjustment of the correct post-processing and may influence the functionality of the component in demanding conditions.

2. Materials and Methods

In this study, simple strip geometry samples were printed with L-PBF and studied with various material characterization methods. The high molybdenum austenitic stainless steel (EN 1.4432) 316L powder was used in AM (Table 1) manufactured by EOS Finland. The AM samples were manufactured to match B-type U-bend samples according to American Society for testing and Materials (ASTM) G30 [32], giving them dimensions of 100 mm × 9 mm × 3 mm, as illustrated in Figure 1. For the reference in the SP and immersion tests, similar samples (100 mm × 9 mm) were cut from the cold-rolled sheet. Low molybdenum variant 316L (EN 1.4404) was used for the shot peening and high molybdenum variant 316L (EN 1.4432) for the immersion tests. The low molybdenum variant was provided in 3 mm thick sheets, while the high molybdenum variant was only available in 1.95 mm thick sheets.

Table 1. Nominal chemical composition of studied 316L variant [33].

Element	Fe	Cr	Ni	Mo	C	Mn	Cu	P	S	Si	N
(wt.%)	balance	17–19	13–15	2.25–3	0.03	2	0.5	0.025	0.01	0.75	0.1

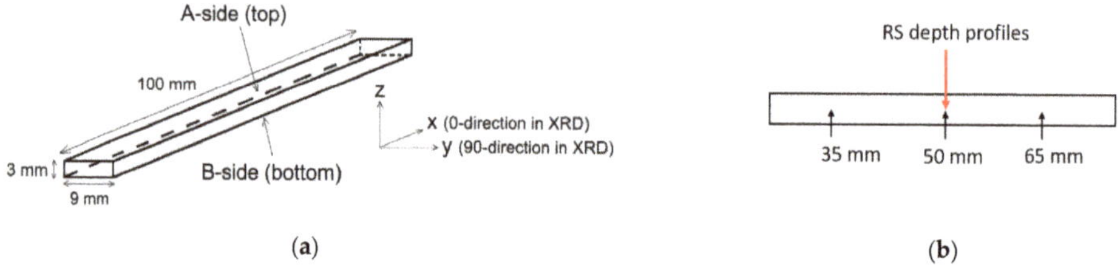

Figure 1. (**a**) Additive manufacturing (AM) strip sample geometry from 316L stainless-steel. (**b**) Locations of residual stress measurements from top and bottom sides (35, 50 and 65 mm from other end) and residual stress depth profiles from middle of sample (50 mm).

The AM strip samples were manufactured by 3DFormtech (Jyväskylä, Finland) with an EOS M290 L-PBF machine (Krailling, Germany) to a solid support structure. In total, 11 samples were printed to the solid structure build plate, at the same time, the 100 mm × 9 mm surface was parallel to the build plate. All samples were oriented in the same direction and they were located on two columns side by side with six or five samples on separate rows, one below another. The printing parameters were the standard recommended settings given by EOS-printing software. Hatch angle rotation of 47° and layer height of 40 µm was used. The edges of samples were printed with 100 W laser power and 900 mm/s scanning speed. The infill was done with 214 W laser power, 928 mm/s scanning speed and 100 µm hatch spacing. The top surfaces of the samples were printed with 150 W laser power, 515 mm/s scanning speed and 100 µm hatch spacing. The bottom sides of the samples were printed with the infill parameters. Eleven strips were treated with four different post-process treatments introduced in the process diagram in Figure 2. After the printing, five as-built samples were cut from the build plate with a band saw. The remaining six on the build plate were heat-treated with a stress relief annealing at 1030 °C for 60 min under vacuum. The samples were cooled using flowing nitrogen gas. After this, the remaining samples were cut from the build plate using a band saw.

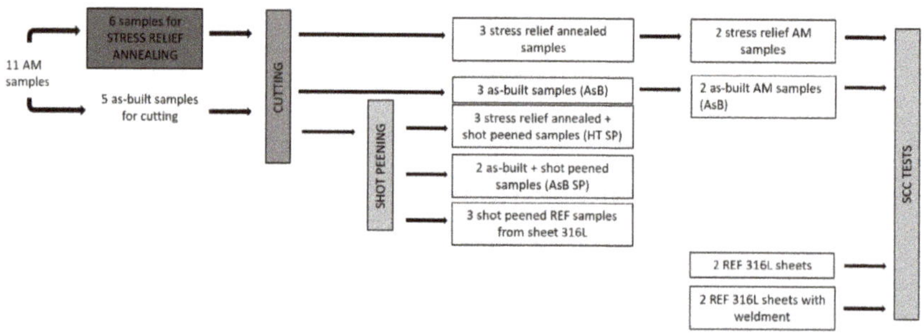

Figure 2. Process chart of samples.

Shot peening was done as a post-treatment for two as-built samples, for three stress relief annealed samples and for three reference samples. The shot peening was carried out in the manual shot peening (SP) cabinet with silicon carbide (SiC) shots with hardness reported at 2000 HV [34], being much harder than the samples (as-built average hardness 220 $HV_{0.05}$). Pressure in the SP system was 7 bar, and the flow rate of the shots were calculated to be 11.4 g/s. The peening was carried out at a 20 cm distance from the sample surface for either 1, 2 or a total of 3 s per surface. Overall, three of the samples were in the as-built (AsB) condition, three were stress relief annealed (AsB HT), three were both annealed and shot peened (HT SP), two samples were as-built and shot peened (AsB SP) and three reference samples were shot peened, as illustrated in Figure 2.

The metallographic cross-sectional samples were prepared by grinding them with 320–4000 SiC papers and then electrolytical polisher (Struers, Ballerup, Denmark) with 10% oxalic acid using 10 s etching time, 20 V as the voltage and a flow rate of 10. The hardness depth profiles were done from the cross-sectional samples using the MMT-X7 (Matsuzawa, Toshima, Japan) microhardness measurement device with a test load of 50 g.

The microstructure of the cross-sectional samples was studied by an optical microscope DM 2500 M (Leica Microsystems, Renens, Switzerland) and a field emission scanning microscope (FESEM) ULTRAplus (Zeiss, Jena, Germany). The microstructure of the samples was also studied by a Symmetry electron backscatter diffraction (EBSD) system (Oxford Instruments, Abingdon, United Kingdom) combined to the FESEM. A step size of 0.15 µm and an acceleration voltage of 15 kV were used. The cross-sectional EBSD samples were prepared by a traditional metallographic method followed by a final polishing using colloidal silica suspension (0.04 µm). The EBSD data were collected by the AZtec software (Oxford Instruments, Abingdon, United Kingdom) and post-processed and analyzed by the CHANNEL 5 software (Oxford Instruments, Abingdon, United Kingdom). In the EBSD results, a band contrast (BC) map represents the quality of the Kikuchi diffraction pattern for each measurement pixel. In the BC map, bright color means that the pattern quality is good, and the crystal orientation can be determined. Black signifies that the pattern quality is poor, and indexing is not possible. The colors in the inverse pole figure (IPF) maps correspond to the crystallographic orientations parallel to the observed plane (z direction), as indicated by a colored stereographic triangle, i.e., an IPF coloring key. Different boundaries were also analyzed and marked on the BC + IPF maps. Low-angle boundaries were considered when the misorientation angle was 2–10°, high-angle boundaries with the angles >10° and twin boundaries (Σ3) were defined by a 60° misorientation along the axis <111>. Phase maps show the distribution of iron and silicon carbide.

The residual stresses (RS) were measured with a portable X-ray diffraction (XRD)-based system XStress 3000 residual stress diffractometer (Stresstech Oy, Jyväskylä, Finland) with the modified Chi method [35]. Measurements from all the samples were done along the longitudinal (0) and transverse directions (90) in relation to the specimen surfaces (illustrated in Figure 1a). The residual stresses from an austenitic steel at lattice plane (311) with a diffraction angle of 148.9° were used with chromium K beta radiation. Measurements were done with 3 mm diameter collimator. In total, six different surface RS measurements were taken from one sample (Figure 1b). The measurement locations were from both surfaces A (top) and B (bottom), from 35, 50 and 65 mm from the other end in the centerline of the sheet sample. One of the locations (50 mm) was in the middle of the samples. Residual stress depth profiles were taken from as-built, HT and reference samples together with certain shot peened samples (one reference sheet, as-built and shot peened sample). At first, the surface residual stress measurement was done. Next, material was removed by electrolytical polishing with the MoviPol-5 electrolytical polisher (Struers, Ballerup, Denmark) with A2 electrolyte containing perchloric acid and ethanol. The removed layer was measured with a dial gage. This routine was repeated up to 100–120 µm depth below the surface.

In order to verify the SCC behavior, the corrosion testing was performed in two test series containing one as-built AM sample, one stress relief annealed AM sample and two reference samples. One reference sample had high Mo content and other was similar, having a weld in the middle. The samples were fixed to specially prepared 4-point bending rigs manufactured from aluminum and painted with epoxy paint. The immersion was performed at 80 °C in a thermal cabinet in a 50% magnesium chloride solution with 75% of yield strength condition with the 4-point bending rigs. The corrosion test series exposure times were 237 and 674 h for two separate test series. After the immersion tests, sample cross-sections were analyzed with an optical microscope in order to reveal possible stress corrosion cracks.

3. Results

3.1. Microstructural Observations

The microstructure of the AM 316L in as-built and after stress relief annealing conditions was studied by the optical microscope and FESEM-EBSD (Figures 3 and 4). The L-PBF selectively melts the powder layers which then solidify, bowl-shaped fusion boundary structures are visible in the as-built structure and the elongated, different-sized grains are oriented parallel to the build direction, where temperature is higher in the upper printed layer (Figure 3a). The overlapping features are resulting from the solidification of the melt pool after each laser scan. The layer thickness in the AM was 40 μm. The surface layer thickness was, however, larger, as seen in the Figure 3a, because the top layer processing parameters (lower laser power, lower speed) were different. In the stress relief annealed AM sample (Figure 3b), the melt pool boundaries were no longer as visible as in the as-built sample (Figure 3a). Figure 4 presents the FESEM images of the AsB 316L sample, showing the cellular structure within the grains. The cell size was measured to be, on average, 0.7 μm in diameter with the intercept technique. The cellular structure changed at the austenite grain boundaries and some of the grains continued across the fusion boundaries.

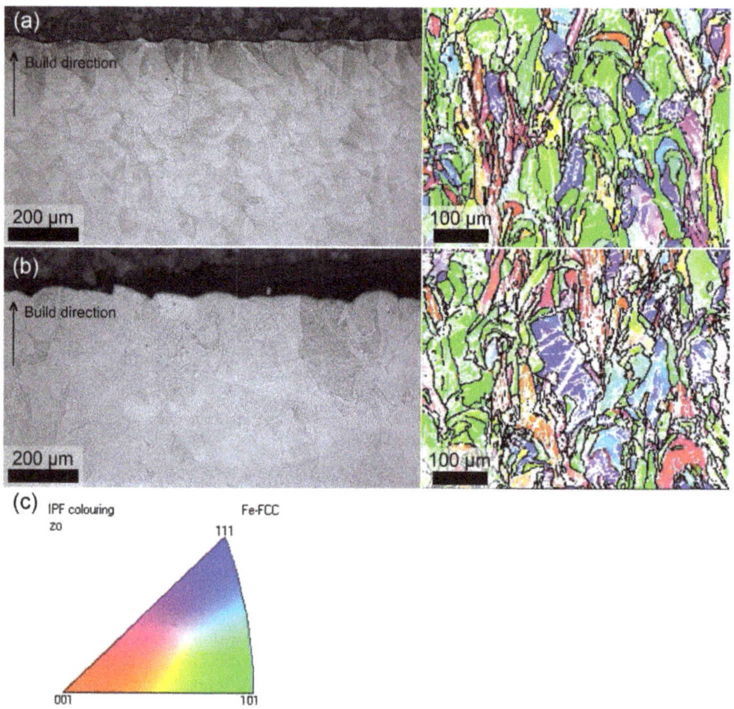

Figure 3. Optical microscope image from surface (left) and inverse pole figure (IPF) map superimposed on band contrast (BC) map from bulk (right) for (**a**) AsB AM sample and (**b**) AsB HT AM sample. The colors in the IPF maps correspond to the orientations (z direction), as indicated by (**c**) IPF coloring key for austenite (iron face-centered cubic structure, Fe-FCC). In IPF + BC maps, low-angle boundaries (2–10°) are marked by white lines and high-angle boundaries (>10°) by black lines.

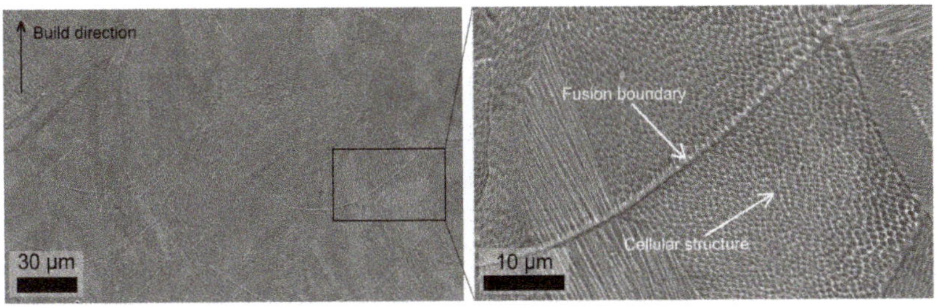

Figure 4. Scanning electron microscopy images of AsB AM 316L sample.

The microstructure of the reference sheet sample and AsB AM 316L was studied by FESEM-EBSD after the shot peening (Figure 5). The AsB HT AM 316L had similar structure as the as-built condition and it is not shown separately. In the bulk of the reference sheet sample, homogenous, round-shaped austinite grains existed, with some twin boundaries (Σ3). All samples studied had a non-indexable or weakly identified 10–20 μm thick surface layer, indicating the formation of extremely refined or heavily deformed crystals during the SP. Below this non- or weakly indexable surface layer, many strain-induced, mainly low-angle, boundaries were formed within the austenitic grains (Figure 5). In the reference sample, also a very small amount of the martensitic structure was observed in the subsurface layer. However, no martensite was observed in the AM samples. The hardness depth profiles of the cross-sectional samples presented in Figure 6 verified the increased hardness values after the SP (from 200 to 250 $HV_{0.05}$ base level to 300–400 $HV_{0.05}$) in the subsurface layer. Based on the phase maps (Figure 5), SiC residues from the SP existed on all sample surfaces.

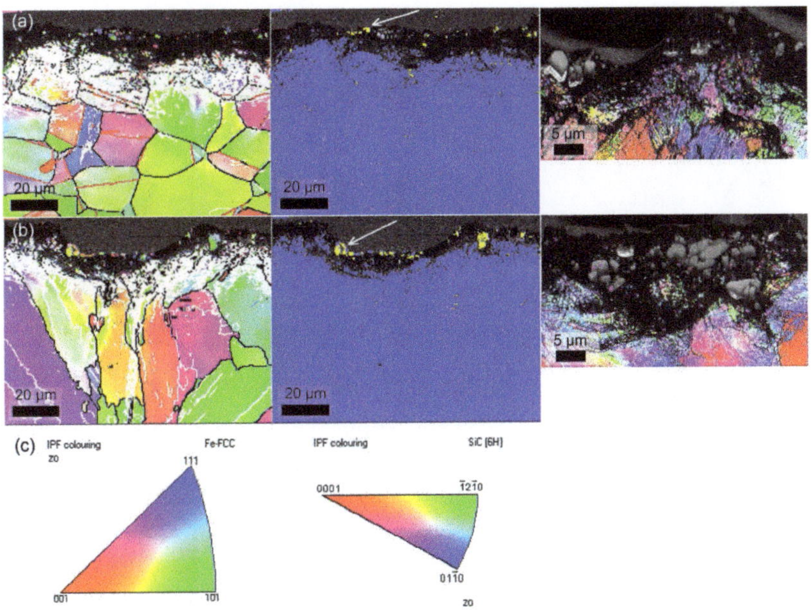

Figure 5. Inverse pole figure (IPF) map superimposed on band contrast (BC) map (with two magnifications, left side and right side) and phase map (middlemost) superimposed on BC map collected after shot peening from (**a**) reference sheet sample (SP, 3 s) and (**b**) AsB AM sample (SP, 2 s). Colors in IPF maps correspond to orientations (z direction), as indicated by (**c**) IPF coloring key for austenite (Fe-FCC) and silicon carbide (SiC). On BC + IPF maps, low-angle boundaries (2–10°) are marked by white lines, high-angle boundaries (>10°) by black lines and twin boundaries (60° <111>) by red lines. In phase maps, iron (austenite) is blue and silicon carbide is yellow, and some of the residual silicon carbides are marked by arrows.

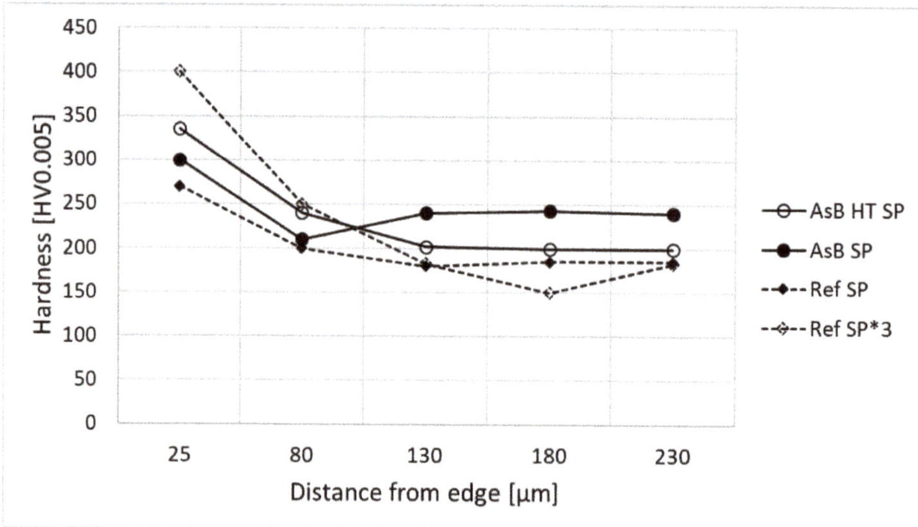

Figure 6. Hardness depth profiles of the cross-sectional samples: AsB HT SP, AsB SP, Ref SP (1 s peening) and REF SP (3 s peening).

3.2. *Residual Stresses*

Surface residual stresses were measured from all different sample categories presented in the Figure 2 process chart. It should be noted that the penetration of the X-rays exceeds only few microns into the material with the XRD method. AsB AM components (numbered as 1, 2, 3) had different residual stress states on the top and bottom surfaces, as presented in Figure 7a. Figure 7b presents the surface residual stress results from three stress relief annealed samples (top and bottom). The annealing treatment decreased the surface residual stresses compared to the AsB samples in top and bottom surfaces (Figure 7a). The effect of the SP is shown in Figure 7c. As expected, all the shot peened samples (AsB SP, HT SP and reference (Ref) sheet SP) had compressive residual stresses on both measurement directions and measured surfaces. The SP created much more compressive stresses to the AM component surface (both as-built and heat-treated) compared to the reference sheet sample (average of −555 MPa on the as-built HT SP and −502 MPa on the as-built SP versus −186 MPa on the Ref SP).

Figure 8 presents the residual stress depth profiles measured from the longitudinal (0) and the transverse (90) directions from the middle of the samples. Three RS profiles were measured for the as-built sample (the top and bottom surfaces) and the stress relief annealed sample (the bottom surface). The other RS profiles were measured for the shot peened samples: the AsB was shot peened for 1 s and the reference sheet samples were shot peened either for 1 s (Ref sheet SP1) or 3 s (Ref sheet SP3). Residual stress depth profiles measured from the shot peened samples (Figure 8) showed a typical profile of compressive residual stress reaching a maximum compression below the surface of the component and then constantly decreasing towards zero. The reference sample shot peened once (1 s) had a maximum of −620 MPa and a minimum of −130 MPa at 100 µm below the surface, and the three times (totally 3 s) shot peened sample had more intense compressive stresses under its surface, ranging from a maximum of −700 to −400 MPa. The shot peened AsB AM sample stands between the shot peened reference samples, at −600 to −280 MPa at 100 µm below the surface. The effect of the SP exceeded over the measured 100 µm layer (Figure 8) in all samples. The SP for 1 s was almost able to transform the 200 MPa tensile residual stress to a maximum of −600 MPa compression below (20 µm) the surface. In addition, as supposed, the SP produced a uniform RS state to the surface and below the surface if the different measurement directions 0 and 90 are compared.

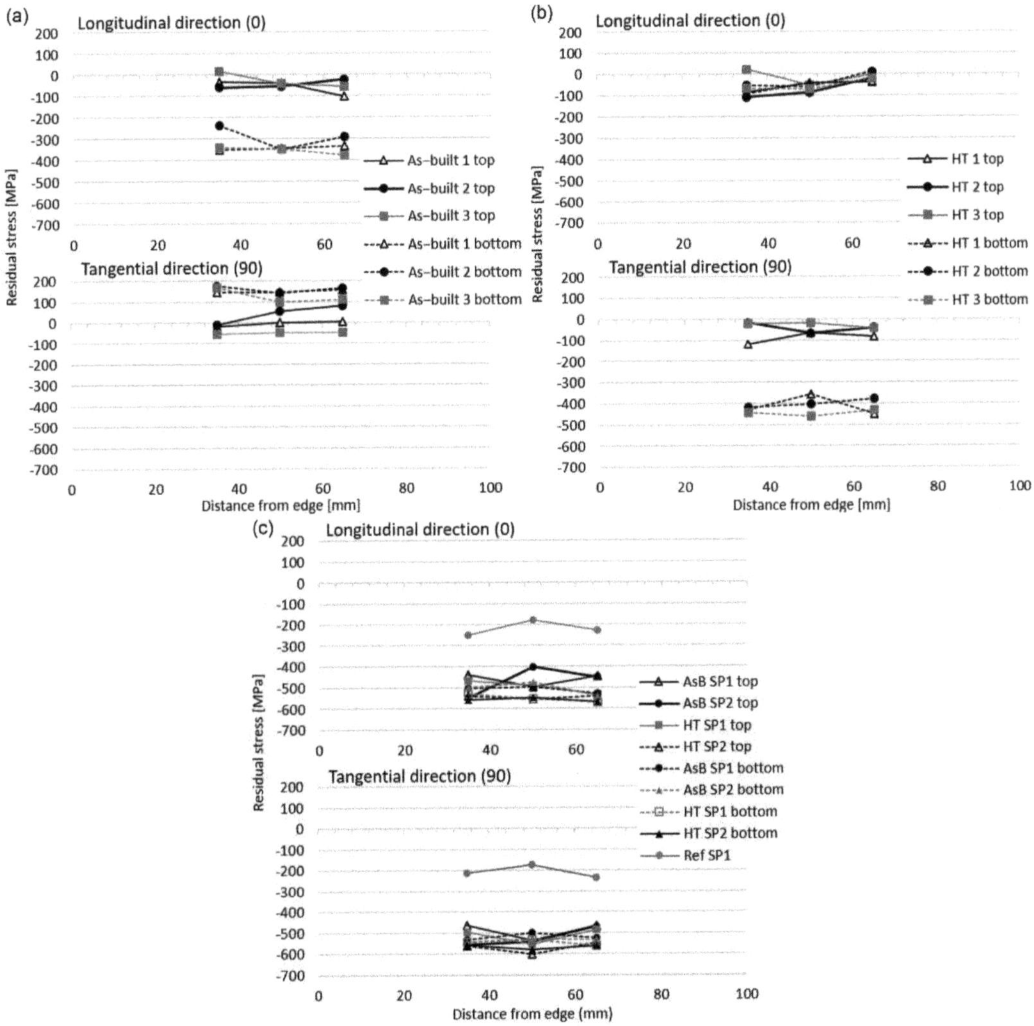

Figure 7. (**a**) Surface residual stresses of three AsB AM samples (top) and (bottom) surfaces in longitudinal and tangential directions. Typical deviation was 20–40 MPa. (**b**) Surface residual stresses of three stress relief annealed AM samples (top) and (bottom) surfaces in longitudinal and tangential directions. Typical deviation was 20–40 MPa. (**c**) Surface residual stresses of shot peened samples AsB, HT and Ref (top), and (bottom) surfaces in longitudinal and tangential directions.

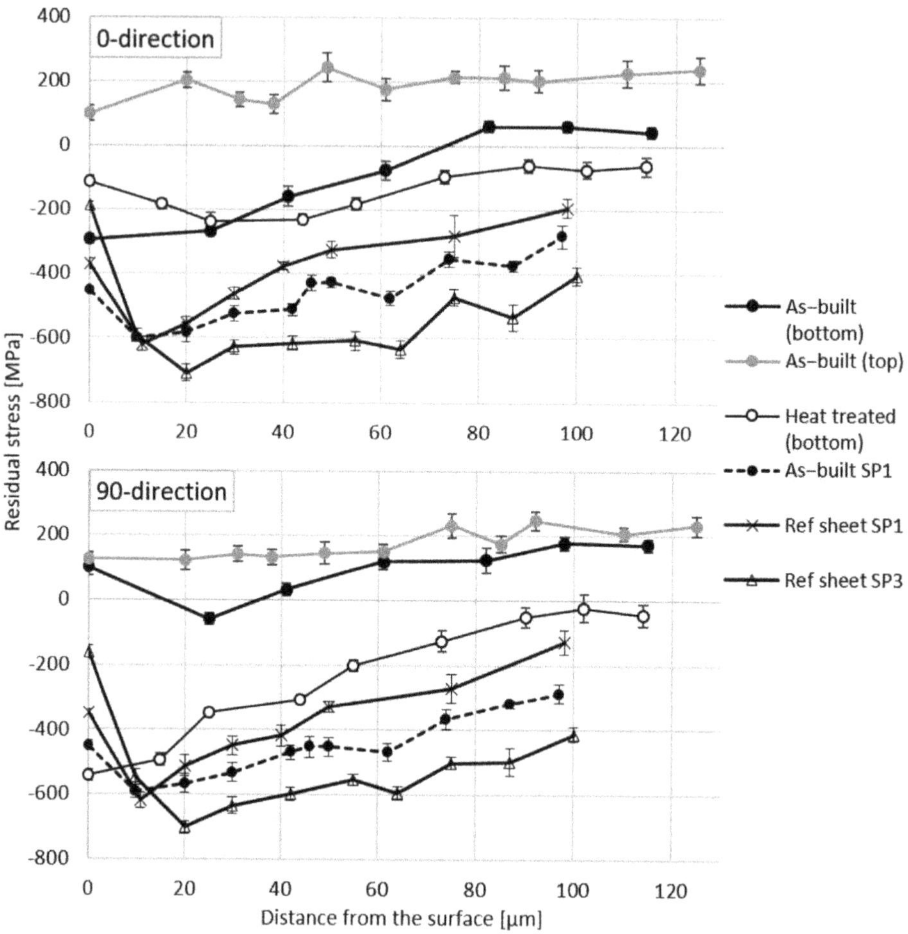

Figure 8. Residual stress depth profiles from 0 and 90 measurement direction for as-built (top and bottom surfaces) and heat-treated samples (bottom surface), and the as-built sample shot peened 3 s and Ref sheet shot peened either 1 or 3 s.

3.3. Immersion Tests in Magnesium Chloride

The possibility of SCC was studied by the 237 and the 674 h immersions in magnesium chloride at 80 °C in four-point bending rigs. The cross-sections of the samples after immersion tests were studied with an optical microscope at two different depths: surface and 0.5 mm depth from the surface. Any signs of corrosion or cracking were not detected in the AM components, AsB, AsB + HT or references (the 316L sheet and the 316L sheet with the weldment). Figure 9 shows examples of optical micrographs of (a) AsB sample and (b) AsB HT material after first corrosion test, and (c) AsB sample and (d) AsB HT material after second corrosion test with no signs of SCC.

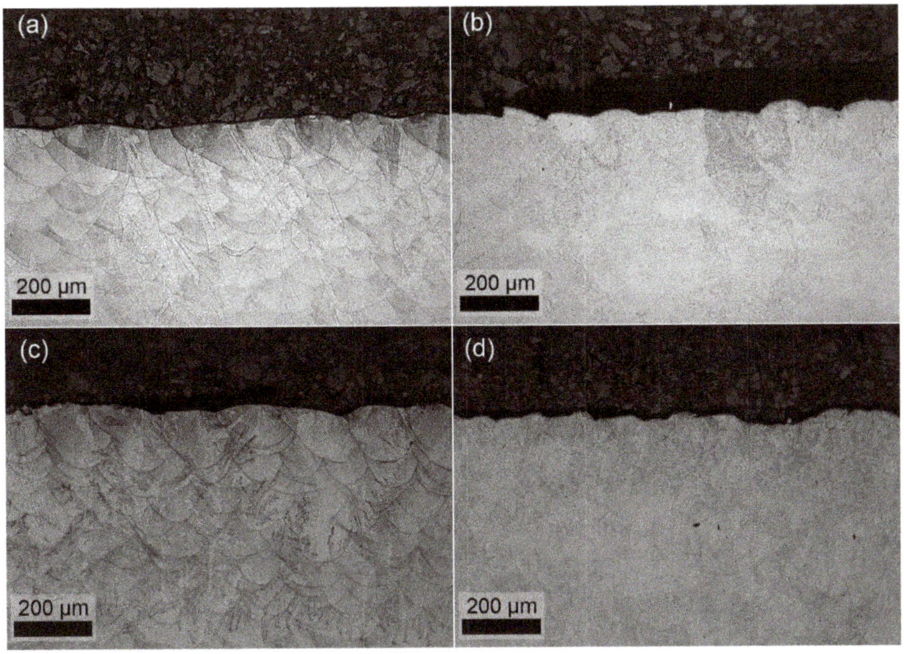

Figure 9. Optical micrographs of (**a**) AsB sample and (**b**) AsB HT after first corrosion test, and (**c**) AsB sample and (**d**) AsB HT after second corrosion test.

4. Discussion

Different post-processing treatments (as-built, stress relief annealed and shot peening) for simple AM 316L samples were investigated and their effect on the residual stress state and microstructure were determined. The microstructures of AsB AM 316L consisted of bowl-shaped fusion boundary structures and the elongated, different-sized grains were oriented parallel to the build direction. In addition, there were many low-angle boundaries within the austenitic grains, agreeing with the literature that a large fraction of low-angle grain boundaries exist in AM 316L, e.g., References [9,14]. The melt pool boundaries were no longer as visible in the stress relief annealed AM sample compared to the as-built sample. The AsB sample contained cellular structure within the grains which changed at the austenite grain boundaries, and some of the grains continued across the fusion boundaries. This kind of hierarchical and inhomogeneous microstructure is typical for AM 316L [9,14,36].

Microstructural studies showed that the bulk reference sheet sample consisted of homogenous, round-shaped austenite grains with some twin boundaries (Σ3), considered as annealing twins. All studied samples had a non-indexable or weakly identified 10–20 μm thick surface layer, indicating the formation of extremely refined or heavily deformed crystals during the SP. Below this surface layer, many strain-induced, mainly low-angle, boundaries were formed within the austenitic grains. Very small amount of martensite was observed in the subsurface layer of the reference sample. However, martensite was not observed from AM samples. The hardness depth profiles of the cross-sectional samples verified the increased hardness values after the SP in the subsurface layer, probably because of the grain refinement and formation of low-angle boundaries. Our findings here agree well with the literature. Based on the studies by Shakhova et al. [37], cold-working of conventional austenitic stainless-steel results in strain-induced boundaries, mainly associated with deformation twin boundaries, and then a martensitic transformation starts primarily at these twins. Both strain-induced martensitic transformation and deformation twins promote the grain refinement [37]. Hong et al. [14] noticed that in the AM 316L sample, no martensite and deformation twins existed after a tensile test at room temperature. They assumed that the high density of low-angle boundaries and a fine cellular microstructure hinder dislocation slip and deformation twinning, decreasing the nucleation sites for martensite. Pham et al. [38] noticed deformation-induced twins (Σ3) in AM 316L after tensile tests at room temperature but no martensite transformation.

AsB AM components had different surface residual stresses on the top and bottom surfaces. One reason for this is the different printing parameters on the top and bottom surfaces. In addition, a sample removal with a band saw from the build plate affected the bottom surface stresses. The difference in residual stress values between measurement directions in the as-built samples can be explained by the fact that they have different orientation on the build plate while printing, or that the sample had different orientation during band sawing. It should be noted that the surface residual stresses on all as-built samples (1, 2, 3) had variations between the different measurement locations on the AM strip and some variations existed also on the band saw bottom surface results. In addition, normally, the components are removed from the build plate after the heat treatment. The causes for the surface residual stress variations might also be the anisotropic heat removal and re-melting of previous layers producing thermal gradients and their effects on the produced residual stresses [16,21]. The surface residual stress results from the top surfaces of the as-built samples are similar to the results reported by Ghasri-Khouzani et al. [18] with 5 mm thick disks of 316L after being removed from the build plate. The varying residual stresses in the different surface regions may be caused by the heterogeneous microstructure and the location-dependent lattice spacings, as suggested in Li et al. [15].

The annealing treatment was noticed to decrease the surface residual stresses compared to the AsB samples in top and bottom surfaces. One reason for the different stress levels for the bottom surface is that the heat-treated samples were cut in a different orientation than the as-built samples, which was verified by visual observations of the cutting marks on the samples. When comparing the top surface results, the stress state was, somewhat unexpectedly, slightly more compressive in the annealed sample than in the as-built sample. The stress state after the heat treatment was expected to be more homogeneous and stress-free. Regardless of the original AM condition (as-built or heat-treated), the post-processing shot peening decreased the surface residual stress results approximately −400 to −600 MPa in both directions. Even if the SP was rather quick, it produced notable compressive residual stresses.

The AsB sample had almost a constant tensile RS depth profile in the top surface in both measured directions. This result is in line with previous studies [18,39]. While the metal powder is added on to the surface of the build plate, it causes the molten metal to cool rapidly and to contract. Tensile residual stresses are developing in the material as the shrinking material is constrained. Correspondingly, compressive residual stresses are developing in the build plate. These stresses are relieved when a new powder is added on the top of the first layer, while layers re-melt and new hot powder effectively anneals it [8]. This continuous process of cooling and annealing generates a residual stress state such that the bottom of the component is in compression while the top is in tension, with a neutral axis somewhere in between.

Based on the measurements here, it is obvious that a stress state with compression at the bottom and tension at the top was present in the as-built samples if they were cut from the build plate prior to the heat treatment. The stress relief annealed samples were cut from the build plate after the heat treatment. The residual stress profile from the heat-treated bottom surface showed compressive residual stresses both in longitudinal and in transverse directions. The transverse direction showed even higher compression on the surface. The band saw might also affect 90 μm below the surface from the bottom surface because the residual stress was closer to zero, compared to the bottom surface of the as-built sample, but it still compressed.

In addition, the stress corrosion cracking resistance of AM 316L was tested in magnesium chloride atmosphere. The magnesium chloride solution was chosen as it is widely known, and it has been shown to be more aggressive than sodium chloride [40]. Due to our testing equipment limitations, the standardized ASTM G36 boiling magnesium chloride test was not possible, but an 80 °C study was conducted. Any signs of corrosion or cracking were not detected in the AM components or reference samples. Therefore, it can be summarized that the AM components had similar tolerance against stress corrosion cracking to the 316L sheet and the sheet with the weldment. It is generally known that SCC might be rapid in sheet 316L, and the failure can occur, in the worst case, also in a short time [41]. Some results of the SCC behavior of austenitic stainless-steels in chloride solutions have been reported. SCC of 316 stainless-steel at 3.5 and 9.35 wt.% NaCl solutions at room temperature under a constant load was studied by Elsariti and Haftirman [41]. They [41] reported that no cracks were detected after 404 h of exposure. After 838 h in the 9.35 wt.% NaCl solution and after 1244 h in 3.5 wt.% NaCl, the samples had cracked [41]. The tests here were performed for AM 316L steel in 50 wt.% $MgCl_2$ solution at 80 °C for up to 674 h. The conditions and materials here and in Reference [41] are quite different, making a comparison hard, but possible. 316L is generally considered to be more corrosion-resistant owing to its added molybdenum content [42]. Elsariti's and Haftirman's [41] observations about SCC in 316 steel in the 9.35 wt.% NaCl solution indicated a relation between chloride concentration and SCC incubation time. Therefore, it is surprising that no SCC was observed in our study on 316L at much higher chloride concentration at an exposure time between the two observation times of 404 and 838 h of Elsariti and Haftirman [41]. The immersion in this study was carried out at 80 °C, while their [41] tests were done at room temperature. The intensity of loading was not mentioned in Reference [41], stating only that a constant load setup was used. The studied materials (316 vs 316L) were also slightly different, although 316L is generally considered more corrosion-resistant only after welding, where the low carbon content of 316L inhibits the precipitation of chromium carbides [42]. The tensioning of the samples might have been influenced and caused the stress to be under the threshold stress required to initiate SCC. In the future, immersion tests will be performed also for the shot peened AM samples to verify the possible effect of SiC residues on the corrosion behavior.

5. Conclusions

Regardless of the benefits of the metal AM, the method still requires several post-processing steps before the component is ready for use. Residual stress generation during the whole processing chain is one important aspect when considering the performance of the part in very demanding atmospheres where, e.g., stress corrosion cracking (SCC) might be possible. In this study, we investigated the effect of different post-treatments on the residual stress state and microstructure of AM 316L and its corrosion resistance in magnesium chloride atmosphere. The main findings of this study are summarized as follows:

- The surface residual stresses varied according to the location due to the complex heating–cooling cycles in AM. The range of surface residual stress variation was 150 MPa in the AsB and 90 MPa in the AsB HT sample. The band sawing produced more compressive surface residual stresses that were also dependent on the sawing orientation.
- Subsequent shot peening levelled out the residual stress differences in different measurement directions, producing compressive residual stresses of more than −400 MPa in the AM samples and the effect of the shot peening exceeded the measured 100 μm layer below the surface.
- The shot peening left residues of SiC on the surfaces, which should be taken into account during the use of the components. In addition, the grain refinement and the formation of low-angle boundaries on the sample surfaces were observed after the shot peening.

- The immersion testing for 273 and 674 h in the magnesium chloride atmosphere at 80 °C with four-point bending did not produce any signs of pitting corrosion nor cracking in the AM components or sheet 316L references.

Author Contributions: Conceptualization, M.K., S.S.-a. and M.H.; methodology, M.K.; software, S.S.-a.; validation, T.J.; formal analysis, S.S.-a.; investigation, M.K., S.S.-a., M.H. and T.J.; resources, S.S.-a.; data curation, S.S.-a.; writing—original draft preparation, M.K., S.S.-a., M.H., T.G. and T.J.; writing—review and editing, S.S.-a., M.H., M.L., T.J., T.G. and M.V.; visualization, T.J. and S.S.-a.; supervision, M.L. and M.V. All authors have read and agreed to the published version of the manuscript.

Funding: This research was partly funded by Association of Finnish Steel and Metal Producers, grant number 750/19.

Institutional Review Board Statement: Not applicable.

Informed Consent Statement: Not applicable.

Data Availability Statement: The data presented in this study are available on request from the corresponding author. The data are not publicly available due the data also forms part of an ongoing study.

Acknowledgments: 3D Formtech Oy (especially Sami Koivulahti and Toni Järvitalo) is thanked for the collaboration with sample preparation. SEM and EBSD studies made use of Tampere Microscopy Center facilities at Tampere University.

Conflicts of Interest: The authors declare no conflict of interest.

References

1. Gausemeier, J.; Echterhoff, N.; Wall, M. *Thinking Ahead the Future of Additive Manufacturing–Innovation Roadmapping of Required Advancements*; Direct Manufacturing Research Center, University of Paderborn: Paderborn, Germany, 2013; p. 110.
2. Attaran, M. The rise of 3-D printing: The advantages of additive manufacturing over traditional manufacturing. *Bus. Horiz.* **2017**, *60*, 677–688. [CrossRef]
3. Bourell, D.; Kruth, J.P.; Leu, M.; Levy, G.; Rosen, D.; Beese, A.M.; Clare, A. Materials for additive manufacturing. *CIRP Ann. Manuf. Technol.* **2017**, *66*, 659–681. [CrossRef]
4. SFS-EN ISO 17296-2: Additive manufacturing. In *General Principles*; Part 2: Overview of process categories and feedstock (ISO 17296-2:2015); Finnish Standards Association SFS: Helsinki, Finland, 2016.
5. Tucho, W.M.; Lysne, V.H.; Austbø, H.; Sjolyst, K.A.; Hansen, V. Investigation of effects of process parameters on microstructure and hardness of SLM manufactured SS316L. *J. Alloys Compd.* **2018**, *740*, 910–925. [CrossRef]
6. Thijs, L.; Verhaeghe, F.; Craeghs, T.; Humbeeck, J.; Kruth, J.P. A study of the microstructural evolution during selective laser melting of Ti-6Al-4V. *Acta Mater.* **2010**, *58*, 3303–3312. [CrossRef]
7. Vilaro, T.; Colin, C.; Bartout, J.D. As-fabricated and heat-treated microstructures of the Ti-6Al-4V alloy processed by selective laser melting. *Metall Mater. Trans. A Phys. Metall Mater. Sci.* **2011**, *42*, 3190–3199. [CrossRef]
8. Gorsse, S.; Hutchinson, C.; Gouné, M.; Banerjee, R. Additive manufacturing of metals: A brief review of the characteristic microstructures and properties of steels, Ti-6Al-4V and high-entropy alloys. *Sci. Technol. Adv. Mater.* **2017**, *18*, 584–610. [CrossRef] [PubMed]
9. Wang, Y.M.; Voisin, T.; Keown, J.T.; Ye, J.; Calta, N.P.; Li, Z.; Zheng, Z.; Zhang, Y.; Chen, W.; Roehling, T.T.; et al. Additively manufactured hierarchical stainless steels with high strength and ductility. *Nat. Mater.* **2018**, *17*, 63–70. [CrossRef] [PubMed]
10. Casati, R.; Lemke, J.; Vedani, M. Microstructure and Fracture Behavior of 316L Austenitic Stainless Steel Produced by Selective Laser Melting. *J. Mater. Sci. Technol.* **2016**, *32*, 738–744. [CrossRef]
11. Montero, M.; Nardone, S.; Hautfenne, C.; Humbeeck, J. Effect of Heat Treatment of 316L Stainless Steel Produced by Selective Laser Melting (SLM). In Proceedings of the 27th Annual International Solid Freeform Fabrication Symposium-An Additive Manufacturing Conference, Austin, TX, USA, 8–10 August 2016; pp. 558–565.
12. Herzog, D.; Seyda, V.; Wycisk, E.; Emmelmann, C. Additive manufacturing of metals. *Acta Mater.* **2016**, *117*, 371–392. [CrossRef]
13. Lou, X.; Song, M.; Wang, M.; Was, G.S.; Rebak, R.B. Radiation damage and irradiation-assisted stress corrosion cracking of additively manufactured 316L stainless steels. *J. Nucl. Mater.* **2018**, *513*, 33–44.
14. Hong, Y.; Zhou, C.; Zheng, Y.; Zhang, L.; Zheng, J.; Chen, X.; An, B. Formation of strain-induced martensite in selective laser melting austenitic stainless steel. *Mater. Sci. Eng. A* **2019**, 740–741. [CrossRef]
15. Li, C.; Liu, Z.Y.; Fang, X.Y.; Guo, Y.B. Residual Stress in Metal Additive Manufacturing. *Procedia CIRP* **2018**, *71*, 348–353. [CrossRef]
16. Robinson, J.H.; Ashton, I.R.T.; Jones, E.; Fox, P.; Sutcliffe, C. The effect of hatch angle rotation on parts manufactured using selective laser melting. *Rapid Prototyp. J.* **2018**, *25*, 289–298. [CrossRef]

17. Humbeeck, J.; Buls, S.; Thijs, L.; Kempen, K.; Vrancken, B.; Kruth, J.P. Selective Laser Melting of Crack-Free High Density M2 High Speed Steel Parts by Baseplate Preheating. *J. Manuf. Sci. Eng.* **2014**, *136*, 061026.
18. Ghasri, K.M.; Peng, H.; Rogge, R.; Attardo, R.; Ostiguy, P.; Neidig, J.; Billo, R.; Hoelzle, D.; Shankar, M.R. Experimental measurement of residual stress and distortion in additively manufactured stainless steel components with various dimensions. *Mater. Sci. Eng. A* **2017**, *707*, 689–700.
19. Kim, H.; Lin, Y.; Tseng, T.L.B. A review on quality control in additive manufacturing. *Rapid Prototyp. J.* **2018**, *24*, 645–669. [CrossRef]
20. Cheng, B.; Shrestha, S.; Chou, K. Stress and deformation evaluations of scanning strategy effect in selective laser melting. *Addit. Manuf.* **2016**, *12*, 240–251.
21. Robinson, J.; Ashton, I.; Fox, P.; Jones, E.D.; Sutcliffe, D. Determination of the effect of scan strategy on residual stress in laser powder bed fusion additive manufacturing. *Addit. Manuf.* **2018**, *23*, 13–24. [CrossRef]
22. Kalentics, N.; Boillat, E.; Peyre, P.; Ciric, K.S.; Bohojevic, N.; Loge, R.E. Tailoring residual stress profile of Selective Laser Melted parts by Laser Shock Peening. *Addit. Manuf.* **2017**, *16*, 90–97. [CrossRef]
23. Wang, Z.D.; Sun, G.F.; Lu, Y.; Chen, M.Z.; Bi, K.D.; Ni, Z.H. Microstructural characterization and mechanical behavior of ultrasonic impact peened and laser shock peened AISI 316L stainless steel. *Surf. Coat. Technol.* **2020**, *385*. [CrossRef]
24. Chen, L.; Richter, B.; Zhang, X.; Ren, X.; Pfefferkorn, F.E. Modification of surface characteristics and electrochemical corrosion behavior of laser powder bed fused stainless-steel 316L after laser polishing. *Addit. Manuf.* **2020**, *32*. [CrossRef]
25. Shiomi, M.; Osakada, K.; Nakamura, K.; Yamashita, T.; Abe, F. Residual stress within metallic model made by selective laser melting process. *CIRP Ann. Manuf. Technol.* **2004**, *53*, 195–198. [CrossRef]
26. Kong, D.; Ni, X.; Dong, C.; Zhang, L.; Man, C.; Yao, J.; Xiao, K.; Li, X. Heat treatment effect on the microstructure and corrosion behavior of 316L stainless steel fabricated by selective laser melting for proton exchange membrane fuel cells. *Electrochim. Acta* **2018**, *276*, 293–303. [CrossRef]
27. Kirka, M.M.; Lee, Y.; Greeley, D.A.; Okello, A.; Goin, M.J.; Prearce, M.T.; Dehoff, R.R. Strategy for Texture Management in Metals Additive Manufacturing. *JOM* **2017**, *69*, 523–531. [CrossRef]
28. Cruz, V.; Chao, Q.; Birbilis, N.; Fabijanic, D.; Hodgson, P.D.; Thomas, S. Electrochemical studies on the effect of residual stress on the corrosion of 316L manufactured by selective laser melting. *Corros. Sci.* **2020**, *164*, 108314. [CrossRef]
29. Sugavaneswaran, M.; Vinoth, J.A.; Barath, K.M.J.; Lokesh, K.; John, R.A. Enhancement of surface characteristics of direct metal laser sintered stainless steel 316L by shot peening. *Surf. Interface Anal.* **2018**, *12*, 31–40. [CrossRef]
30. Lodhi, M.J.K.; Deen, K.M.; Greenlee, W.M.C.; Haider, W. Additively manufactured 316L stainless steel with improved corrosion resistance and biological response for biomedical applications. *Addit. Manuf.* **2019**, *27*, 8–19. [CrossRef]
31. Lou, X.; Andersen, P.L.; Rebak, R.B. Oxide inclusions in laser additive manufactured stainless steel and their effects on impact toughness and stress corrosion cracking behaviour. *J. Nucl. Mater.* **2019**, *499*, 182–190. [CrossRef]
32. ASTM G30-97. *Standard Practice for Making and Using U-Bend Stress-Corrosion Test Specimens*; ASTM International: West Conshohocken, PA, USA, 2016; p. 7.
33. EOS Stainless Steel 316L. Material Data Sheet. Available online: https://www.eos.info/en/additive-manufacturing/3d-printing-metal/dmls-metal-materials/stainless-steel (accessed on 22 December 2020).
34. Ćurković, L.; Rede, V.; Grilec, K. Utjecaj opterećenja na izmjerenu tvrdoću silicijeve karbidne keramike. *Kem. Ind. Chem. Eng.* **2010**, *59*, 483–488.
35. SFS-EN 15305. *Non-Destructive Testing-Test Method for Residual Stress Analysis by X-ray Diffraction*; Finnish Standards Association SFS: Helsinki, Finland, 2008.
36. Valente, E.H.; Christiansen, T.L.; Somers, M.A.J. High-temperature solution nitriding and low-temperature surface nitriding of 3D printed stainless steel. In Proceedings of the European Conference on Heat Treatment (ECHT), Friedrichshafen, Germany, 12–13 April 2018.
37. Shakhova, I.; Dudko, V.; Belyakov, A.; Tsuzaki, K.; Kaibyshev, R. Effect of large strain cold rolling and subsequent annealing on microstructure and mechanical properties of an austenitic stainless steel. *Mat. Sci. Eng. A* **2012**, *545*, 176–186. [CrossRef]
38. Pham, M.; Dovgyy, B.; Hooper, P.A. Twinning induced plasticity in austenitic stainless steel 316L made by additive manufacturing. *Mat. Sci. Eng. A* **2017**, *704*, 102–111. [CrossRef]
39. Brown, D.W.; Bernardin, J.D.; Carpenter, J.S.; Clausen, B.; Spernjak, D.; Thompson, J.M. Neutron diffraction measurements of residual stress in additively manufactured stainless steel. *Mater. Sci. Eng. A* **2016**, *678*, 291–298. [CrossRef]
40. Prosek, T.; Iversen, A.; Taxén, C.; Thierry, D. Low-Temperature Stress Corrosion Cracking of Stainless Steels in the Atmosphere in the Presence of Chloride Deposits. *CORROSION* **2009**, *65*, 105–117. [CrossRef]
41. Elsariti, S.M.; Haftirman, W.H. Behaviour of stress corrosion cracking of austenitic stainless steels in sodium chloride solutions. *Proc. Eng.* **2013**, *53*, 650–654. [CrossRef]
42. Outokumpu. Stainless steels for highly corrosive environments Outokumpu Supra range datasheet. Available online: https://www.outokumpu.com/fi-fi/products/product-ranges/supra (accessed on 12 November 2020).

metals

Article

Process Window for Electron Beam Melting of 316LN Stainless Steel

Stefan Roos * and Lars-Erik Rännar

Sports Tech Research Centre, Department of Quality Management and Mechanical Engineering, Mid Sweden University, Akademigatan 1, SE-83125 Östersund, Sweden; Lars-Erik.Rannar@miun.se
* Correspondence: stefan.roos@miun.se; Tel.: +46-10-142-85-32

Abstract: Electron beam melting (EBM) is currently hampered by the low number of materials available for processing. This work presents an experimental study of process parameter development related to EBM processing of stainless steel alloy 316LN. Area energy (AE) input and beam deflection rate were varied to produce a wide array of samples in order to determine which combination of process parameters produced dense (>99%) material. Both microstructure and tensile properties were studied. The aim was to determine a process window which results in dense material. The range of AE which produced dense materials was found to be wider for 316LN than for many other reported materials, especially at lower beam deflection rates. Tensile and microstructural analysis showed that increasing the beam deflection rate, and consequently lowering the AE, resulted in material with a smaller grain size, lower ductility, lower yield strength, and a narrower window for producing material that is neither porous nor swelling.

Keywords: powder bed fusion; electron beam melting (EBM); process window; stainless steel; 316LN

Citation: Roos, S.; Rännar, L.-E. Process Window for Electron Beam Melting of 316LN Stainless Steel. *Metals* **2021**, *11*, 137. https://doi.org/10.3390/met11010137

Received: 21 December 2020
Accepted: 8 January 2021
Published: 12 January 2021

1. Introduction

One advantage that laser-based powder bed fusion (laser-PBF) methods still have over the electron beam melting (EBM) method is the sheer number of different materials available for processing. The dominant EBM machine supplier for industrial applications, Arcam (Arcam AB, Gothenburg, Sweden), currently supports six different materials, while the numerous suppliers of laser-based equipment offer more than 20 different materials. EBM process windows for pure copper [1–3], Inconel 718 [4], and Ti64 [5,6] have previously been reported. Wang et al. [7] presented work on EBM process development for 316L, and other researchers [8–11] have also presented work on EBM of 316LN, which is a nitrogen-enriched version of 316L, although without disclosing their process development methods. Stainless steel alloy 316L is a versatile engineering material that features outstanding corrosion resistance, good weldability, high strength, and high ductility [10]; it is readily available in a powder form suitable for additive manufacturing and offers a lower material cost than, for instance, the commonly used alloy Ti6Al4V. Therefore, 316L is a desirable choice for industrial adaptation for EBM. Using 316L(N) for additive manufacturing is not novel in itself; laser-PBF [12–14] and directed energy deposition [15–17] methods have been used to process 316L feedstock into solid parts. The novelty is found with the adaptation of the EBM process to the 316LN powder feedstock, and where previously mentioned research has demonstrated the feasibility of using EBM for 316LN for producing parts, this work explores the processing parameter space for which good, solid parts can be manufactured.

When using EBM to produce materials with high relative density and low top surface roughness, it is vital to find the correct process parameters [18,19]. The present work focuses on managing two of the critical process parameters: electron beam deflection rate and electron beam power input. The combination of said parameters determines the amount of beam energy applied to each part of the powder bed, often described as area energy

(AE) (J/mm^2). The parameters also determine the build temperature and solidification rate, which influence microstructure [20,21], as well as the build time, which relates to the overall productivity of the machine.

This study presents a well-defined process window for 316LN stainless steel, along with an investigation into the relationship between varying process parameters. The resulting microstructure and mechanical properties are investigated for a selection of process parameter settings. From an industrial standpoint, the important productivity aspect is presented, and one can relate the productivity to material properties.

2. Materials and Methods

2.1. Sample Manufacturing and Process Parameters

All samples were manufactured using a modified Arcam S20 EBM machine (GE Additive, Mölnlycke, Sweden) facilitating a 3 kW tungsten filament electron gun with an approximately 200 µm powder bed interaction diameter. Each build consisted of 30 pieces of 15 mm × 15 mm × 15 mm cubes arranged on a 170 mm × 170 mm stainless steel substrate (Figure 1a), with each cube assigned a unique set of process parameters. The process area energy (AE), calculated by Equation (1)

$$\mathbf{AE}\left(\frac{\mathbf{J}}{\mathbf{mm^2}}\right) = \frac{60 \times \mathbf{Beam\ Current\ (mA)}}{\mathbf{Line\ spaceing\ (mm)} \times \mathbf{Beam\ deflection\ rate}\ \left(\frac{\mathbf{mm}}{\mathbf{s}}\right)} \quad (1)$$

was varied by changing beam deflection rate and beam current while layer thickness (70 µm), line spacing (0.1 mm), process temperature (initially 810–820 °C, measured below start plate), hatch scanning strategy (snake pattern with layer-wise directional change), and focus offset (5 mA) remained constant. Equations (2) and (3) were used for calculating volumetric processing rate and volumetric energy input.

$$\mathbf{Volumetric\ energy\ input}\left(\frac{\mathbf{J}}{\mathbf{mm^3}}\right) = \frac{\mathbf{AE}\left(\frac{\mathbf{J}}{\mathbf{mm^2}}\right)}{\mathbf{layer\ thickness\ (mm)}} \quad (2)$$

$$\mathbf{Volumetric\ processing\ rate}\left(\frac{\mathbf{mm^3}}{\mathbf{s}}\right) = \frac{\mathbf{Processing\ rate}\left(\frac{\mathbf{mm^2}}{\mathbf{s}}\right)}{\mathbf{layer\ thickness\ (mm)}} \quad (3)$$

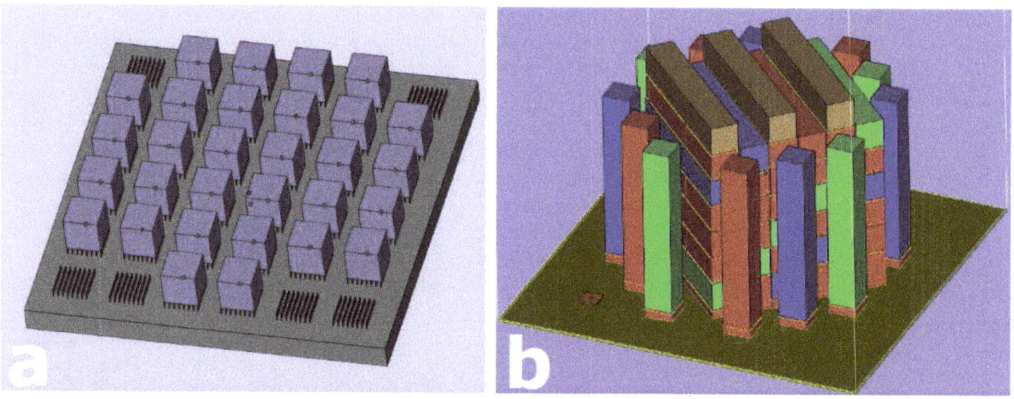

Figure 1. Layouts of builds for sample manufacturing: (**a**) cubes for evaluation of different process parameters; (**b**) layout for tensile bar build.

Automatic temperature control features were deactivated, and process temperature was regulated by altering preheat theme parameters during manufacturing. Each sample top surface was visually analyzed using light microscopy (Nikon Instruments, Tokyo, Japan) to determine porosity or swelling tendencies, and these observations were later quantified by Archimedes-type density measurements. A total of 120 sample cubes were manufactured over 4 builds. Tensile specimens were manufactured in the form of 15 mm × 15 mm × 90 mm rectangular blocks. Specimens arranged lengthwise in the xy-plane were angled at 45° with respect to the x- and y-axes in order to maintain a constant scan length between layers (Figure 1b). All specimens were built on 3.5 mm supports to facilitate ease of post-build removal and assurance of initial process stability. A total of 30 tensile bars were manufactured.

2.2. *Powder Material*

A gas-atomized 316LN precursor powder (Carpenter, Torshälla, Sweden) with 53–150 μm grain size range was used for this study. The chemical composition of the powder was 17.6 wt.% Cr, 12.3 wt.% Ni, 2.46 wt.% Mo, 1.7 wt.% Mn, 0.5 wt.% Si, 0.075 wt.% N, 0.013 wt.% C, and Fe for balance. The morphology of the powder was analyzed using scanning electron microscopy, flowability using a Hall flow funnel, and packing ratio using a scale with 10^{-4} g resolution. The powder used for this study had been recycled in the standard Arcam powder recovery system several times.

2.3. *Density Measurement*

Sample relative density was determined using Archimedes' method. From Archimedes' principle, sample density (ρ_{sample}) was calculated via Equation (4)

$$\rho_{\mathbf{sample}} = \frac{W_{\mathbf{in\ air}} \times \rho_{\mathbf{liq.}}}{W_{\mathbf{in\ air}} - W_{\mathbf{in\ liq.}}} \tag{4}$$

where $W_{\text{in air}}$ and $W_{\text{in liq.}}$ are the measured sample weight in air and distilled water, respectively, and $\rho_{\text{liq.}}$ is the density of the liquid for submersion. Three repetitive measurements were taken for each sample.

2.4. *Testing of Mechanical Properties*

Tensile specimens were machined from 15 mm × 15 mm × 90 mm rectangular blocks into circular cross-section rods of dimensions compliant with the "Small size specimen 3" of ASTM E8/E8M-13a (Ø6 mm and length 56 mm of reduced section). Tensile tests were carried out at room temperature using an Instron 5969 Universal Testing System (Instron, Norwood, MA, USA) with a clip-on extensometer. Four tensile specimens were tested for each selected process parameter setting: 1000/2.7, 3000/1.8, and 5000/1.5 (beam deflection rate (mm/s)/AE (J/mm^2)). The process parameters were selected based on AE location being in the center of their approved respective beam deflection rate in the process window. Rockwell B hardness was determined using a Mitutoyo HR200 with a type B ball indenter.

2.5. *Microstructural Characterization*

Samples were encapsulated in acrylic resin and mechanically polished using colloidal silica as final polishing medium. Microstructural features were revealed via electrochemical etching with oxalic acid (saturated solution + 10% H20) for 30–60 s at 2 V with approximately 0.03 A (no current limit). Scanning electron microscopy and energy-dispersive X-ray spectroscopy (EDS) (Tescan Maia 3, Tescan, Brno, Czech Republic) were used to analyze the elemental composition of the processed material and to produce imagery for grain size analysis. EDS data was acquired and analyzed using AZtec software (Oxford Instruments, Abingdon, Oxfordshire, UK).

Optical spectroscopy was used to analyze and retrieve images of melt tracks revealed by the etching process. Image analysis was used to quantify the melt track width.

The grain dimensions were examined using particle analysis in ImageJ software (National Institutes of Health, Bethesda, Maryland, MD, USA).

2.6. *Statistical Analysis*

A null hypothesis stating that group arithmetic means are equal was used. Alpha of 0.05 ($p \leq 0.05$) was used for rejection of the null hypothesis. Group means were compared with a *t*-test method. Since the expectation of the results was unknown, the *t*-test results were evaluated for double-sided variance. The arithmetic mean and standard deviation were used to present the test data.

3. Results and Discussion

The powder was mainly spherical in shape with some satellites attached, although some oddly shaped powder grains were observed (Figure 2). Particles deviating from spherical shape did not influence flowability (14.7 s/50 g Hall flow) or spreadability (visually determined) to the point of affecting the quality of powder distribution during processing. The bulk density of powder was approximately 4.98 g/cm^3, or 62.3%.

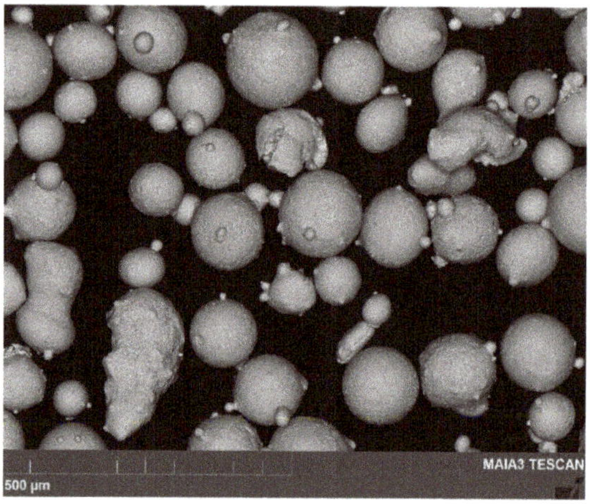

Figure 2. Scanning electron microscope image of 316LN precursor powder, showing mainly spherical powder grains with some satellites attached.

At lower beam deflection rates, there was a wide AE spectrum for producing dense material, but the acceptable AE span decreased as the deflection rate increased (Figure 3). Three process settings of varying beam deflection rates and AE were selected from the process window: samples "a" (1000 mm/s, 2.7 J/mm^2), "b" (3000 mm/s, 1.8 J/mm^2), and "c" (5000 mm/s, 1.5 J/mm^2).

Density measurements showed a well-defined point of AE below which the samples had decreasing density (Figure 4). In addition, the density of the samples increased as the AE increased, even when entering the swelling region of the process window.

The impact of different process parameters on surface topography was visually clear. While density measurement showed that lower beam deflection rates resulted in higher relative density, the visual surface quality in terms of apparent surface roughness improved with higher deflection rates (Figure 3). One could hypothesize that this was an effect of melt pool dynamics, where the lower beam deflection rate caused the material to completely solidify between beam passes. The higher beam deflection rate may have entered a dynamic regime where the adjacent melt track was at least partly in a liquid state, thus combining the two melt tracks more seamlessly. This would also explain why the melt tracks appeared narrower at lower beam deflection rates, even though the opposite situation would be a reasonable assumption given the higher area energy input. This hypothesis is supported by the cross sections of samples a, b, and c (see Figure 5), with melt track widths of average (SD): 254 μm (38.3), 442 μm (78.8), and 649 μm (27.5), respectively.

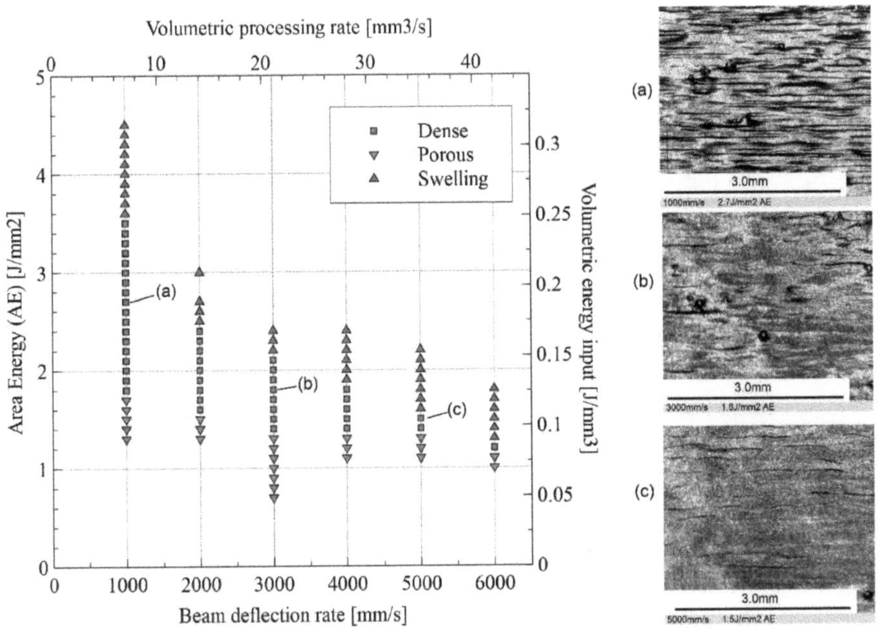

Figure 3. Process window for electron beam melting (EBM)-processed 316LN stainless steel at 70 um layer thickness and 0.1 mm line spacing; (**a**–**c**) the top surfaces of the specimens and their respective positions in the process window. For ease of comparison to other process development studies, volumetric energy (accounting for layer thickness), and volumetric processing rate have been added as secondary axes.

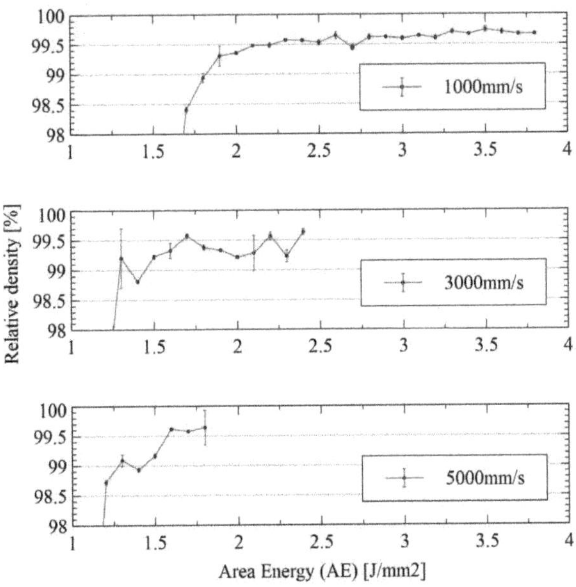

Figure 4. Relative density measurement results (arithmetic mean ± SD) for different beam deflection rates and area energies.

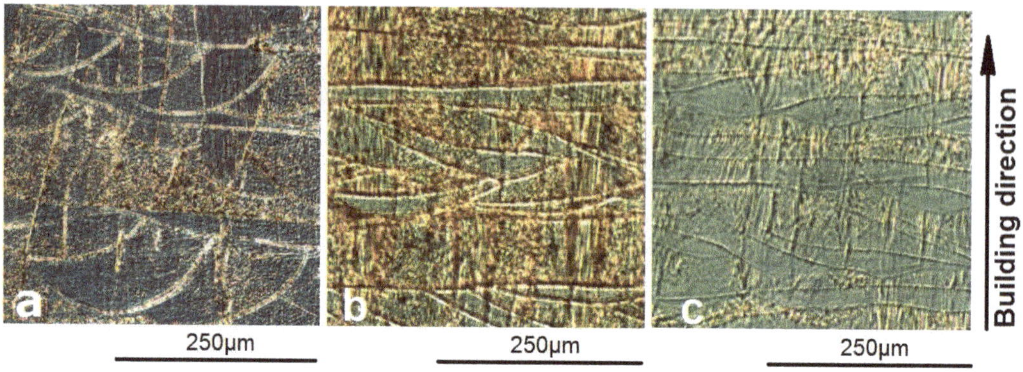

Figure 5. Melt tracks from samples with different process settings: (**a**) sample a (1000 mm/s), (**b**) sample b (3000 mm/s), (**c**) sample c (5000 mm/s).

All z-axis-oriented samples showed material properties that were inferior in all aspects to their xy-oriented counterparts. Figure 6 shows the difference in breaking behavior for z-oriented bars at 5000 mm/s, which appeared more brittle than all other samples.

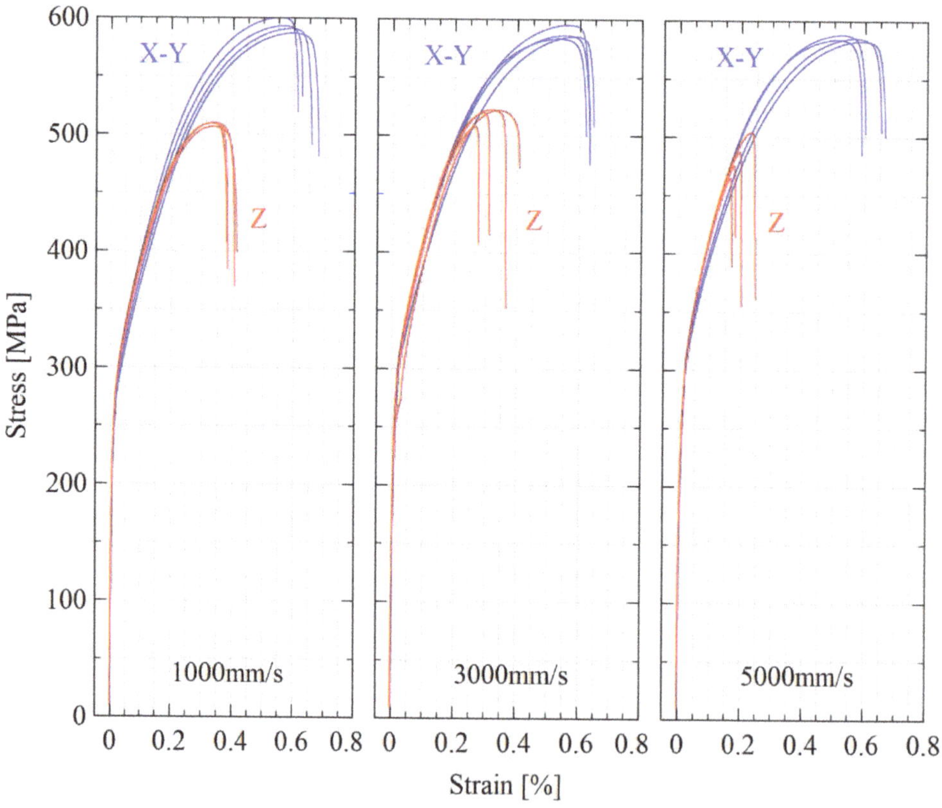

Figure 6. Results from tensile testing of EBM-processed 316LN stainless steel.

All of the xy-oriented tensile bars exceeded the specifications for both cast and wrought 316LN described by the ASTM standards ASTM240 (cast), ASTM A276 (wrought), and ASTM F3184 (316L AM) (Table 1). The z-oriented tensile bars fulfilled the yield strength requirement for all tested process settings but failed to comply with the ultimate tensile strength and elongation requirements.

Table 1. Tensile test results of 316LN stainless steel processed via EBM (arithmetic mean ± standard deviation (n-1)).

EBM-Built Samples	YS (MPa)	UTS (MPa)	Elongation at Break (%) *
xy 1000 mm/s	271.3 ± 7.8	592.7 ± 5.5	63.2 ± 2.2/72.4 ± 4.3
xy 3000 mm/s	284.2 ± 4.6	587.4 ± 5.2	63.5 ± 1.7/75.0 ± 0.0
xy 5000 mm/s	290.1 ± 5.9	585.2 ± 2.4	62.2 ± 3.9/78.1 ± 2.1
z 1000 mm/s	262.4 ± 2.5	508.9 ± 1.7	40.2 ± 2.6/39.6 ± 8.8
z 3000 mm/s	273.0 ± 3.3	518.1 ± 7.0	34.5 ± 5.5/35.9 ± 9.2
z 5000 mm/s	279.4 ± 1.6	483.2 ± 16.7	20.0 ± 4.0/16.7 ± 3.8
ASTM A743M (cast)	205	515	40 ***
ASTM A276 (wrought)	255	515 **	35 ***
ASTM F3184 (316L AM)	205	515	30

* Values show post-test measurements for gauge lengths 50/24 mm respectively; ** Annealed, thickness > 12.7 mm (1/2″); *** Gauge length = 50 mm.

Anisotropic mechanical properties in additively manufactured components are a common occurrence and are known to originate in the microstructure [17,22,23]. Thermal treatment throughout the entire build cycle due to elevated build temperatures in the EBM-build chamber have a history of producing as-built materials with good strength, high ductility, and low residual stress [24], which explains the high elongation at break recorded for the xy-oriented tensile bars. Microstructural heterogeneity, such as elongated grains, tends to yield inferior strength and ductility in the building direction, an effect which was also present in this study and explains the difference between xy-oriented tensile bars and z-oriented tensile bars. Lack of fusion between layers due to melting errors can also play a significant role if such a defect is present in the tensile bar. The fact that all tensile bars within each parameter set displayed a similar tensile behavior makes the presence of lack of fusion defects improbable.

Hardness tests showed that the Rockwell B hardness (HRB) value did not differ significantly between the samples, with HRB values of 78.3 ± 3.6, 77.5 ± 2.3, and 78.6 ± 2.0 for samples with process settings of 1000 mm/s, 3000 mm/s, and 5000 mm/s, respectively—all well below the maximum of 95 specified by ASTM A240.

Microstructural analysis revealed that the grains and sub-grains were non-equiaxed. The grains showed elongation along the build direction for all examined specimens, while elongated sub-grains were oriented in multiple directions (Figure 7).

Grain size measurements in the xy-plane showed grain areas of 1847 ± 2509, 1485 ± 1977, and 1101 ± 1483 μm^2 for samples a, b, and c, respectively. When a t-test was used to evaluate the null hypothesis of equal arithmetic means, the probability of accepting the null hypothesis was $p = 0.50$ for a and b, $p = 0.19$ for b and c, and $p = 0.14$ for a and c. Despite a trend where a faster beam deflection rate and lower AE led to finer grains, the statistical analysis did not support any conclusive statements.

EDS analysis showed that there were no statistically significant differences in elemental composition between the samples resulting from different sets of process parameters. Typical area scan location, contained within a single grain, can be seen in Figure 8.

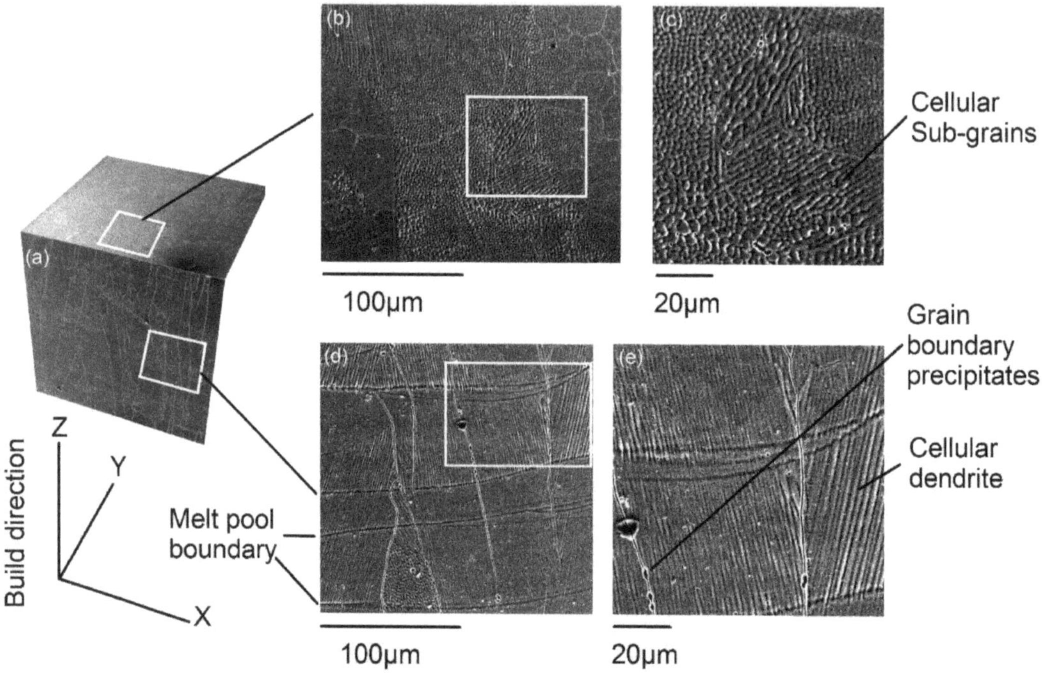

Figure 7. Microstructure of EBM-built 316LN stainless steel samples: (**a**) overview of magnified areas of interest, (**b**) low and (**c**) high magnification scanning electron microscope image of cross section perpendicular to build direction, highlighting cellular sub-grains, (**d**) low and (**e**) high magnification of cross section parallel to build direction showing boundaries of elongated grains, grain boundary precipitates, and cellular dendrites.

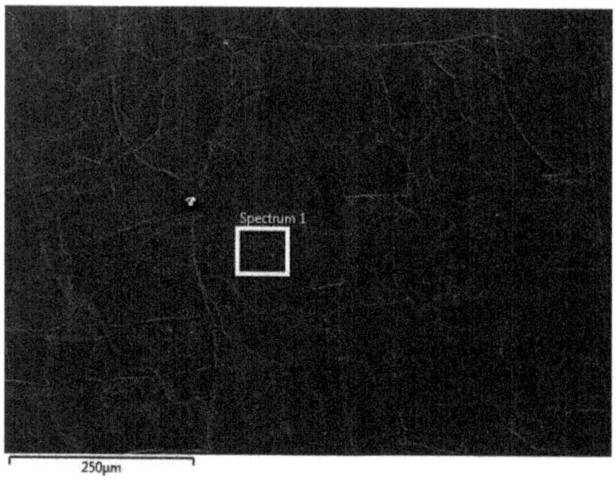

Figure 8. Representative SEM image of cross section oriented in the *xy*-plane used for EDS analysis. EDS spectrum collected from area marked "Spectrum 1" contained within a single microstructural grain.

The scan was performed in the *xy* cross sections (building direction towards reader) at approximately 50% build height.

Levels of chromium and molybdenum were consistently slightly higher than manufacturer specification in all samples. One possible explanation for this is that the EDS scanning was performed within the boundaries of individual grains, and so any migration of elements towards the grain surfaces would not be included in the results. The precipitate scan should be seen as comparative rather than absolute. The interaction volume of the EDS scanning was of the same size magnitude as the precipitates themselves, causing the surrounding material to influence readings. Grain boundary precipitate scans did show a minor elevation in chromium levels, while nickel, molybdenum, and silicon all showed slightly lower levels than bulk material scans (Table 2). No manganese was detected in the precipitates.

Table 2. EBM-manufactured 316LN stainless steel sample elemental weight percentage and (SD) acquired via EDS. Sample designation a, b, and c referenced from process window presented in Figure 3.

Element	Cr	Ni	Mo	Si	Mn	Fe
Sample a	18.18 (0.05)	12.24 (0.08)	2.82 (0.05)	0.62 (0.02)	1.6 (0.04)	64.44 (0.09)
Sample b	18.17 (0.07)	12.30 (0.10)	2.78 (0.07)	0.64 (0.02)	1.66 (0.06)	64.35 (0.11)
Sample c	18.23 (0.05)	12.35 (0.08)	2.78 (0.05)	0.67 (0.02)	1.75 (0.05)	64.12 (0.09)
Grain boundary inclusions	19.25 (0.07)	11.96 (0.10)	1.97 (0.06)	0.45 (0.02)	-	66.28 (0.11)

The statistical analysis (Table 3) revealed several cases where the null hypothesis was rejected, with a significant difference between mean values for samples a, b, and c.

Table 3. *p*-values for a true null hypothesis of equal means between groups. Alpha of 0.05 used for determination of statistical significance. Data from tests of samples with process settings a, b, and c (Figure 3). *p*-values less than 0.05 are highlighted.

Compared Data	a and b	b and c	a and c
Grain size	0.5	0.194	0.135
Melt track width	0.003	0.002	3.09×10^{-7}
Ultimate tensile strength *xy*	0.215	0.477	0.068
Ultimate tensile strength *z*	0.085	0.018	0.054
Yield strength *xy*	0.036	0.169	0.008
Yield strength *z*	0.002	0.025	8.73×10^{-5}
Elongation *xy*	0.865	0.586	0.672
Elongation *z*	0.133	0.008	0.0004
Hardness	0.582	0.439	0.822

Statistical analysis of the test data showed that changing the beam deflection rate and AE significantly changed the melt track width and yield strength in the *z*-direction for all tested process settings. Comparing properties for samples a and c meant a greater change in beam deflection rate and beam power than comparing a and b, or b and c. The comparison between a and c also revealed statistically significant differences in yield strength in the *xy*-direction and elongation in the *z*-direction, but the impact was not of sufficient magnitude to show statistical relevance for each change in process settings. The analysis did not show any statistically significant differences between samples in terms of grain size, ultimate tensile strength in the *xy*-direction, elongation in the *xy*-direction, or hardness.

Future research should include the optimization of process parameters to process the precursor powder into solid material that is compliant with the current ASTM material specifications and corresponds to or exceeds the material properties of laser-based powder bed fusion-processed material. In order to keep the layer time to a minimum, the process window width for the higher beam deflection rates should, if possible, be increased by altering the parameters that were kept fixed in this study. Developing an understanding of how the dynamic melting characteristics of high deflection rates affect the thermal history of the processed material is the key to success in producing high-quality parts in the shortest possible timeframe.

4. Conclusions

Increasing the beam deflection rate and lowering the area energy counterintuitively led to a wider melt track. It also led to lower ductility and ultimate tensile strength in the *z*-direction.

Changing the beam deflection rate and area energy produced no significant differences in material grain size, ultimate tensile strength in the *xy*-direction, elongation in the *xy*-direction, or hardness.

Changes in the beam deflection rate and area energy had a greater influence on the *z*-direction properties than the *xy*-direction properties, with the sole exception of yield strength.

Even though the elemental composition was retained, the EBM-processed 316LN in this study failed to comply with existing ASTM standards for 316 L and 316 LN material. The main issue was the *z*-direction properties.

Author Contributions: S.R. has performed the research design and experiments, also performed the writing of this paper. L.-E.R. has proofread this paper and facilitated the use of equipment and resources. All authors have read and agreed to the published version of the manuscript.

Funding: Research is conducted with financial support from the European Regional Development Fund through the Interreg Sweden–Norway program.

Institutional Review Board Statement: Not Applicable.

Informed Consent Statement: Not Applicable.

Data Availability Statement: The raw/processed data required to reproduce these findings cannot be shared at this time as the data also forms part of an ongoing study.

Conflicts of Interest: The authors declare no conflict of interest.

References

1. Frigola, P.; Harrysson, O.A.; West, H.A.; Aman, R.L.; Rigsbee, J.M.; Ramirez, D.A.; Murr, L.E.; Medina, F.; Wicker, R.B.; Rodriguez, E. Fabricating copper components with electron beam melting. *Adv. Mater. Process.* **2014**, *172*, 20–24.
2. Lodes, M.A.; Guschlbauer, R.; Körner, C. Process development for the manufacturing of 99.94% pure copper via selective electron beam melting. *Mater. Lett.* **2015**, *143*, 298–301. [CrossRef]
3. Guschlbauer, R.; Momeni, S.; Osmanlic, F.; Körner, C. Process development of 99.95% pure copper processed via selective electron beam melting and its mechanical and physical properties. *Mater. Charact.* **2018**, *143*, 163–170. [CrossRef]
4. Helmer, H.E.; Körner, C.; Singer, R.F. Additive manufacturing of nickel-based superalloy Inconel 718 by selective electron beam melting: Processing window and microstructure. *J. Mater. Res.* **2014**, *29*, 1987–1996. [CrossRef]
5. Kirchner, A.; Klöden, B.; Luft, J.; Weißgärber, T.; Kieback, B. Process window for electron beam melting of Ti-6Al-4V. *Powder Metall.* **2015**, *58*, 246–249. [CrossRef]
6. Pobel, C.R.; Osmanlic, F.; Lodes, M.A.; Wachter, S.; Körner, C. Processing windows for Ti-6Al-4V fabricated by selective electron beam melting with improved beam focus and different scan line spacings. *Rapid Prototyp. J.* **2019**, *25*, 665–671. [CrossRef]
7. Wang, C.; Tan, X.; Liu, E.; Tor, S.B. Process parameter optimization and mechanical properties for additively manufactured stainless steel 316L parts by selective electron beam melting. *Mater. Des.* **2018**, *147*, 157–166. [CrossRef]
8. Rännar, L.-E.; Koptyug, A.; Olsén, J.; Saeidi, K.; Shen, Z. Hierarchical structures of stainless steel 316L manufactured by electron beam melting. *Addit. Manuf.* **2017**, *17*, 106–112. [CrossRef]
9. Olsén, J.; Shen, Z.; Liu, L.; Koptyug, A.; Rännar, L.-E. Micro- and macro-structural heterogeneities in 316L stainless steel prepared by electron-beam melting. *Mater. Charact.* **2018**, *141*, 1–7. [CrossRef]
10. Zhong, Y.; Rännar, L.-E.; Liu, L.; Koptyug, A.; Wikman, S.; Olsen, J.; Cui, D.; Shen, Z. Additive manufacturing of 316L stainless steel by electron beam melting for nuclear fusion applications. *J. Nucl. Mater.* **2017**, *486*, 234–245. [CrossRef]
11. Zhong, Y.; Rännar, L.-E.; Wikman, S.; Koptyug, A.; Liu, L.; Cui, D.; Shen, Z. Additive manufacturing of ITER first wall panel parts by two approaches: Selective laser melting and electron beam melting. *Fusion Eng. Des.* **2017**, *116*, 24–33. [CrossRef]
12. Zhang, M.; Sun, C.-N.; Zhang, X.; Goh, P.C.; Wei, J.; Hardacre, D.; Li, H. Fatigue and fracture behaviour of laser powder bed fusion stainless steel 316L: Influence of processing parameters. *Mater. Sci. Eng. A* **2017**, *703*, 251–261. [CrossRef]

13. Li, Z.; Voisin, T.; McKeown, J.T.; Ye, J.; Braun, T.; Kamath, C.; King, W.E.; Wang, Y.M. Tensile properties, strain rate sensitivity, and activation volume of additively manufactured 316L stainless steels. *Int. J. Plast.* **2019**, *120*, 395–410. [CrossRef]
14. Choo, H.; Sham, K.-L.; Bohling, J.; Ngo, A.; Xiao, X.; Ren, Y.; Depond, P.J.; Matthews, M.J.; Garlea, E. Effect of laser power on defect, texture, and microstructure of a laser powder bed fusion processed 316L stainless steel. *Mater. Des.* **2019**, *164*, 107534. [CrossRef]
15. Feenstra, D.R.; Cruz, V.; Gao, X.; Molotnikov, A.; Birbilis, N. Effect of build height on the properties of large format stainless steel 316L fabricated via directed energy deposition. *Addit. Manuf.* **2020**, *34*, 101205. [CrossRef]
16. Azinpour, E.; Darabi, R.; de Sa, J.C.; Santos, A.; Hodek, J.; Dzugan, J. Fracture analysis in directed energy deposition (DED) manufactured 316L stainless steel using a phase-field approach. *Finite Elem. Anal. Des.* **2020**, *177*, 103417. [CrossRef]
17. Saboori, A.; Aversa, A.; Marchese, G.; Biamino, S.; Lombardi, M.; Fino, P. Microstructure and mechanical properties of AISI 316L produced by directed energy deposition-based additive manufacturing: A review. *Appl. Sci.* **2020**, *10*, 3310. [CrossRef]
18. Safdar, A.; He, H.Z.; Wei, L.Y.; Snis, A.; de Paz, L.E.C. Effect of process parameters settings and thickness on surface roughness of EBM produced Ti-6Al-4V. *Rapid Prototyp. J.* **2012**, *18*, 401–408. [CrossRef]
19. Klingvall, E.R.; Rännar, L.E.; Bäckstöm, M.; Carlsson, P. The effect of EBM process parameters upon surface roughness. *Rapid Prototyp. J.* **2016**, *22*, 495–503. [CrossRef]
20. Pushilina, N.; Syrtanov, M.; Kashkarov, E.; Murashkina, T.; Kudiiarov, V.; Laptev, R.; Lider, A.; Koptyug, A. Influence of manufacturing parameters on microstructure and hydrogen sorption behavior of electron beam melted titanium Ti-6Al-4V alloy. *Materials* **2018**, *11*, 763. [CrossRef]
21. Hrabe, N.; Quinn, T. Effects of processing on microstructure and mechanical properties of a titanium alloy (Ti–6Al–4V) fabricated using electron beam melting (EBM), Part 2: Energy input, orientation, and location. *Mater. Sci. Eng. A* **2013**, *573*, 271–277. [CrossRef]
22. Kok, Y.; Tan, X.P.; Wang, P.; Nai, M.L.S.; Loh, N.H.; Liu, E.; Tor, S.B. Anisotropy and heterogeneity of microstructure and mechanical properties in metal additive manufacturing: A critical review. *Mater. Des.* **2018**, *139*, 565–586. [CrossRef]
23. Carroll, B.E.; Palmer, T.A.; Beese, A.M. Anisotropic tensile behavior of Ti–6Al–4V components fabricated with directed energy deposition additive manufacturing. *Acta Mater.* **2015**, *87*, 309–320. [CrossRef]
24. Zhao, X.; Li, S.; Zhang, M.; Liu, Y.; Sercombe, T.B.; Wang, S.; Hao, Y.; Yang, R.; Murr, L.E. Comparison of the microstructures and mechanical properties of Ti–6Al–4V fabricated by selective laser melting and electron beam melting. *Mater. Des.* **2016**, *95*, 21–31. [CrossRef]

metals

MDPI

Review

A Review of Heat Treatments on Improving the Quality and Residual Stresses of the Ti–6Al–4V Parts Produced by Additive Manufacturing

Óscar Teixeira [1], Francisco J. G. Silva [1,*], Luís P. Ferreira [1] and Eleonora Atzeni [2]

1 ISEP—School of Engineering, Polytechnic of Porto, Rua Dr. António Bernardino de Almeida, 431, 4249-015 Porto, Portugal; oscarfrmteixeira96@gmail.com (Ó.T.); luispintoferreira@eu.ipp.pt (L.P.F.)

2 Department of Management and Production Engineering (DIGEP), Politecnico di Torino, Corso Duca degli Abruzzi, 24, 10129 Torino (TO), Italy; eleonora.atzeni@polito.it

* Correspondence: fgs@isep.ipp.pt; Tel.: +351-228-340-500

Received: 4 July 2020; Accepted: 24 July 2020; Published: 27 July 2020

Abstract: Additive manufacturing (AM) can be seen as a disruptive process that builds complex components layer upon layer. Two of its distinct technologies are Selective Laser Melting (SLM) and Electron Beam Melting (EBM), which are powder bed fusion processes that create metallic parts with the aid of a beam source. One of the most studied and manufactured superalloys in metal AM is the Ti–6Al–4V, which can be applied in the aerospace field due to its low density and high melting point, and in the biomedical area owing to its high corrosion resistance and excellent biocompatibility when in contact with tissues or bones of the human body. The research novelty of this work is the aggregation of all kinds of data from the last 20 years of investigation about Ti–6Al–4V parts manufactured via SLM and EBM, namely information related to residual stresses (RS), as well as the influence played by different heat treatments in reducing porosity and increasing mechanical properties. Throughout the report, it can be seen that the expected microstructure of the Ti–6Al–4V alloy is different in both manufacturing processes, mainly due to the distinct cooling rates. However, heat treatments can modify the microstructure, reduce RS, and increase the ductility, fatigue life, and hardness of the components. Furthermore, distinct post-treatments can induce compressive RS on the part's surface, consequently enhancing the fatigue life.

Keywords: additive manufacturing; metal powder bed fusion; selective laser melting; electron beam melting; Ti–6Al–4V; residual stresses; heat treatments; microstructure

1. Introduction

Additive manufacturing (AM), also known by many as three-dimensional printing (3DP), was born in 1981 through Hideo Kodama [1] and can be described, according to the International Organization for Standardization (ISO) 52900-15 standard [2], as a manufacturing process that creates 3D objects from scratch, typically layer after layer. As a matter of fact, this technology can be seen as a disruptive one, as it completely changed the creation of parts through the addition of material instead of the common subtraction of it, something that is characteristic of subtractive manufacturing [3]. Furthermore, it can be classified as a sustainable process as it enables the production of complex geometry components with minimal waste of material [4]. One of the process categories associated with this new manufacturing process is called Powder Bed Fusion, which agglomerates three distinct technologies, namely Selective Laser Sintering (SLS), Selective Laser Melting (SLM), and Electron Beam Melting (EBM) (Figure 1) [5]. However, if one desires to produce exclusively metal components, then it must employ the last two mentioned technologies. Thus, from now on, this paper review will solely focus on SLM and EBM, as they are seen as metal Powder Bed Fusion technologies.

In fact, SLM appeared in 1996 through Wilhelm Meiners, Konrad Wissenbach, and Andres Gasser [6], and it is a process that must be done in a closed chamber surrounded by an inert gas, such as argon or nitrogen, to avoid oxidation of the material after melting, remove spatter, metal vapor, and plasma plumes [7–10]. In addition, the parts are usually connected to a substrate plate by support structures, in order to reduce deformations when manufacturing. The operating method is done by firstly heating up the building platform in order to reduce deformations caused by temperature differences (only on some machines); then, the powder cartridge rises, thanks to an actuator, and dispenses a predefined amount of powder. After that, the leveling roller/blade creates a uniform layer of powder on the construction platform and places the substrate excess in another cartridge opposite the first one, after which a high-energy density laser melts the powder, forming a melt pool of approximately 100 μm width [11]. Finally, the construction platform lowers according to the thickness of each layer, which usually varies between 20 and 100 μm depending on the material, and the process is repeated until the final part is completed [12].

ARCAM was founded in 1997 and is the only enterprise that manufactures EBM machines, as they own the patent [13]. This technology uses accelerated electrons in the order of 0.1 to 0.4 times the speed of light [14] as a beam source, and the process is done in a vacuum chamber [15]. Furthermore, the additive manufacturing steps are the same as the SLM technology; however, the temperature of the build chamber is much higher (500–800 °C) [16–19]. Hence, one should expect fewer residual stresses (RS) because of the lower temperature gradient during manufacturing (Tmelt–Tambient). In addition, the typical layer thickness of the parts is between 50 and 200 μm, due to the deeper penetration and higher energy of the beam, making the manufacturing process quicker than SLM [16].

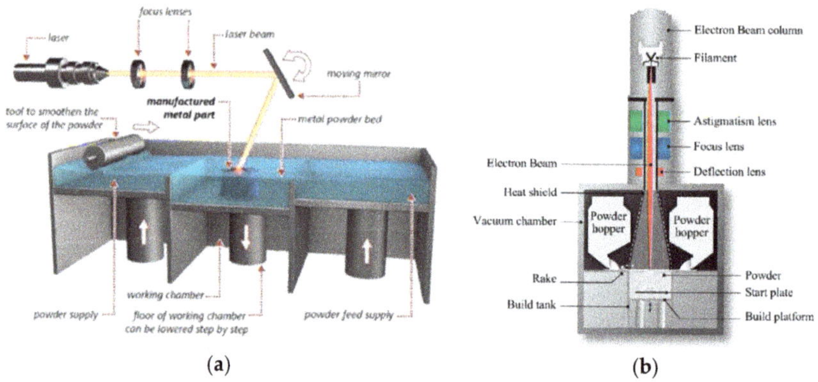

Figure 1. Main components of (**a**) Selective Laser Melting (SLM) machine, (**b**) Electron Beam Melting (EBM) machine. Adapted from [20], under CC BY 4.0. Original sources: (**a**) EMPA, (**b**) ARCAM.

Titanium alloys have been receiving special attention from the aerospace industry, thanks to the low density and the high melting point of this material (around 1650 °C) [21]. In other words, several companies turn to titanium when they need a light component that is capable of safely supporting high loads and/or when working temperatures are high [22]. In addition, it has also a high corrosion resistance [23–25], and it is often used in medical implants due to its excellent biocompatibility when in contact with tissues or bones of the human body [26]. Note that this material is quite difficult and expensive to machine, so its use and diffusion in AM is of great interest [27,28]. In fact, the titanium alloy most studied by the scientific community and used to manufacture final components is Ti–6Al–4V (UNS R56400) [29]. The crystalline structure of this alloy is composed of a compact hexagonal structure (phase α) and body-centered cubic (phase β). The existence of aluminum and vanadium causes the α and β phases to stabilize, respectively, maintaining the dual phase at room temperature [22]. In general, the microstructure, when depositing material, presents a martensitic phase (α′) due to the rapid cooling of the manufacturing process. In addition to the quick cooling, the α' phase will only form if the build temperature is kept below the martensite start temperature (Ms) [30]. The aforementioned temperature varies from 575 °C up to 800 °C [31–33] as the impure elements along with the initial microstructure and composition homogeneity affect the Ms temperature [34]. In Figure 2, it is possible to analyze how the phase transformation is affected by the distinct cooling rates, i.e., in the region of 525–410 °C/s the α' phase will be formed; however, if the cooling rate is kept under 20 °C/s, it will lead to the formation of the phase α.

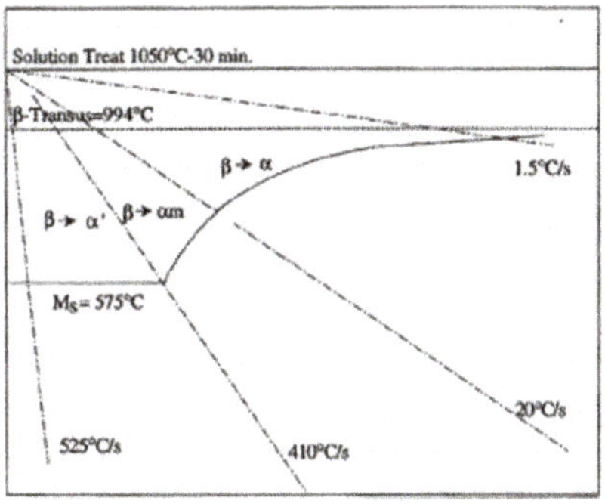

Figure 2. Diagram that explains how the cooling rate affects the phase transformation. Reproduced from [35], under CC BY 4.0.

In laser-based additive manufacturing processes, RS are induced by thermal principles, as the gradient temperature is high because of the heating and thermal expansion upon the deposition of a new layer, and its subsequent cooling and contraction [36]. Therefore, production errors in the components might occur due to geometrical distortion. Thus, it is mandatory to understand the expected orientation and magnitude of the RS in order to do accurate predictions of the final part properties. Moreover, high magnitude (100–500 MPa) RS are more prone to occur in the SLM rather than in the EBM process [37].

As a matter of fact, residual stresses can be divided into three different types, Type I, II, and III, depending on the operational length scale, as shown in Figure 3. Type I are macroscopic stresses that act on the scale of component geometry, Type II, often called intergranular stresses, are microstresses acting at the individual grain scale, and Type III act at the atomic scale [38]. Nowadays, multiple techniques are employed to measure residual stresses, and they can be divided into two major groups: destructive, and non-destructive testing. On one hand, if one seeks to find residual stresses located on the surface level in a non-destructive way, X-ray Diffraction (XRD) is one of the most suitable techniques for that effect due to its great accuracy (±20 MPa) and the large variety of equipment available [39]. On the other hand, if the goal is to investigate internal residual stresses in a destructive way, one can use the stripping method, crack compliance method, or the contour method. Furthermore, there is also an economic semi-destructive way to discover surface residual stresses called the hole-drilling strain gage method [40]. It is considered a semi-destructive method as it does not affect, in a significant way, the structural integrity of the part. Currently, an ISO standard does not exist; however, according to American Society for Testing and Materials (ASTM) E837-08 [41] standard, the aforementioned method can be described as a technique that computes residual stresses near the surface of a material by sticking a strain rosette to the part's surface and then drilling a hole at the center of the rosette, which later determines the relieved strains.

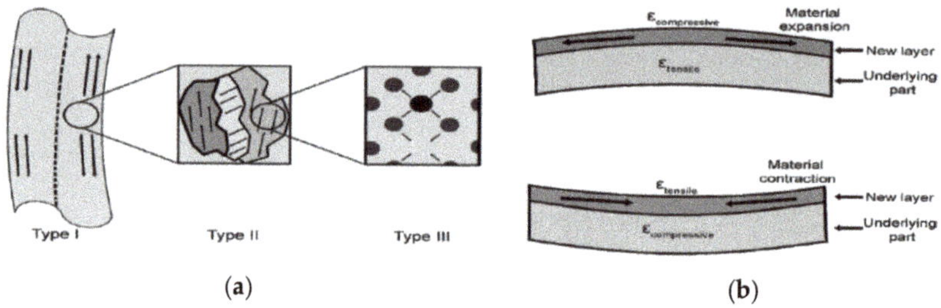

Figure 3. (**a**) Residual stresses classification and (**b**) plastic deformation development. Reproduced from [36], with permission of Elsevier, 1996.

The research novelty of this work is the aggregation of all kinds of information from the last 20 years of research about the Ti–6Al–4V alloy in a structured way, namely information related to residual stresses, as well as the influence played by different heat treatments in reducing porosity, increasing mechanical properties, and the stress relief of additive manufactured components. The authors strongly believe that the compilation of the aforementioned data is essential because it allows understanding the research evolution of this titanium alloy, and it provides a good starting point for those who want to learn about additive manufacturing. Furthermore, the reading of this review paper can also be useful to understand some key aspects of the Ti–6Al–4V alloy, thus improving the quality and performance of the manufactured parts.

2. Methodology

The methodology followed in this article started with the selection of the theme, which is heat treatments and residual stresses on metal powder bed fusion Ti–6Al–4V parts. In this way, the authors tried to analyze the amount of information available on the ScienceDirect platform in order to get a general idea of the number of articles published on this topic. When performing an advanced search by title, abstract, and keywords on the topics "Additive Manufacturing + Ti-6Al-4V", 537 articles were found, of which only 11 were review articles, as shown in Figure 4.

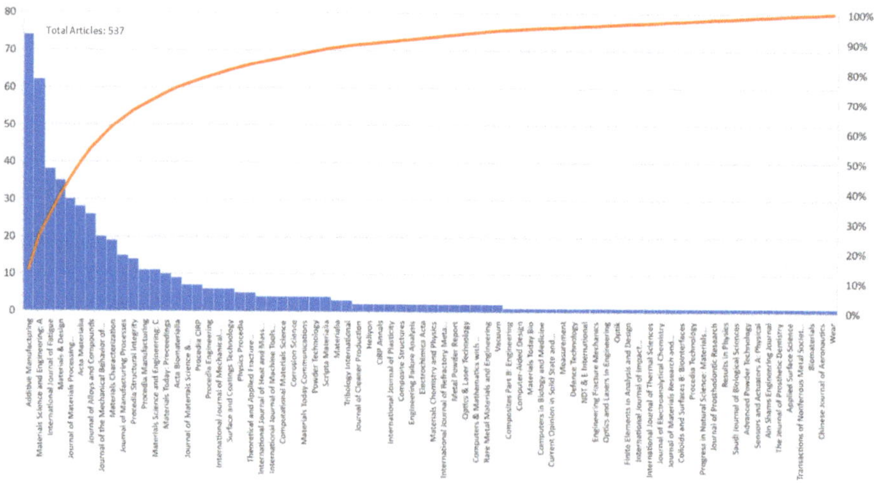

Figure 4. Pareto's chart of the research "Additive Manufacturing + Ti–6Al–4V" on ScienceDirect.

Then, after a thorough research for the topics "Heat Treatments + Selective Laser Melting + Ti-6Al-4V", as shown in Figure 5a, it was concluded that the number of results had decreased considerably, with a total of 61 research articles. After that, the authors searched for "Residual Stresses + Selective Laser Melting + Ti–6Al–4V", as shown in Figure 5b, and they found that there were only 31 research articles. Lastly, the authors removed the word "Selective Laser Melting" on the last two searches and added the word "Electron Beam Melting", and they found out 18 articles about heat treatments and 10 about residual stresses, as shown in Figure 6. Note that if the research had not been done in an advanced way, the results would have been much higher; however, more than 50% of the results would not be relevant for this review article.

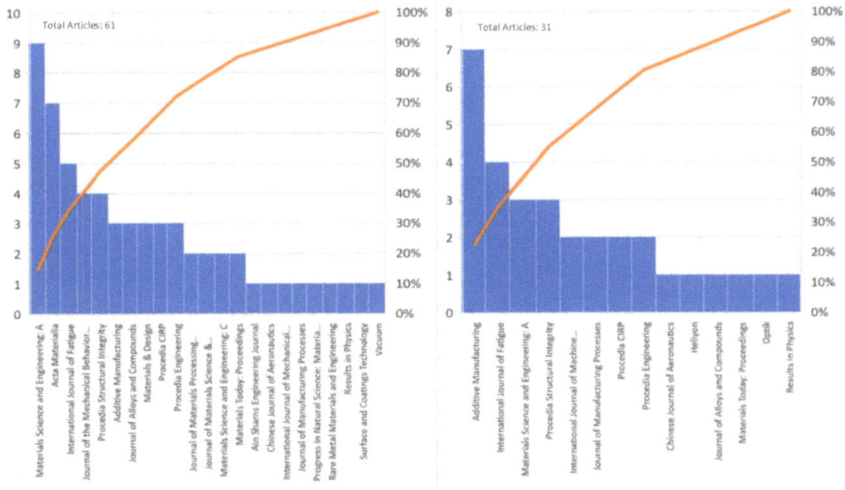

(**a**) (**b**)

Figure 5. Pareto's charts of the research, (**a**) "Heat Treatments + Selective Laser Melting + Ti–6Al–4V", and (**b**) "Residual Stresses + Selective Laser Melting + Ti–6Al–4V", on ScienceDirect.

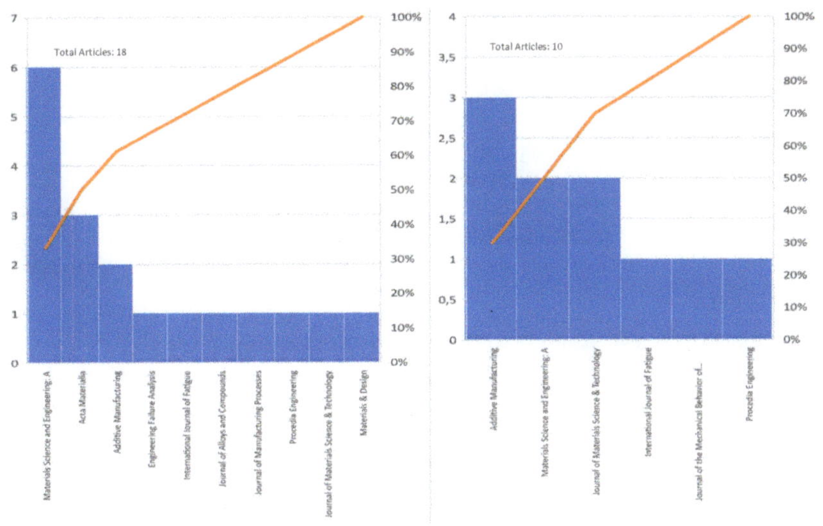

(**a**) (**b**)

Figure 6. Pareto's charts of the research, (**a**) "Heat Treatments + Electron Beam Melting + Ti–6Al–4V", and (**b**) "Residual Stresses + Electron Beam Melting + Ti–6Al–4V", on ScienceDirect.

3. Literature Review

Before heading to the first subsection of the literature review, the authors thought that the SLMed and EMBed as-built microstructures of the titanium alloy should be compared as they are heavily dependent on the manufacturing process. Furthermore, the microstructure also plays a crucial role in governing the material properties of the components. Therefore, several investigators studied the as-built microstructure of SLMed titanium samples [42–46]. Thijs et al. [42] reported that all the as-built specimens produced via SLM showed a microstructure with fine acicular α′ precipitates in the columnar original β grains, which was due to the rapid cooling of the manufacturing process, as shown in Figure 7a. Similarly, Facchini et al. [43] and Gong et al. [44] found the same microstructure results as the aforementioned study, while Benedetti et al. [45] reported that the as-built titanium specimens had a very fine acicular α′ microstructure. On the contrary, Xu et al. [46] managed to accomplish an ultrafine lamellar α + β structure of SLMed titanium as-built specimens without having notable grain coarsening. The authors used the following process variables: layer thickness of 60 μm, focal offset distance of 2 mm, and energy density of 50.62 J/mm^3, they and reported that a careful selection of the printing parameters was the key for such microstructure.

Several other authors studied the microstructure of as-built titanium specimens produced via EBM [47–51]. Xu et al. [47] reported that an α + β lamellar microstructure, as shown in Figure 7b, was achieved for the following processing parameters: accelerated voltage of 60 kV, beam current of 10 mA, filament current of 432 mA, focus current of 670 mA, gun vacuum pressure of 3.4×10^{-3} Pa, chamber vacuum pressure of 7×10^{-2} Pa, and work distance of 150 mm. Likewise, Galarraga et al. [48], Zhao et al. [49], Chern et al. [50], and Murr et al. [51] found an α + β lamellar microstructure in the EBMed as-built specimens.

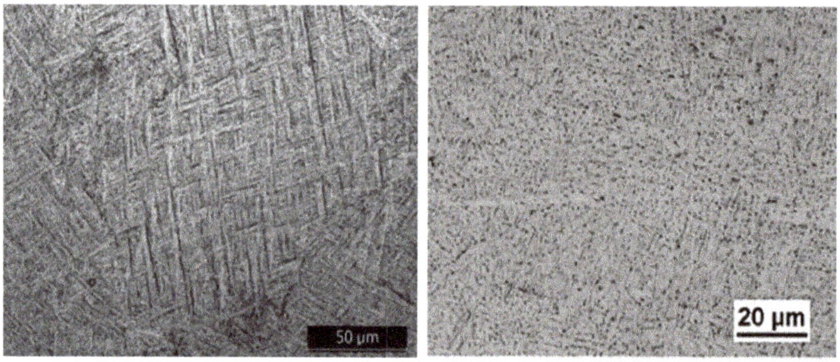

(a) (b)

Figure 7. Characteristic microstructure of (**a**) SLM process (fine acicular α′ in columnar original β grains), and (**b**) EBM process (α + β lamellar). Reproduced from [45,48] respectively, with permission of Elsevier 2017, 2016.

Furthermore, it is also relevant to understand the influence played by distinct additive manufacturing processes on the mechanical properties, profile surface roughness (Ra), and porosity of Ti–6Al–4V components. Thus, in Table 1, it is possible to analyze the distinct as-built mechanical properties of titanium samples. From the aforementioned table, it is worth noting that the SLMed samples, when compared to the EBMed ones, exhibit a 20% to 30% increase of the yield strength (YS) and ultimate tensile strength (UTS); however, the ductility is much lower due to the inherent rapid cooling of the SLM process which promotes an α′ martensitic microstructure within long columnar prior β grains. Regarding the Ra, it is known that different scan speeds, powder sizes, and layer thicknesses of the EBM and SLM manufacturing processes promote distinct surface roughness [35]. Edwards et al. [52] stated that the titanium specimens produced via SLM had an Ra between 32 and 38.5 μm, depending on the build orientation. Moreover, Chan et al. [53] compared the surface roughness of Ti–6Al–4V samples manufactured via SLM and EBM and reported that the latter manufacturing process revealed an Ra of 131 μm, while the former presented 38.5 μm. The authors associated the high surface roughness of the EBMed sample to the presence of unfilled gaps and unmelted powder attached to the surface. Fousová et al. [54] studied the effect of internal defects and surface roughness on the mechanical properties of Ti–6Al–4V samples produced via SLM and EBM. In order to achieve a similar microstructure as the EBM as-built specimens, the SLM ones underwent a heat treatment at 820 °C for 5.5 h. After analyzing the microstructure and conducting a uniaxial tension test and a fatigue test, the authors reported that both EBMed and SLMed specimens exhibited a two-phase lamellar microstructure; however, internal defects were distinct due the manufacturing process. EBMed samples showed spherical pores resulting from gas entrapment, while SLMed ones revealed insufficient melting defects. Despite that, the researchers concluded that the static properties in the tension of both samples reached comparable values, but the fatigue test demonstrated that the SLMed specimens exhibited greater fatigue strength (220 ± 24 MPa) than the EBMed ones (115 ± 13 MPa) due to the lower surface roughness. Besides, Qiu et al. [55] studied the effect of laser power and layer thickness on surface roughness and porosity level of SLMed titanium as-built samples. The researchers reported that a low layer thickness combined with a high laser power leads to the production of specimens with a porosity level inferior to 1 vol%. In fact, when it comes to part porosity, several authors [44,49,56] reported that in the SLM process, one should expect porosities between 0.1 and 0.5 vol%, while others [44,57] stated that components produced through EBM present an inherent porosity between 0.1 and 0.3 vol%. Recently, Ning et al. [58] created an analytical model based on the relationship between porosity evolution and the thermal behavior of the material, which predicted part porosity in metal powder bed additive manufacturing without resorting to any iteration-based simulations or Finite Element Modeling (FEM). For that, the authors followed the algorithm that can be seen in Figure 8. The molten pool dimensions were computed through the comparison of the material's melting temperature with the expected temperatures using an analytical thermal model. The porosity evaluation was computed by subtracting the volume fraction of the powder bed void from the post-processed porosity of the part. The volume fraction of the powder was investigated by image analysis and through an advancing front approach. Then, after determining the analytical results of the expected porosity, the authors compared the values with additively manufactured Ti–6Al–4V specimens from the literature and concluded that their model could predict porosity with a high amount of precision. Furthermore, the authors stated that the computational time of the model was short, allowing a quick and precise prediction for distinct processing parameters.

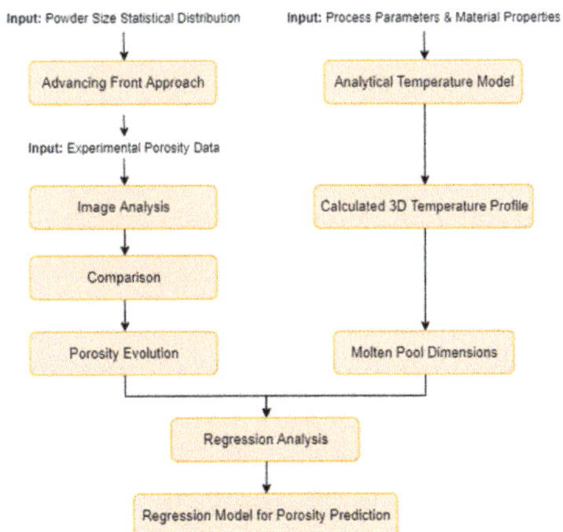

Figure 8. Schematic illustration of the algorithm used in the porosity prediction model. Adapted from [58], with permission of Elsevier, 2020.

In another report, Zhang et al. [59] studied and compared the wear resistance of as-built SLM, as-built EBM, and forged Ti–6Al–4V samples. Then, the authors performed friction and wear tests to the three types of samples and concluded that when compared to the forged specimens, the additively manufactured ones exhibited a better wear resistance owing to their weaker delamination and higher hardness. Moreover, the SLMed samples showed a higher wear rate than the EBMed ones due to the higher number of horizontal cracks of the former, as shown in Figure 9.

Table 1. Mechanical properties of as-built Ti–6Al–4V specimens in tensile testing. UTS: ultimate tensile strength, YS: yield strength.

Manufacturing Process	YS [MPa]	UTS [MPa]	ε [%]	Ref.
EBM	881 ± 12.5	978.5 ± 11.5	10.7 ± 1.5	[60]
EBM	851.8 ± 5.8	964.5 ± 0.3	16.3 ± 0.8	[61]
SLM	1056 ± 29	1351 ± 34	5.5 ± 0.8	[62]
SLM	1143 ± 30	1219 ± 20	4.9 ± 0.6	[63]

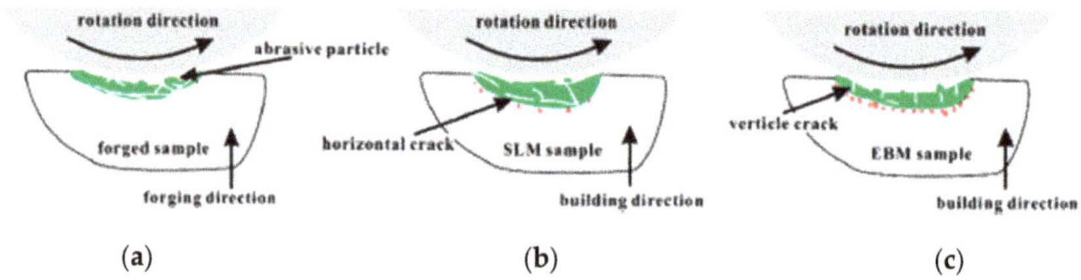

Figure 9. Wear mechanisms in the Ti–6Al–4V samples built by (**a**) forging, (**b**) SLM, and (**c**) EBM. Reproduced from [59], under CC BY 4.0.

3.1. Heat Treatments

Heat treatments are known for reducing residual stresses, improving mechanical properties, and changing the microstructure of titanium components. Balyakin et al. [64] studied the heat treatment effect on the surface roughness of titanium specimens manufactured via SLM. The researchers conducted distinct heat treatments followed by mechanical brushing and chemical polishing and reported that after heating the samples to 950 °C for 30 min followed by oil quenching and subsequent aging at 650 °C for 120 min, the specimens exhibited the worst surface roughness (Ra = 4.41 μm) measured just before undergoing mechanical polishing and chemical polishing, whilst the best final surface roughness (Ra = 0.27 μm) was achieved through the following preliminary heat treatments: annealing at 650 °C for 120 min followed by furnace cooling (FC) plus heating up to 950 °C for 30 min followed by water quenching (WQ). Vayssette et al. [65] investigated the roughness influence on multiaxial high-cycle fatigue of titanium samples manufactured via SLM and EBM. Firstly, when producing the specimens, the authors used processing parameters that minimized the RS formation due to high-thermal gradients. Secondly, the samples were post-treated with a stress relief heat treatment followed by Hot Isostatic Pressing (HIP), which aimed to reduce each part's porosity. After that, the researchers analyzed the microstructure and reported that both samples exhibited a lamellar microstructure with a mean 2 μm lamella size. Then, the authors measured the Ra values of both samples before and after chemical polishing, which for the SLM samples were 18.9 μm and 11.2 μm, and for the EBM, they were 38.9 μm and 17.7 μm, respectively. Finally, the researchers conducted fatigue tests and reported that despite the roughness improvement induced by the chemical etching, the fatigue properties did not improve significantly. Similarly, Yuan et al. [66] investigated the heat treatment effect on the compressive fatigue properties of titanium samples produced via EBM. For that, after manufacturing the samples, the authors conducted an annealing treatment with three different temperatures (750 °C, 850 °C, and 950 °C) for 1.5 h followed by FC to room temperature. Then, they compared the as-built microstructure to the post-treated one and concluded that the heat treatment promoted the decomposition of the acicular α' martensite into an α + β phase. After conducting fatigue tests, the researchers concluded that when annealing at 950 °C, near the phase boundary, the width of the α lamellae coarsens up to 20 μm and promotes a ductility improvement of the struts, an increase of the fatigue strength, and the fatigue endurance ratio raises to nearly 0.6, which is almost three times more than that reported in another work [67].

Several authors [68–73] studied the fatigue properties of the titanium alloy manufactured via metal powder bed fusion. Chastand et al. [68] investigated the influence of different manufacturing processes, build orientation, surface roughness, and HIP treatment on the fatigue life of titanium samples. To do that, the researchers manufactured several specimens in SLM and EBM machines with different build orientations, namely XY and Z. Then, all the SLMed specimens underwent a stress relief treatment at 640 °C for 4 h, while others undertook a stress relief + HIP treatment at 920 °C under 1020 bar for 2 h. After that, the authors conducted the same HIP treatment to 12 EBMed specimens and then machined and polished the surface of several SLMed and EBMed samples. After conducting the fatigue tests, Chastand et al. [68] reported that the manufacturing processes (SLM and EBM) effect is negligible on the fatigue properties, as a similar type of defects was found in both cases. Moreover, machining and polishing improved the fatigue properties of the titanium alloy because it lowered the surface roughness and removed superficial defects. Furthermore, the HIP treatment led to the size decrease of the internal defects, hence improving the fatigue properties. To conclude, the authors also reported that the effect of the build direction highly depends on the presence of unmelted zones. In a similar study, Wu et al. [71] investigated the porosity and HIP treatment effect on the fatigue endurance ratio of SLMed titanium samples. For that, the researchers purposely manufactured cellular structures with three different porosities (33 vol%, 50 vol%, and 84 vol%) and then subjected some of them to an HIP treatment at 1000 °C and a pressure of 150 MPa for 1 h. After studying the microstructure, the authors reported that the heat treatment promoted the transformation of the martensitic microstructure into a lamellar α + β phase. Then, the researchers conducted fatigue tests and concluded that the porosity increase in the as-built samples led to a decrease in the fatigue endurance ratio, as shown in Figure 10, due to the low ductility of α′. However, in the HIPed samples, as the porosity increased from 33 vol% to 50 vol%, the fatigue endurance ratio at 10^6 cycles increased from 0.5 up to nearly 0.55, Figure 10. Moreover, when compared to the as-built specimens, the HIPed ones exhibited far better fatigue endurance ratios as the porosity increased. Similarly, Vispoli et al. [72] investigated the surface roughness effect on the fatigue behaviour of Ti–6Al–4V samples produced via SLM. The authors reported that the fatigue resistance of the as-built specimens, when compared to the machined ones, evidenced a drastic reduction due to the elevated roughness.

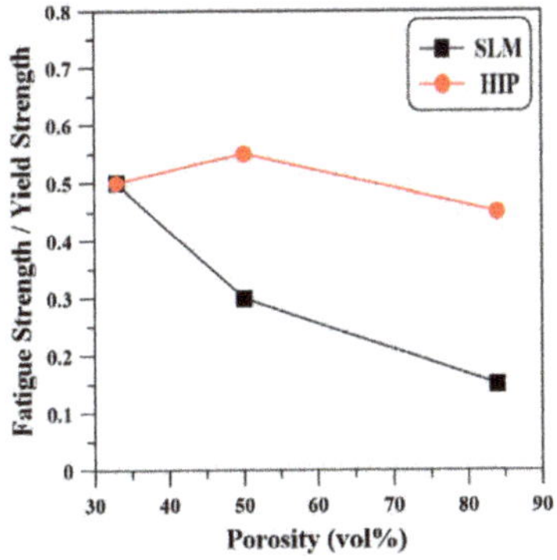

Figure 10. Porosity influence on the fatigue endurance ratios at 10^6 cycles of as-built and specimens subject to Hot Isostatic Pressing (HIP). Reproduced from [71], with permission of Elsevier, 2020.

Likewise, Yu et al. [73] studied the influence of distinct heat treatments and surface finishing processes on the fatigue performance of titanium samples produced via SLM. The authors employed four different finishing processes, namely turning, 400 mesh sandpaper grinding, 400 mesh sandpaper grinding + sandblasting, and polishing. Furthermore, the researchers also conducted three distinct heat treatments, which can be seen in Table 2. After the fatigue tests, the authors concluded that all the surface-finishing processes reduced the as-built arithmetic mean roughness (Ra = 12.16 µm), hence significantly improving the fatigue performance of the samples. Furthermore, as is possible to analyze in Figure 11a, when compared to the as-built specimens, the HIPed and HT-920 ones showed lower YS and UTS; however, the ductility was greatly improved. On the contrary, the HT-850-550 specimens showed no obvious reduction in the YS and UTS when compared to the as-built ones, and the ductility of the former specimens was also better than the latter. Finally, the heat treatments promoted the increase of the fatigue limit, as the HT-920 and HT-850-550 samples exhibited a stress amplitude of approximtaely 400 MPa for 10^6 cycles, and the HIPed samples showed a stress amplitude of approximtaely 500 MPa for the same number of cycles, as shown in Figure 11b.

Table 2. Distinct heat treatments performed by Yu et al. [73]. FC: furnace cooling, WQ: water quenching.

Code Name in Figure 11a	Condition
SLMed	As-built
HIPed	HIP at 920 °C and a pressure of 100 MPa for 2 h
HT-920	Heat treatment at 920 °C for 2 h followed by FC
HT-850-550	Double heat treatment, firstly done at 850 °C for 2 h followed by WQ and then 550 °C for 4 h followed by AC

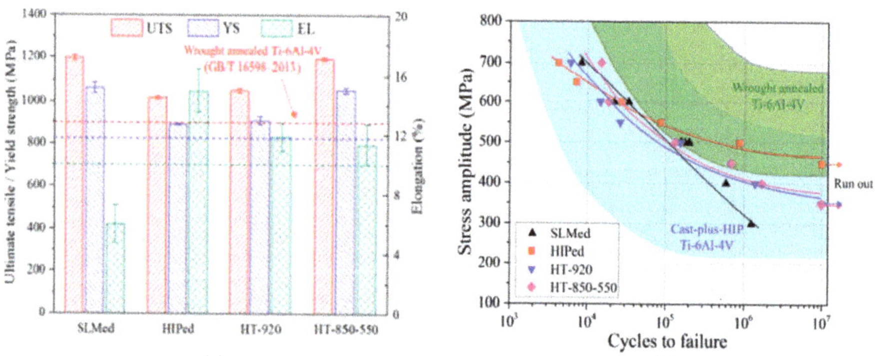

Figure 11. (**a**) Tensile properties and elongation of the different specimens, (**b**) Fatigue performances of the distinct specimens. Adapted from [73], with permission of Elsevier, 2019.

Kim et al. [74] investigated the effect of microstructure control through heat treatments on the high-temperature creep behavior of Ti–6Al–4V specimens manufactured via SLM. The authors conducted a heat treatment at 1040 °C for 1 h and then analyzed the microstructure of the as-built and heat-treated specimens. The former exhibit a typical martensitic microstructure, while the latter showed a β phase at the interface of a large α colony area, meaning that a Widmanstätten's structure was formed. Subsequently, the researchers did two different mechanical tests, namely a compressive test at 500 °C and a high-temperature creep test, and they reported that the compressive yield strength of the as-built specimens was approximately 930 MPa and that in the heat-treated specimens was 557 MPa. Nevertheless, the high-temperature creep test results confirmed that the heat-treated samples exhibited a lower steady-state creep rate, proving that the martensitic microstructure has a lower creep performance than the Widmanstätten one.

In a different study, Ettefagh et al. [75] investigated the influence of the annealing heat treatment on the corrosion resistance of SLMed titanium components. For that, the authors conducted two distinct post-treatments at 600 °C and 800 °C for 2 h. The microstructure analysis revealed that the heat treatment done at 600 °C did not provide the proper condition for the vanadium diffusion, hence the microstructure remained similar to the as-built one (α' needle-like). However, the 800 °C heat treatment led to the formation of the β phase and the stress relief of the α' phase. Then, the researchers performed the corrosion tests and ascertained that the corrosion rate of the as-built specimens was nearly 16 times worse than the cold-rolled commercial titanium samples, which were used as a benchmark. Moreover, the martensitic phase stress relief and the formation of the β phase on the 800 °C post-treated samples greatly improved the corrosion behavior of the material, which exhibited results comparable to the benchmark.

Wang et al. [76] studied the heat treatment effect on the biocompatibility and osseointegration of SLMed titanium samples. For that, the researchers employed an annealing treatment at a temperature of 820 °C for 4 h to half of the manufactured specimens. Then, all the specimens were ultrasonic degreased in acetone and anhydrous ethanol baths for 1 h, respectively. Later, the authors applied human bone mesenchymal stem cells (hBMSCs) to both types of specimens, measured cell proliferation on the first, fourth, and seventh day of incubation, and reported that the heat-treated specimens showed a significant enhancement of hBMSCs adhesion and proliferation. The authors concluded that the annealing treatment promoted better mechanical properties, a more homogeneous, rougher, and more hydrophilic surface, as well as the formation of a crystalline rutile (TiO2) layer, which elevated the corrosion resistance and biocompatibility of the implant. Similarly, Pazhanivel et al. [77] investigated the post-treatment effect on the corrosion resistance of titanium specimens produced via SLM. Firstly, the authors produced several specimens on an EOS M290 machine using different laser power and scan speed parameters. After performing a tensile test, the researchers reported that the specimen that achieved the best elongation (10.88%) was manufactured with a laser power of 310 W and a scan speed of 1340 mm/s. Then, the researchers subjected some specimens to a heat treatment at 850 °C for 2 h and continued for 2 h more to go through isothermal heating. After that, the furnace temperature was set below 200 °C to reduce the specimen's temperature to 200 °C within 1 h, and, besides, the samples were also shot-peened. Later, the authors studied the microstructure of the as-built and post-treated Ti–6Al–4V samples and reported that the former exhibited the α' as a major phase, while the latter was composed of ultrafine grain lamellar $\alpha + \beta$ phases. Subsequently, corrosive tests were performed in a NaCl and phosphate buffer corrosion media, and the researchers concluded that due to the microstructural change, the post-treated samples demonstrated lower corrosion rates. Likewise, Leon et al. [78] investigated the HIP effect on the corrosion performance of Ti–6Al–4V samples manufactured by EBM. After producing the specimens, the authors conducted a thermal treatment at 925 °C with a pressure of 100 MPa for 3 h and then performed electrochemical measurements and stress-corrosion examination by means of slow strain-rate testing. After analyzing the results, the authors concluded that the heat treatment had a slightly advantageous effect on the mechanochemical performance and corrosion resistance of the titanium alloy due to the reduction of α/β interphase area, the increased amount of β phase, as well as the amplified passivation film stability.

Despite not using any thermal treatment, Sharma et al. [79] investigated the corrosion resistance of as-built Ti–6Al–4V bioimplants manufactured through SLM, having in mind that their dissolution inside the human body usually leads to inflammations in the internal organs. The authors used three types of corrosion media (NaOH, NaCl, and H2SO4), which varied from alkaline to acidic and also used a physiological simulated body fluid (SBF) medium as a comparison. Then, after manufacturing the samples, the researchers studied the microstructure of the SLMed and commercial cast titanium, which was used as a reference, and determined that the former exhibited a martensitic microstructure, while the latter showed a uniform distribution of $\alpha + \beta$ phases. Subsequently, the authors conducted several corrosion tests and reported that the SLMed samples had a worse corrosion resistance than the reference because of the α' phase and the higher surface porosity. Nevertheless, the charge transfer resistance, which measures the difficulty faced when an electron is moved from one atom/compound to another, and the polarization of both samples were noticeably high in NaOH, NaCl, and SBF, meaning that the samples exhibited a good corrosion resistance in the aforementioned media. However, in a more acidic and aggressive solution (H2SO4), they showed an enormously poor corrosion resistance.

Nalli et al. [80] investigated the barrel finishing and heat treatment effect on the mechanical properties of Ti–6Al–4V samples manufactured through SLM. For that, some specimens were heat-treated with distinct temperatures (482 °C, 704 °C, 788 °C, and 800 °C) and holding times (3 h, 4 h, 5 h, 6 h, 8 h, and 10 h). Furthermore, others were barrel finished at 30 and 42 rpm, and as a reference, the authors left some specimens in the as-built state. Then, mechanical tests were performed, and after analyzing the results, the researchers reported that as the temperature of the heat treatment and the intensity of the barrel finishing increased, the elongation at break of the specimens improved as well. However, as the elongation increased, the YS and UTS decreased; nevertheless, the authors concluded that the heat treatment temperature of 788 °C and barrel finishing speed of 30 rpm provided the best results in terms of ductility versus strength.

In order to have a better understanding of the effects caused by distinct post-treatments on the mechanical properties of Ti–6Al–4V components, the authors of the current work agglomerated distinct information from diverse scientific reports related to YS, UTS, and elongation at break (ε), which can be analyzed in Table 3.

Table 3. Manufacturing process and post-treatment effect on the mechanical properties of the Ti–6Al–4V alloy.

Mnf. Process	Condition	Orientation	YS [MPa]	UTS [MPa]	ε [%]	Ref.
Wrought	-	Vertical	836 ± 9	942 ± 8	12.5 ± 1.2	[81]
Wrought	-	Horizontal	832 ± 10	933 ± 7	13.0 ± 1.5	[81]
Forged	Mill Annealed	-	970	1030	16	[82]
Forged	Mill Annealed	-	960 ± 10	1006 ± 10	18.37 ± 0.88	[83]
Cast	-	-	865	980	13.5	[84]
Cast	-	-	750 ± 2	875 ± 10	4.5 ± 0.2	[85]
ASTM F136	-	-	≥795	≥860	≥10	[86]
EBM	As-built	Vertical	812 ± 12	851 ± 19	3.6 ± 0.9	[87]
EBM	As-built	Horizontal	783 ± 15	833 ± 22	2.7 ± 0.4	[87]
EBM	As-built	Vertical	869 ± 7.2	928 ± 9.8	9.9 ± 1.7	[63]
EBM	As-built	Horizontal	899 ± 4.7	978 ± 3.2	9.5 ± 1.2	[63]
EBM	As-built	Vertical	1001–1051	1073–1116	11–15	[82]
EBM	As-built	Horizontal	973–1006	1032–1066	12–15	[82]
EBM	HIPed	-	723–817	817–918	3–9	[88]
EBM	Two-stage HIPed	-	885 ± 6	985 ± 12	-	[89]
EBM	Annealed	-	741–842	837–918	3–9	[88]
EBM	Stress Relieved	-	778–943	885–1015	3–9	[88]
SLM	As-built	Vertical	1150 ± 67	1246 ± 134	1.4 ± 0.5	[81]
SLM	As-built	Horizontal	1273 ± 53	1421 ± 120	3.2 ± 0.5	[81]
SLM	As-built	Vertical	1143 ± 30	1219 ± 20	4.89 ± 0.6	[63]
SLM	As-built	Horizontal	1195 ± 19	1269 ± 9	5.0 ± 0.5	[63]
SLM	HIPed	Vertical	883–888	973 – 974	18.5–19.4	[90]
SLM	HIPed	-	912 ± 30	1005 ± 30	8.3 ± 2	[91]
SLM	Annealed	Vertical	1045–1054	1115–1116	9.5–12.4	[90]
SLM	Annealed	Vertical	905–911	987–989	7.4–12.5	[90]
SLM	Annealed	Horizontal	913 ± 7	1019 ± 11	8.9 ± 1	[84]
SLM	Annealed	Horizontal	944 ± 8	1036 ± 30	8.5 ± 1	[84]
SLM	Stress Relieved	Vertical	937 ± 9	1052 ± 11	9.6 ± 0.9	[92]

Similarly, several authors [92–100] also studied the heat treatment effect on the mechanical properties and microstructure of Ti–6Al–4V components. For instance, Benzing et al. [92] demonstrated the efficiency of a novel HIP treatment strategy employed to EBMed specimens. The heat treatment consisted of a two-stage HIP, which can be described as follows: 1050 °C with a pressure of 100 MPa for 2 h with a 12 °C/min heating rate and a nearly 1600 °C/min cooling rate done between 1050 and 500 °C + temper HIP at 800 °C with a pressure of 30 MPa for 2 h and 12 °C/min heating and cooling rates. After studying the microstructure and performing mechanical tests, the authors reported that the new heat treatment maintained the α lath thickness of 1.20 μm ± 0.32 μm, created prior-β grains, sealed the internal porosity, removed microstructural heterogeneities, and maintained the YS and UTS identical to the as-built specimens. In another report, Sabban et al. [96] applied an innovative heat treatment strategy that achieves a bimodal globularized microstructure without the need for a previous plastic deformation process. After manufacturing the specimens in an SLM machine, the researchers employed a heat treatment, as shown in Figure 12, with a thermal cycle between 875 and 975 °C, a heating rate of 2.5 °C/min, and a cooling rate of 1 °C/min. The authors reported that the numerous heating cycles transformed the lamellar α phase into the β phase, which later transformed into globular α upon slow cooling. Moreover, the Electron Backscatter Diffraction (EBSD) micrograph showed that the heating cycles also aided to break the α lamella into smaller segments, which subsequently were converted into spatially separated globular α. Besides, the ductility of the heat-treated specimens, when compared to the as-built ones, was enhanced, the YS and UTS suffered a small decrease, the toughness increased from 98 to 163 MJ/m^3, and beneficial compressive residual stresses were created at the surface level. Likewise, Galarraga et al. [98] submitted EBMed titanium specimens to different solution heat treatments with distinct cooling rates and reported that the water-cooled samples, when compared to the furnace-cooled ones, exhibited an utterly α' martensitic microstructure which promoted an increase of 31 % of the UTS; however, they lowered the ductility by 86%. Haar et al. [100] developed a novel duplex anneal treatment that was firstly done at 910 °C for 8 h followed by water-quenching and then 750 °C for 4 h followed by furnace-cooling. The first step of the heat treatment created a bimodal microstructure of α in the α' matrix, while the second step promoted the decomposition of α' into $\alpha + \beta$ lamellar. Subsequently, the authors conducted tensile tests and verified that when compared to the as-built SLMed and conventionally annealed samples, the duplex annealed ones demonstrated a far greater ductility.

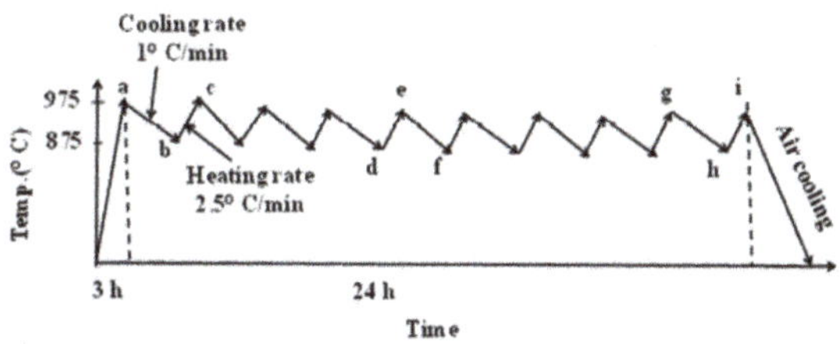

Figure 12. Schematic of the heat treatment employed for a period of 24 h. Reproduced from [96], with permission of Elsevier, 2019.

3.2. Residual Stresses

In additive manufacturing, residual stresses are spatially nonuniform and can lead to worse fatigue life and mechanical properties of the built components, as well as layer delamination and part distortion [101]. Moreover, the length of the scan vectors used when producing the components can also influence the distribution and magnitude of RS, as a longer scan vector results in a higher temperature gradient, which is opposite to shorter scan vectors that are preferable to create lower residual stresses [102]. Hence, one of the most usual scanning strategies employed is the chessboard scanning [103]. Ali et al. [104] investigated the effect of the scan strategy and vector length on SLMed Ti–6Al–4V specimens. For that, the authors used distinct scan strategies, namely 45° alternating, 90° alternating, 2 mm × 2 mm chessboard, 3 mm × 3 mm chessboard, 5 mm × 5 mm chessboard, 5 mm × 5 mm chessboard with scan vectors rotated 45° in adjacent blocks, and 5 mm × 5 mm chessboard with scan vectors rotated 90° in adjacent blocks. After analyzing the finite element simulation results, the researchers reported that a 90° alternating scan strategy caused the lowest residual stresses. Furthermore, the growth of the scan vector length in the chessboard scanning strategy resulted in increased residual stresses. Likewise, Song et al. [105] used finite element simulation and experimental verification to ascertain the effect of scanning strategies on the residual stresses of SLMed Ti–6Al–4V samples. The authors employed three distinct scan strategies, which were line scan, 15° alternating scan, and 90° alternating scan. After conducting several tests, they reported that the scan strategy had little to no influence on the molten pool size, and the 15° alternating scan strategy produced the lowest residual stresses. Lu et al. [106] investigated the RS evolution on distinct chessboard scanning areas through Vickers micro-indentation. After performing the aforementioned test, the authors reported that the lowest RS were found in the 2×2 mm^2 sample; however, that only happened because cracks were formed, and stresses were released. Thus, the authors concluded that the 5×5 mm^2 scanning area was the best to produce end parts with the lowest RS.

Laser Shock Peening (LSP) has been widely studied by several authors [107–114], which state that the aforementioned post-treatment refines the microstructure and induces compressive residual stresses on the surface level, thus enhancing the fatigue lifetime of the components made from distinct titanium alloys. Zhang et al. [110] conducted a residual stress analysis through XRD and a micro-hardness test through Vickers indentation and reported that the LSP treatment induced compressive residual stresses and enhanced the micro-hardness of the Ti–6Al–4V samples. Plus, multiple laser shocks had a positive effect on the surface hardening. Furthermore, the authors also ascertained that the number of overlapped laser spots, which were meant to act as the laser peening intensity, provided better fatigue life properties when set to 3. In another study, Ren et al. [111] reported that the LSP treatment was able to reduce the surface roughness and repair surface details of Ti–6Al–4V samples if done with a lower laser pulse energy (6.3 J). Besides, when the authors employed a higher laser pulse energy (7.9 J), the surface hardness significantly increased. Likewise, Jin et al. [113] investigated the influence of LSP on the fatigue behavior and microstructure of Ti–6Al–4V samples manufactured via EBM. For the post-treatment, the authors used a Nd:YAG laser with a frequency of 1 Hz, a wavelength of 1064 nm, a beam diameter of 2.5 mm, and a pulse duration of 12 ns. Then, the microstructure was studied, and the researchers reported that the EBMed specimens showed a microstructure composed of $\alpha + \beta$ phases, as shown in Figure 13a. Besides, the LSP treatment caused grain refinement of the α phase via deformation twinning, as shown in Figure 13b. Subsequently, the authors performed mechanical tests and concluded that the LSPed samples, when compared to the as-built ones, exhibited a micro-hardness increase of 11%, as well as a run-out fatigue strength (2×10^6 cycles) increase of 100 MPa. Moreover, the LSP treatment inhibited the crack initiation and enhanced the required work of fatigue fracture, which, according to the literature reviewed by the authors, was due to the formation of compressive residual stresses and the grain refinement. In a similar study, Lan et al. [114] investigated the effect of the LSP treatment on the microstructural evolution and mechanical properties of Ti–6Al–4V specimens. The authors employed the same LSP parameters as Jin et al. [113] and then studied the microstructural evolution of the as-built and post-treated samples. They reported that when compared to the as-built samples, the post-treated ones exhibited a refined α lamella into sub-micro-equiaxed grains and nano-equiaxed grains. Later, the researchers measured the residual stresses of both types of samples through XRD and concluded that the LSP treatment changed the surface stress state from nearly 45 MPa to −380 MPa. Finally, the authors conducted a tensile test and ascertained that the LSP-treated samples possessed a higher tensile strength and elongation, due to the grain refinement and the induced compressive RS.

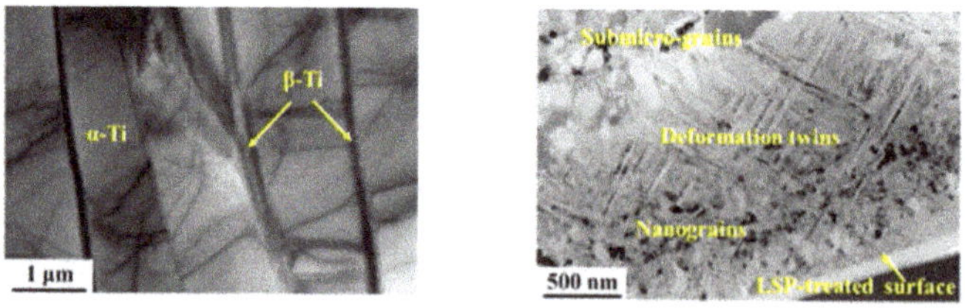

(a) (b)

Figure 13. Transmission Electron Microscopy (TEM) image of (**a**) EBM as-built specimens, and (**b**) Laser Shock Peening (LSP)-treated specimens. Adapted from [113], with permission of Elsevier, 2020.

Eyzat et al. [115] studied the effect of distinct post-treatments on the mechanical properties, residual stresses, and roughness of Ti–6Al–4V samples manufactured via SLM. For that, the researchers employed a stress relief heat treatment at 400 °C for 2 h to some specimens and the previous heat treatment + Surface Mechanical Attrition Treatment (SMAT) to the others. The latter post-treatment uses a high frequency (20 kHz) ultrasonic generator, which leads to the vibration and subsequent impact of spherical balls into the specimen's surface and promotes plastic deformation. After performing mechanical tests, investigating the specimen's surface roughness, and analyzing the residual stresses through the XRD method, the authors concluded that SMAT treatment changed the RS from tensile (approximately 100 MPa) to compressive (approximately −300 MPa) as it can be seen in Figure 14, reducing by 80% the high surface roughness, and boosting the mechanical properties (YS, UTS, and micro-hardness) by 10% to 15%. Yakout et al. [116] studied the correlation between residual stresses and process parameters of SLMed Ti–6Al–4V samples. For that, the authors used the Renishaw AM 400 machine and the titanium grade 23 powder from the same enterprise. Then, several specimens were manufactured using distinct laser power (P), scanning speed (v), and hatch spacing (h), while the layer thickness (t) was kept constant at 40 μm. Note that the aforementioned process parameters are all correlated in the Volumetric Energy Density (Ev) formula, which is:

$$E_v = \frac{P}{v \times h \times t} \tag{1}$$

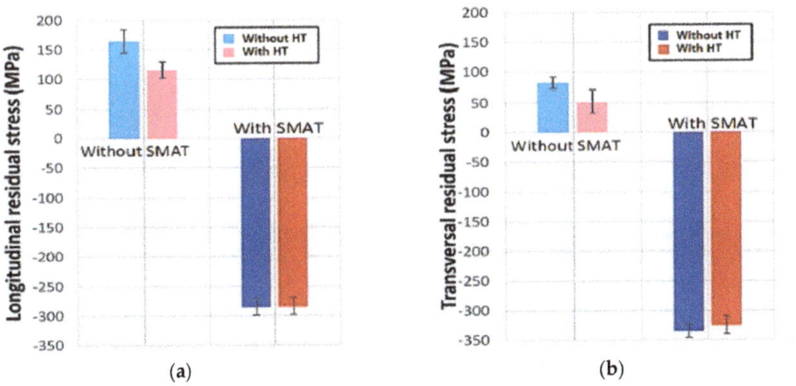

Figure 14. Post-treatment effects on (**a**) longitudinal residual stresses and (**b**) transversal residual stresses. Adapted from [115], under CC BY 4.0.

Later, after conducting distinct studies of the produced samples, the researchers concluded that the *Ev* increase led to a growth in the part's thermal expansion and density up to a maximum point, which was considered the critical energy density (*Ev* = 86.8 J/mm^3). Furthermore, the components manufactured with a volumetric energy density inferior to the critical one exhibited less residual stresses, as the XRD analysis exposed.

In a similar study, Vastola et al. [117] investigated, through FEM, the effect of beam scan speed, beam power density, beam size, and chamber bed temperature on the distribution and magnitude of residual stresses on EBMed Ti–6Al–4V samples. The authors reported that lower scan speeds deepened the Heat-Affected Zone (HAZ), while higher speeds reduce the HAZ. Furthermore, a 20% increase of the beam power caused an approximate 15% growth of the HAZ. Moreover, the researchers also stated that small beam sizes provided higher stresses within a smaller Heat-Affected Zone. Finally, it was concluded that the bed pre-heating temperature was the most crucial parameter for reducing the residual stresses, as a 50 °C increase of the pre-heating temperature promoted an approximate 20% decrease of residual stresses, as shown in Figure 15.

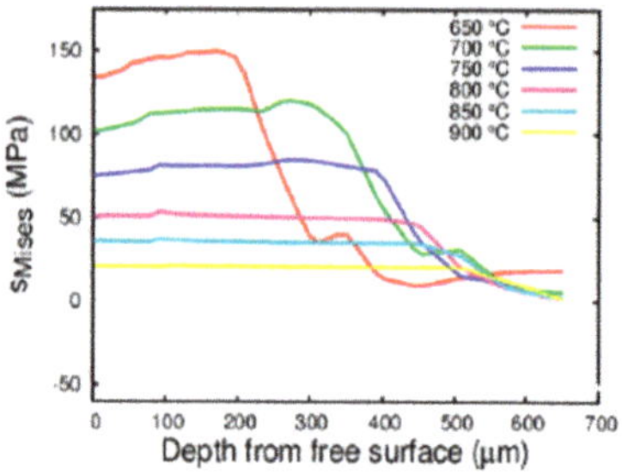

Figure 15. Quantitative effect of bed pre-heating temperature on residual stresses, where SMises is the Von Mises Stress. Adapted from [117], with permission of Elsevier, 2016.

In another study, Ning et al. [118] developed an analytical model that predicts RS and part distortion without the need for any iteration-based simulations or FEM. The inputs for the aforementioned model were the laser parameters, scan parameters, and material parameters, and the thermal model was calculated by a moving point heat source solution along with a heat sink solution. Then, the thermal stress model was calculated from the thermal model and the RS were computed from the thermal stress model by applying an elastoplastic relaxation method. After that, part distortion was computed using a surface displacement model, which finally outputted the distortion at distinct locations. The authors validated the accuracy of their analytical model by manufacturing several cantilever-like Ti–6Al–4V specimens on an SLM machine and measuring the part distortion on a Coordinate Measuring Machine (CMM). After comparing the experimental results with the analytical ones, the researchers ascertained that similar agreements were seen in terms of part distortion at distinct locations. Furthermore, the authors also emphasized that the computational time of the analytical model was only 10 s on a personal computer, thus allowing a high computational efficiency.

Finally, in Figures 16 and 17, it is possible to analyze two similar diagrams that explain and summarize the expected mechanical properties and microstructures of as-built and heat-treated EBMed and SLMed Ti–6Al–4V components, respectively.

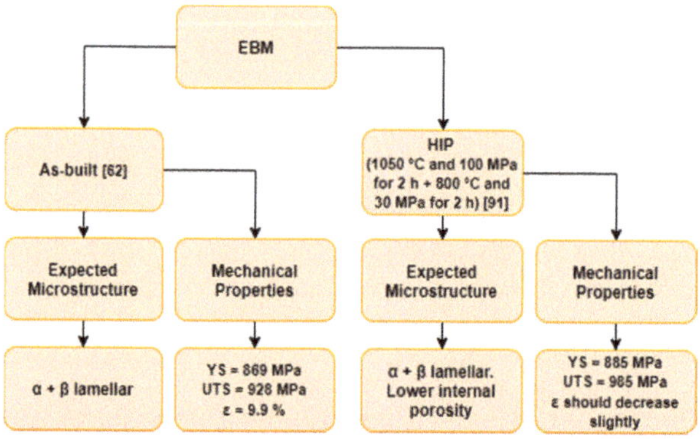

Figure 16. Diagram of the expected microstructure and mechanical properties of as-built and heat-treated EBMed Ti–6Al–4V samples.

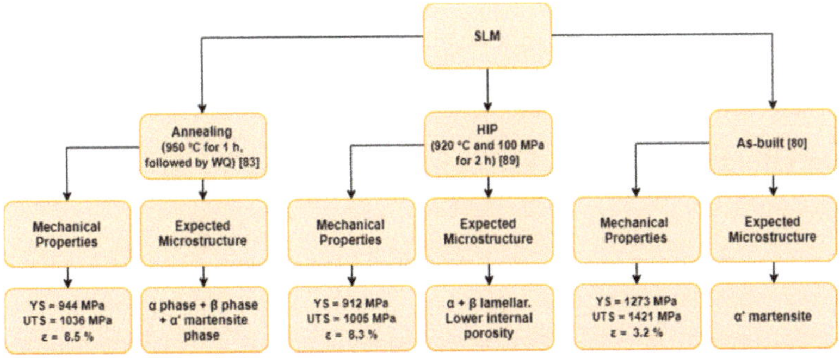

Figure 17. Diagram of the expected microstructure and mechanical properties of as-built and heat-treated SLMed Ti–6Al–4V samples.

4. Concluding Remarks and Outlook

One of the most used and studied superalloys in additive manufacturing is the Ti–6Al–4V, which can be applied in several distinct areas, namely in the aerospace field due to its low density and high melting point (around 1650 °C), and the biomedical area owing to its high corrosion resistance and excellent biocompatibility when in contact with tissues or bones of the human body. For this review work, the authors analyzed roughly 22% of the 537 different publications seen in Figure 4, and more than half of those 22% were about heat treatments and residual stresses of Ti–6Al–4V components produced via SLM and/or EBM in order to provide a solid and unbiased overview. The main ideas can be described as follows:

- Owing to the inherent characteristics of the SLM process, i.e., high cooling rates and quick solidification, the expected microstructure in the as-built state is characterized by fine acicular α' martensitic precipitates in columnar original β grains, opposed to an $\alpha + \beta$ lamellar microstructure obtained through EBM. The latter manufacturing process achieves superior building chamber temperatures; therefore, it reduces the part's cooling rate and forms a bimodal microstructure.
- If one seeks to improve the mechanical properties, such as ductility and hardness, and decrease the residual stresses of SLMed Ti–6Al–4V parts, then heat treatments should be conducted. A duplex anneal treatment can be firstly done at 910 °C for 8 h followed by water-quenching and then 750 °C for 4 h followed by furnace-cooling. The first step of the heat treatment creates a bimodal microstructure of α in the α' matrix, while the second step promotes the decomposition of α' into $\alpha + \beta$ lamellar.
- Hot Isostatic Pressing must be done if one desires to increase the density and fatigue life of the manufactured component, as the aforementioned treatment is responsible for decreasing the size of the internal defects. Besides, one can also predict the expected porosity of the final pieces due to recent developments of trustworthy and efficient computer analytical models. Furthermore, the surface roughness also plays a crucial role in the fatigue life of the final part and is heavily dependent of the manufacturing process, i.e., SLM leads to lower Ra values than EBM; nevertheless, post-machining, sandblasting, and polishing can greatly decrease the Ra values, consequently weakening the crack initiation and improving the fatigue life.
- It has been seen that residual stresses can be lowered through heat treatments as well as optimized printing parameters, such as the bed pre-heating temperature in EBM or the correlation of distinct process parameters in the Volumetric Energy Density formula on SLMed components. Besides, high scanning speeds coupled with shorter scan vectors, i.e., chessboard scanning, has been demonstrated to lower the residual stresses. Furthermore, LSP and SMAT post-treatments can also enhance the fatigue life as they induce compressive residual stresses on the surface and increase the microhardness of the components.

Despite the tremendous research over the last years about additively manufactured Ti–6Al–4V parts, there are still some problems to overcome—namely, the mechanical properties variability when subjected to different heat treatments and process parameters. Furthermore, heat treatments should be further investigated and improved, because often one sacrifices yield and ultimate tensile strength for ductility. In addition, there is still room to improve regarding the residual stresses of Ti–6Al–4V components, i.e., researching the RS evolution under different heat treatments and printing parameters.

Lastly, the main limitations felt by the authors upon writing this review paper were the restricted access of several scientific articles as well as the comprehension of a small number of results exhibited by distinct authors. Moreover, as one can analyze in the Pareto charts (Figures 5 and 6), the scientific community has studied more SLMed Ti–6Al–4V components than EBMed ones, and only a few works focus on residual stresses, which ultimately was a big limitation of information.

Author Contributions: Conceptualization: Ó.T. and F.J.G.S.; methodology: Ó.T., L.P.F., E.A. and F.J.G.S.; investigation: Ó.T.; formal analysis: E.A., L.P.F. and F.J.G.S.; supervision: F.J.G.S. and E.A.; validation: F.J.G.S., L.P.F. and E.A.; writing original draft: Ó.T.; writing, reviewing and editing: F.J.G.S.; resources: E.A.; project administration: F.J.G.S., L.P.F. and E.A. All authors have read and agreed to the published version of the manuscript.

Funding: This research received no external funding.

Acknowledgments: The main author would like to acknowledge his mentor for all the help and support given through the writing of this report.

Conflicts of Interest: The authors declare no conflict of interest.

List of Symbols and Abbreviations

3DP	Three-Dimensional Printing
AM	Additive Manufacturing
ASTM	American Society for Testing and Materials
CMM	Coordinate Measuring Machine
EBM	Electron Beam Melting
EBSD	Electron Backscatter Diffraction
Ev	Volumetric Energy Density
FC	Furnace Cooling
FEM	Finite Element Modeling
h	Hatch Spacing
H2SO4	Sulfuric Acid
HAZ	Heat-Affected Zone
hBMSCs	Human Bone Mesenchymal Stem Cells
HIP	Hot Isostatic Pressing
ISO	International Organization for Standardization
LSP	Laser Shock Peening
Ms	Martensite Start Temperature
NaCl	Sodium Chloride
NaOH	Sodium Hydroxide
P	Laser Power
Ra	Surface Roughness
Ref.	Reference
RS	Residual Stresses
SBF	Simulated Body Fluid
SLM	Selective Laser Melting
SLS	Selective Laser Sintering
SMAT	Surface Mechanical Attrition Treatment
t	Layer Thickness
$T_{ambient}$	Ambient Temperature
T_{melt}	Melting Temperature

UTS	Ultimate Tensile Strength
v	Scanning Speed
WQ	Water Quenching
XRD	X-ray Diffraction
YS	Yield Strength
ε	Elongation at Break

References

1. Goldberg, D. History of 3D Printing: It's Older than You Are (That Is, If You're Under 30). AutoDesk. 2018. Available online: https://www.autodesk.com/redshift/history-of-3d-printing/ (accessed on 4 December 2019).
2. ISO/ASTM 52900-15. *Standard Terminology for Additive Manufacturing–General Principles–Terminology*; ASTM International: West Conshohocken, PA, USA, 2015.
3. Silva, F.J.G.; Campilho, R.D.S.G.; Gouveia, R.M.; Pinto, G.; Baptista, A. A Novel Approach to Optimize the Design of Parts for Additive Manufacturing. *Procedia Manuf.* **2018**, *17*, 53–61. [CrossRef]
4. Gouveia, R.M.; Silva, F.J.G.; Atzeni, E.; Sormaz, D.; Alves, J.L.; Pereira, A.B. Effect of Scan Strategies and Use of Support Structures on Surface Quality and Hardness of L-PBF AlSi10Mg Parts. *Materials* **2020**, *13*, 2248. [CrossRef]
5. ASTM F2792-12a. *Standard Terminology for Additive Manufacturing Technologies (Withdrawn 2015)*; ASTM International: West Conshohocken, PA, USA, 2012; pp. 1–3.
6. Meiners, W. Shaped Body Especially Prototype or Replacement Part Production. German Patent 19649865, 12 February 1998.
7. Giganto, S.; Zapico, P.; Castro-Sastre, M.Á.; Martínez-Pellitero, S.; Leo, P.; Perulli, P. Influence of the scanning strategy parameters upon the quality of the SLM parts. *Procedia Manuf.* **2019**, *41*, 698–705. [CrossRef]
8. Anwar, A.B.; Ibrahim, I.H.; Pham, Q.C. Spatter transport by inert gas flow in selective laser melting: A simulation study. *Powder Technol.* **2019**, *352*, 103–116. [CrossRef]
9. Ferrar, B.; Mullen, L.; Jones, E.; Stamp, R.; Sutcliffe, C.J. Gas flow effects on selective laser melting (SLM) manufacturing performance. *J. Mater. Process. Technol.* **2012**, *212*, 355–364. [CrossRef]
10. Konečná, R.; Nicoletto, G.; Riva, E. Notch fatigue behavior of Inconel 718 produced by selective laser melting. *Procedia Struct. Integr.* **2019**, *17*, 138–145. [CrossRef]
11. Contuzzi, N.; Campanelli, S.L.; Ludovico, A.D. 3D finite element analysis in the Selective Laser Melting process. *Int. J. Simul. Model.* **2011**, *10*, 113–121. [CrossRef]
12. Nguyen, Q.B.; Luu, D.N.; Nai, S.M.L.; Zhu, Z.; Chen, Z.; Wei, J. The role of powder layer thickness on the quality of SLM printed parts. *Arch. Civ. Mech. Eng.* **2018**, *18*, 948–955. [CrossRef]
13. Arcam, A.B. Arrangement for the Production of a Three-Dimensional Product. U.S. Patent 20060141089, 29 June 2006.
14. Körner, C.; Attar, E.; Heinl, P. Mesoscopic simulation of selective beam melting processes. *J. Mater. Process. Technol.* **2011**, *211*, 978–987. [CrossRef]
15. Galati, M.; Iuliano, L. A literature review of powder-based electron beam melting focusing on numerical simulations. *Addit. Manuf.* **2018**, *19*, 1–20. [CrossRef]
16. Li, P.; Warner, D.H.; Fatemi, A.; Phan, N. Critical assessment of the fatigue performance of additively manufactured Ti-6Al-4V and perspective for future research. *Int. J. Fatigue* **2016**, *85*, 130–143. [CrossRef]
17. Prabhakar, P.; Sames, W.J.; Dehoff, R.; Babu, S.S. Computational modeling of residual stress formation during the electron beam melting process for Inconel 718. *Addit. Manuf.* **2015**, *7*, 83–91. [CrossRef]
18. Smith, C.J.; Derguti, F.; Hernandez Nava, E.; Thomas, M.; Tammas-Williams, S.; Gulizia, S.; Fraser, D.; Todd, I. Dimensional accuracy of Electron Beam Melting (EBM) additive manufacture with regard to weight optimized truss structures. *J. Mater. Process. Technol.* **2016**, *229*, 128–138. [CrossRef]
19. Gokuldoss, P.K.; Kolla, S.; Eckert, J. Additive manufacturing processes: Selective laser melting, electron beam melting and binder jetting-selection guidelines. *Materials* **2017**, *10*, 672. [CrossRef]
20. Azam, F.I.; Abdul Rani, A.M.; Altaf, K.; Rao, T.V.V.L.N.; Zaharin, H.A. An In-Depth Review on Direct Additive Manufacturing of Metals. *IOP Conf. Ser. Mater. Sci. Eng.* **2018**, *328*, 012005. [CrossRef]
21. Properties of Titanium Ti-6Al-4V (Grade 5). Matweb. Available online: http://www.matweb.com/search/datasheet.aspx?MatGUID=10d463eb3d3d4ff48fc57e0ad1037434 (accessed on 6 December 2019).

22. Zuback, J.S.; DebRoy, T. The hardness of additively manufactured alloys. *Materials* **2018**, *11*, 2070. [CrossRef]
23. Dai, N.; Zhang, L.C.; Zhang, J.; Zhang, X.; Ni, Q.; Chen, Y.; Wu, M.; Yang, C. Distinction in corrosion resistance of selective laser melted Ti-6Al-4V alloy on different planes. *Corros. Sci.* **2016**, *111*, 703–710. [CrossRef]
24. Lin, J.; Lv, Y.; Liu, Y.; Sun, Z.; Wang, K.; Li, Z.; Wu, Y.; Xu, B. Microstructural evolution and mechanical property of Ti-6Al-4V wall deposited by continuous plasma arc additive manufacturing without post heat treatment. *J. Mech. Behav. Biomed. Mater.* **2017**, *69*, 19–29. [CrossRef]
25. Tamilselvi, S.; Raman, V.; Rajendran, N. Corrosion behaviour of Ti–6Al–7Nb and Ti–6Al–4V ELI alloys in the simulated body fluid solution by electrochemical impedance spectroscopy. *Electrochim. Acta* **2006**, *52*, 839–846. [CrossRef]
26. Castellanos, S.D.; Alves, J.L.; Neto, R.J. A comparative study of manufacturing processes of complex surface parts in Titanium Ti6Al4V. *Ciência Tecnol. Dos Mater.* **2017**, *29*, 73–78. [CrossRef]
27. Lütjering, G.; Williams, J.C. *Titanium*, 2nd ed.; Springer: New York, NY, USA, 2007. [CrossRef]
28. Huang, R.; Riddle, M.; Graziano, D.; Warren, J.; Das, S.; Nimbalkar, S.; Cresko, J.; Masanet, E. Energy and emissions saving potential of additive manufacturing: The case of lightweight aircraft components. *J. Clean. Prod.* **2016**, *135*, 1559–1570. [CrossRef]
29. Ngo, T.D.; Kashani, A.; Imbalzano, G.; Nguyen, K.T.Q.; Hui, D. Additive manufacturing (3D printing): A review of materials, methods, applications and challenges. *Compos. Part B Eng.* **2018**, *143*, 172–196. [CrossRef]
30. Ducato, A.; Fratini, L.; La Cascia, M.; Mazzola, G. An automated visual inspection system for the classification of the phases of Ti-6Al-4V titanium alloy. *Lect. Notes Comput.* **2013**, *8048*, 362–369. [CrossRef]
31. Ahmed, T.; Rack, H.J. Phase transformations during cooling in $\alpha+\beta$ titanium alloys. *Mater. Sci. Eng. A* **1998**, *243*, 206–211. [CrossRef]
32. Boyer, R.; Collings, E.W.; Welsch, G. *Materials Properties Handbook: Titanium Alloys*; ASM International: Cleveland, OH, USA, 1994.
33. Gil Mur, F.X.; Rodríguez, D.; Planell, J.A. Influence of tempering temperature and time on the α'-Ti-6Al-4V martensite. *J. Alloys Compd.* **1996**, *234*, 287–289. [CrossRef]
34. Reisgen, U.; Olschok, S.; Sharma, R.; Gach, S. Influence on martensite-start-temperature and volume expansion of low-transformation-temperature materials used for residual stress relief in beam welding. *Mater. Sci. Eng. Technol.* **2017**, *48*, 1276–1282. [CrossRef]
35. Liu, S.; Shin, Y.C. Additive manufacturing of Ti6Al4V alloy: A review. *Mater. Des.* **2019**, *164*, 107552. [CrossRef]
36. Bartlett, J.L.; Li, X. An overview of residual stresses in metal powder bed fusion. *Addit. Manuf.* **2019**, *27*, 131–149. [CrossRef]
37. Vayssette, B.; Saintier, N.; Brugger, C.; Elmay, M.; Pessard, E. Surface roughness of Ti-6Al-4V parts obtained by SLM and EBM: Effect on the High Cycle Fatigue life. *Procedia Eng.* **2018**, *213*, 89–97. [CrossRef]
38. Liu, D.; Flewitt, P.E.J. Raman measurements of stress in films and coatings. *Spectrosc. Prop. Inorg. Organomet. Compd.* **2014**, *45*, 141–177. [CrossRef]
39. Guo, J.; Fu, H.; Pan, B.; Kang, R. Recent progress of residual stress measurement methods: A review. *Chin. J. Aeronaut.* **2020**. [CrossRef]
40. Barros, R.; Silva, F.J.G.; Gouveia, R.M.; Saboori, A.; Marchese, G.; Biamino, S.; Salmi, A.; Atzeni, E. Laser Powder Bed Fusion of Inconel 718: Residual Stress Analysis Before and After Heat Treatment. *Metals* **2019**, *9*, 1290. [CrossRef]
41. ASTM E 837-08. *Standard Test Method for Determining Residual Stresses by the Hole-Drilling Strain-Gages*; ASTM International: West Conshohocken, PA, USA, 2008; Volume 1, pp. 1–17.
42. Thijs, L.; Verhaeghe, F.; Craeghs, T.; Humbeeck, J.V.; Kruth, J.P. A study of the microstructural evolution during selective laser melting of Ti-6Al-4V. *Acta Mater.* **2010**, *58*, 3303–3312. [CrossRef]
43. Facchini, L.; Magalini, E.; Robotti, P.; Molinari, A.; Höges, S.; Wissenbach, K. Ductility of a Ti-6Al-4V alloy produced by selective laser melting of prealloyed powders. *Rapid Prototyp. J.* **2010**, *16*, 450–459. [CrossRef]
44. Gong, H.; Rafi, K.; Gu, H.; Janaki Ram, G.D.; Starr, T.; Stucker, B. Influence of defects on mechanical properties of Ti-6Al-4V components produced by selective laser melting and electron beam melting. *Mater. Des.* **2015**, *86*, 545–554. [CrossRef]

45. Benedetti, M.; Torresani, E.; Leoni, M.; Fontanari, V.; Bandini, M.; Pederzolli, C.; Potrich, C. The effect of post-sintering treatments on the fatigue and biological behavior of Ti-6Al-4V ELI parts made by selective laser melting. *J. Mech. Behav. Biomed. Mater.* **2017**, *71*, 295–306. [CrossRef]
46. Xu, W.; Sun, S.; Elambasseril, J.; Liu, Q.; Brandt, M.; Qian, M. Ti-6Al-4V Additively Manufactured by Selective Laser Melting with Superior Mechanical Properties. *JOM* **2015**, *67*, 668–673. [CrossRef]
47. Xu, J.; Zhu, J.; Fan, J.; Zhou, Q.; Peng, Y.; Guo, S. Microstructure and mechanical properties of Ti–6Al–4V alloy fabricated using electron beam freeform fabrication. *Vacuum* **2019**, *167*, 364–373. [CrossRef]
48. Galarraga, H.; Lados, D.A.; Dehoff, R.R.; Kirka, M.M.; Nandwana, P. Effects of the microstructure and porosity on properties of Ti-6Al-4V ELI alloy fabricated by electron beam melting (EBM). *Addit. Manuf.* **2016**, *10*, 47–57. [CrossRef]
49. Zhao, X.; Li, S.; Zhang, M.; Liu, Y.; Sercombe, T.B.; Wang, S.; Hao, Y.; Yang, R.; Murr, L.E. Comparison of the microstructures and mechanical properties of Ti-6Al-4V fabricated by selective laser melting and electron beam melting. *Mater. Des.* **2016**, *95*, 21–31. [CrossRef]
50. Chern, A.H.; Nandwana, P.; McDaniels, R.; Dehoff, R.R.; Liaw, P.K.; Tryon, R.; Duty, C.E. Build orientation, surface roughness, and scan path influence on the microstructure, mechanical properties, and flexural fatigue behavior of Ti–6Al–4V fabricated by electron beam melting. *Mater. Sci. Eng. A* **2020**, *772*, 138740. [CrossRef]
51. Murr, L.E.; Esquivel, E.V.; Quinones, S.A.; Gaytan, S.M.; Lopez, M.I.; Martinez, E.Y.; Medina, F.; Hernandez, D.H.; Martinez, E.; Martinez, J.L.; et al. Microstructures and mechanical properties of electron beam-rapid manufactured Ti-6Al-4V biomedical prototypes compared to wrought Ti-6Al-4V. *Mater. Charact.* **2009**, *60*, 96–105. [CrossRef]
52. Edwards, P.; Ramulu, M. Fatigue performance evaluation of selective laser melted Ti-6Al-4V. *Mater. Sci. Eng. A* **2014**, *598*, 327–337. [CrossRef]
53. Chan, K.S.; Koike, M.; Mason, R.L.; Okabe, T. Fatigue life of titanium alloys fabricated by additive layer manufacturing techniques for dental implants. *Metall. Mater. Trans. A Phys. Metall. Mater. Sci.* **2013**, *44*, 1010–1022. [CrossRef]
54. Fousová, M.; Vojtěch, D.; Doubrava, K.; Daniel, M.; Lin, C.F. Influence of inherent surface and internal defects on mechanical properties of additively manufactured Ti6Al4V alloy: Comparison between selective laser melting and electron beam melting. *Materials* **2018**, *11*, 537. [CrossRef] [PubMed]
55. Qiu, C.; Panwisawas, C.; Ward, M.; Basoalto, H.C.; Brooks, J.W.; Attallah, M.M. On the role of melt flow into the surface structure and porosity development during selective laser melting. *Acta Mater.* **2015**, *96*, 72–79. [CrossRef]
56. Stef, J.; Poulon-Quintin, A.; Redjaimia, A.; Ghanbaja, J.; Ferry, O.; De Sousa, M.; Gouné, M. Mechanism of porosity formation and influence on mechanical properties in selective laser melting of Ti-6Al-4V parts. *Mater. Des.* **2018**, *156*, 480–493. [CrossRef]
57. Tammas-Williams, S.; Zhao, H.; Léonard, F.; Derguti, F.; Todd, I.; Prangnell, P.B. XCT analysis of the influence of melt strategies on defect population in Ti-6Al-4V components manufactured by Selective Electron Beam Melting. *Mater. Charact.* **2015**, *102*, 47–61. [CrossRef]
58. Ning, J.; Sievers, D.E.; Garmestani, H.; Liang, S.Y. Analytical modeling of part porosity in metal additive manufacturing. *Int. J. Mech. Sci.* **2020**, *172*, 105428. [CrossRef]
59. Zhang, W.; Qin, P.; Wang, Z.; Yang, C.; Kollo, L.; Grzesiak, D.; Prashanth, K.G. Superior wear resistance in EBM-Processed TC4 alloy compared with SLM and forged samples. *Materials* **2019**, *12*, 782. [CrossRef]
60. Mohammadhosseini, A.; Fraser, D.; Masood, S.H.; Jahedi, M. Microstructure and mechanical properties of Ti–6Al–4V manufactured by electron beam melting process. *Mater. Res. Innov.* **2013**, *17*, 106–112. [CrossRef]
61. Tan, X.; Kok, Y.; Tan, Y.J.; Descoins, M.; Mangelinck, D.; Tor, S.B.; Leong, K.F.; Chua, C.K. Graded microstructure and mechanical properties of additive manufactured Ti-6Al-4V via electron beam melting. *Acta Mater.* **2015**, *97*, 1–16. [CrossRef]
62. Losertová, M.; Kubeš, V. Microstructure and mechanical properties of selective laser melted Ti6Al4V alloy. *IOP Conf. Ser. Mater. Sci. Eng.* **2017**, *266*, 012009. [CrossRef]
63. Rafi, H.K.; Karthik, N.V.; Gong, H.; Starr, T.L.; Stucker, B.E. Microstructures and mechanical properties of Ti6Al4V parts fabricated by selective laser melting and electron beam melting. *J. Mater. Eng. Perform.* **2013**, *22*, 3872–3883. [CrossRef]

64. Balyakin, A.; Zhuchenko, E.; Nosova, E. Study of heat treatment impact on the surface defects appearance on samples obtained by selective laser melting of Ti-6Al-4V during chemical polishing. *Mater. Today Proc.* **2019**, *19*, 2307–2311. [CrossRef]
65. Vayssette, B.; Saintier, N.; Brugger, C.; El May, M. Surface roughness effect of SLM and EBM Ti-6Al-4V on multiaxial high cycle fatigue. *Theor. Appl. Fract. Mech.* **2020**, *108*, 102581. [CrossRef]
66. Yuan, W.; Hou, W.; Li, S.; Hao, Y.; Yang, R.; Zhang, L.C.; Zhu, Y. Heat treatment enhancing the compressive fatigue properties of open-cellular Ti-6Al-4V alloy prototypes fabricated by electron beam melting. *J. Mater. Sci. Technol.* **2018**, *34*, 1127–1131. [CrossRef]
67. Zhao, S.; Li, S.J.; Hou, W.T.; Hao, Y.L.; Yang, R.; Misra, R.D.K. The influence of cell morphology on the compressive fatigue behavior of Ti-6Al-4V meshes fabricated by electron beam melting. *J. Mech. Behav. Biomed. Mater.* **2016**, *59*, 251–264. [CrossRef]
68. Chastand, V.; Quaegebeur, P.; Maia, W.; Charkaluk, E. Comparative study of fatigue properties of Ti-6Al-4V specimens built by electron beam melting (EBM) and selective laser melting (SLM). *Mater. Charact.* **2018**, *143*, 76–81. [CrossRef]
69. Pegues, J.W.; Shao, S.; Shamsaei, N.; Sanaei, N.; Fatemi, A.; Warner, D.H.; Li, P.; Phan, N. Fatigue of additive manufactured Ti-6Al-4V, Part I: The effects of powder feedstock, manufacturing, and post-process conditions on the resulting microstructure and defects. *Int. J. Fatigue* **2020**, *132*, 105358. [CrossRef]
70. Molaei, R.; Fatemi, A.; Sanaei, N.; Pegues, J.; Shamsaei, N.; Shao, S.; Li, P.; Warner, D.H.; Phan, N. Fatigue of additive manufactured Ti-6Al-4V, Part II: The relationship between microstructure, material cyclic properties, and component performance. *Int. J. Fatigue* **2020**, *132*, 105363. [CrossRef]
71. Wu, M.-W.; Chen, J.-K.; Lin, B.-H.; Chiang, P.-H.; Tsai, M.-K. Compressive fatigue properties of additive-manufactured Ti-6Al-4V cellular material with different porosities. *Mater. Sci. Eng. A* **2020**, *790*, 139695. [CrossRef]
72. Viespoli, L.M.; Bressan, S.; Itoh, T.; Hiyoshi, N.; Prashanth, K.G.; Berto, F. Creep and high temperature fatigue performance of as build selective laser melted Ti-based 6Al-4V titanium alloy. *Eng. Fail. Anal.* **2020**, *111*, 104477. [CrossRef]
73. Yu, H.; Li, F.; Wang, Z.; Zeng, X. Fatigue performances of selective laser melted Ti-6Al-4V alloy: Influence of surface finishing, hot isostatic pressing and heat treatments. *Int. J. Fatigue* **2019**, *120*, 175–183. [CrossRef]
74. Kim, Y.K.; Park, S.H.; Yu, J.H.; AlMangour, B.; Lee, K.A. Improvement in the high-temperature creep properties via heat treatment of Ti-6Al-4V alloy manufactured by selective laser melting. *Mater. Sci. Eng. A* **2018**, *715*, 33–40. [CrossRef]
75. Hemmasian Ettefagh, A.; Zeng, C.; Guo, S.; Raush, J. Corrosion behavior of additively manufactured Ti-6Al-4V parts and the effect of post annealing. *Addit. Manuf.* **2019**, *28*, 252–258. [CrossRef]
76. Wang, M.; Wu, Y.; Lu, S.; Chen, T.; Zhao, Y.; Chen, H.; Tang, Z. Fabrication and characterization of selective laser melting printed Ti–6Al–4V alloys subjected to heat treatment for customized implants design. *Prog. Nat. Sci. Mater. Int.* **2016**, *26*, 671–677. [CrossRef]
77. Pazhanivel, B.; Sathiya, P.; Sozhan, G. Ultra-fine bimodal ($\alpha + \beta$) microstructure induced mechanical strength and corrosion resistance of Ti-6Al-4V alloy produced via laser powder bed fusion process. *Opt. Laser Technol.* **2020**, *125*, 106017. [CrossRef]
78. Leon, A.; Levy, G.K.; Ron, T.; Shirizly, A.; Aghion, E. The effect of hot isostatic pressure on the corrosion performance of Ti-6Al-4 V produced by an electron-beam melting additive manufacturing process. *Addit. Manuf.* **2020**, *33*, 101039. [CrossRef]
79. Sharma, A.; Oh, M.C.; Kim, J.T.; Srivastava, A.K.; Ahn, B. Investigation of electrochemical corrosion behavior of additive manufactured Ti–6Al–4V alloy for medical implants in different electrolytes. *J. Alloys Compd.* **2020**, *830*, 154620. [CrossRef]
80. Nalli, F.; Bottini, L.; Boschetto, A.; Cortese, L.; Veniali, F. Effect of industrial heat treatment and barrel finishing on the mechanical performance of Ti6AL4V processed by selective laser melting. *Appl. Sci.* **2020**, *10*, 2280. [CrossRef]
81. Wysocki, B.; Maj, P.; Sitek, R.; Buhagiar, J.; Kurzydłowski, K.J.; Świeszkowski, W. Laser and electron beam additive manufacturing methods of fabricating titanium bone implants. *Appl. Sci.* **2017**, *7*, 657. [CrossRef]
82. Zhai, Y.; Galarraga, H.; Lados, D.A. Microstructure, static properties, and fatigue crack growth mechanisms in Ti-6Al-4V fabricated by additive manufacturing: LENS and EBM. *Eng. Fail. Anal.* **2016**, *69*, 3–14. [CrossRef]

83. Vrancken, B.; Thijs, L.; Kruth, J.P.; Van Humbeeck, J. Heat treatment of Ti6Al4V produced by Selective Laser Melting: Microstructure and mechanical properties. *J. Alloys Compd.* **2012**, *541*, 177–185. [CrossRef]
84. Vilaro, T.; Colin, C.; Bartout, J.D. As-fabricated and heat-treated microstructures of the Ti-6Al-4V alloy processed by selective laser melting. *Metall. Mater. Trans. A Phys. Metall. Mater. Sci.* **2011**, *42*, 3190–3199. [CrossRef]
85. Koike, M.; Greer, P.; Owen, K.; Lilly, G.; Murr, L.E.; Gaytan, S.M.; Martinez, E.; Okabe, T. Evaluation of titanium alloys fabricated using rapid prototyping technologies-electron beam melting and laser beam melting. *Materials* **2011**, *4*, 1776–1792. [CrossRef]
86. ASTM F136-08. *Standard Specification for Wrought Titanium-6Aluminum-4Vanadium ELI (Extra Low Interstitial) Alloy for Surgical Implant Applications (UNS R56401)*; ASTM International: West Conshohocken, PA, USA, 2008. Available online: www.astm.org (accessed on 24 June 2020).
87. Edwards, P.; O'Conner, A.; Ramulu, M. Electron Beam Additive Manufacturing of Titanium Components: Properties and Performance. *J. Manuf. Sci. Eng.* **2013**, *135*, 061016. [CrossRef]
88. Hayes, B.J.; Martin, B.W.; Welk, B.; Kuhr, S.J.; Ales, T.K.; Brice, D.A.; Ghamarian, I.; Baker, A.H.; Haden, C.V.; Harlow, D.G.; et al. Predicting tensile properties of Ti-6Al-4V produced via directed energy deposition. *Acta Mater.* **2017**, *133*, 120–133. [CrossRef]
89. Benzing, J.; Hrabe, N.; Quinn, T.; White, R.; Rentz, R.; Ahlfors, M. Hot isostatic pressing (HIP) to achieve isotropic microstructure and retain as-built strength in an additive manufacturing titanium alloy (Ti-6Al-4V). *Mater. Lett.* **2019**, *257*, 126690. [CrossRef]
90. Kasperovich, G.; Hausmann, J. Improvement of fatigue resistance and ductility of TiAl6V4 processed by selective laser melting. *J. Mater. Process. Technol.* **2015**, *220*, 202–214. [CrossRef]
91. Leuders, S.; Thöne, M.; Riemer, A.; Niendorf, T.; Tröster, T.; Richard, H.A.; Maier, H.J. On the mechanical behaviour of titanium alloy TiAl6V4 manufactured by selective laser melting: Fatigue resistance and crack growth performance. *Int. J. Fatigue* **2013**, *48*, 300–307. [CrossRef]
92. Simonelli, M.; Tse, Y.Y.; Tuck, C. Effect of the build orientation on the mechanical properties and fracture modes of SLM Ti-6Al-4V. *Mater. Sci. Eng. A* **2014**, *616*, 1–11. [CrossRef]
93. Galati, M.; Saboori, A.; Biamino, S.; Calignano, F.; Lombardi, M.; Marchiandi, G.; Minetola, P.; Fino, P.; Iuliano, L. Ti-6Al-4V lattice structures produced by EBM: Heat treatment and mechanical properties. *Procedia CIRP* **2020**, *88*, 411–416. [CrossRef]
94. Morita, T.; Tsuda, C.; Nakano, T. Influences of scanning speed and short-time heat treatment on fundamental properties of Ti-6Al-4V alloy produced by EBM method. *Mater. Sci. Eng. A* **2017**, *704*, 246–251. [CrossRef]
95. Zhang, X.Y.; Fang, G.; Leeflang, S.; Böttger, A.J.; Zadpoor, A.A.; Zhou, J. Effect of subtransus heat treatment on the microstructure and mechanical properties of additively manufactured Ti-6Al-4V alloy. *J. Alloys Compd.* **2018**, *735*, 1562–1575. [CrossRef]
96. Sabban, R.; Bahl, S.; Chatterjee, K.; Suwas, S. Globularization using heat treatment in additively manufactured Ti-6Al-4V for high strength and toughness. *Acta Mater.* **2019**, *162*, 239–254. [CrossRef]
97. Tsai, M.T.; Chen, Y.W.; Chao, C.Y.; Jang, J.S.C.; Tsai, C.C.; Su, Y.L.; Kuo, C.N. Heat-treatment effects on mechanical properties and microstructure evolution of Ti-6Al-4V alloy fabricated by laser powder bed fusion. *J. Alloys Compd.* **2020**, *816*, 152615. [CrossRef]
98. Galarraga, H.; Warren, R.J.; Lados, D.A.; Dehoff, R.R.; Kirka, M.M.; Nandwana, P. Effects of heat treatments on microstructure and properties of Ti-6Al-4V ELI alloy fabricated by electron beam melting (EBM). *Mater. Sci. Eng. A* **2017**, *685*, 417–428. [CrossRef]
99. Zhang, D.; Wang, L.; Zhang, H.; Maldar, A.; Zhu, G.; Chen, W.; Park, J.S.; Wang, J.; Zeng, X. Effect of heat treatment on the tensile behavior of selective laser melted Ti-6Al-4V by in situ X-ray characterization. *Acta Mater.* **2020**, *189*, 93–104. [CrossRef]
100. Ter Haar, G.M.; Becker, T.H. Selective laser melting produced Ti-6Al-4V: Post-process heat treatments to achieve superior tensile properties. *Materials* **2018**, *11*, 146. [CrossRef]
101. Mercelis, P.; Kruth, J.P. Residual stresses in selective laser sintering and selective laser melting. *Rapid Prototyp. J.* **2006**, *12*, 254–265. [CrossRef]
102. Kruth, J.P.; Deckers, J.; Yasa, E.; Wauthlé, R. Assessing and comparing influencing factors of residual stresses in selective laser melting using a novel analysis method. *Proc. Inst. Mech. Eng. Part B J. Eng. Manuf.* **2012**, *226*, 980–991. [CrossRef]

103. Zaeh, M.F.; Branner, G. Investigations on residual stresses and deformations in selective laser melting. *Prod. Eng.* **2010**, *4*, 35–45. [CrossRef]
104. Ali, H.; Ghadbeigi, H.; Mumtaz, K. Effect of scanning strategies on residual stress and mechanical properties of Selective Laser Melted Ti6Al4V. *Mater. Sci. Eng. A* **2018**, *712*, 175–187. [CrossRef]
105. Song, J.; Wu, W.; Zhang, L.; He, B.; Lu, L.; Ni, X.; Long, Q.; Zhu, G. Role of scanning strategy on residual stress distribution in Ti-6Al-4V alloy prepared by selective laser melting. *Optik* **2018**, *170*, 342–352. [CrossRef]
106. Lu, Y.; Wu, S.; Gan, Y.; Huang, T.; Yang, C.; Junjie, L.; Lin, J. Study on the microstructure, mechanical property and residual stress of SLM Inconel-718 alloy manufactured by differing island scanning strategy. *Opt. Laser Technol.* **2015**, *75*, 197–206. [CrossRef]
107. Maawad, E.; Sano, Y.; Wagner, L.; Brokmeier, H.G.; Genzel, C. Investigation of laser shock peening effects on residual stress state and fatigue performance of titanium alloys. *Mater. Sci. Eng. A* **2012**, *536*, 82–91. [CrossRef]
108. Tong, Z.; Ren, X.; Ren, Y.; Dai, F.; Ye, Y.; Zhou, W.; Chen, L.; Ye, Z. Effect of laser shock peening on microstructure and hot corrosion of TC11 alloy. *Surf. Coat. Technol.* **2018**, *335*, 32–40. [CrossRef]
109. Huang, S.; Zhu, Y.; Guo, W.; Peng, P.; Diao, X. Impact toughness and microstructural response of Ti-17 titanium alloy subjected to laser shock peening. *Surf. Coat. Technol.* **2017**, *327*, 32–41. [CrossRef]
110. Zhang, X.C.; Zhang, Y.K.; Lu, J.Z.; Xuan, F.Z.; Wang, Z.D.; Tu, S.T. Improvement of fatigue life of Ti-6Al-4V alloy by laser shock peening. *Mater. Sci. Eng. A* **2010**, *527*, 3411–3415. [CrossRef]
111. Ren, X.D.; Zhou, W.F.; Liu, F.F.; Ren, Y.P.; Yuan, S.Q.; Ren, N.F.; Xu, S.D.; Yang, T. Microstructure evolution and grain refinement of Ti-6Al-4V alloy by laser shock processing. *Appl. Surf. Sci.* **2016**, *363*, 44–49. [CrossRef]
112. Guo, W.; Sun, R.; Song, B.; Zhu, Y.; Li, F.; Che, Z.; Li, B.; Guo, C.; Liu, L.; Peng, P. Laser shock peening of laser additive manufactured Ti6Al4V titanium alloy. *Surf. Coat. Technol.* **2018**, *349*, 503–510. [CrossRef]
113. Jin, X.; Lan, L.; Gao, S.; He, B.; Rong, Y. Effects of laser shock peening on microstructure and fatigue behavior of Ti–6Al–4V alloy fabricated via electron beam melting. *Mater. Sci. Eng. A* **2020**, *780*, 139199. [CrossRef]
114. Lan, L.; Jin, X.; Gao, S.; He, B.; Rong, Y. Microstructural evolution and stress state related to mechanical properties of electron beam melted Ti-6Al-4V alloy modified by laser shock peening. *J. Mater. Sci. Technol.* **2020**, *50*, 153–161. [CrossRef]
115. Eyzat, Y.; Chemkhi, M.; Portella, Q.; Gardan, J.; Remond, J.; Retraint, D. Characterization and mechanical properties of As-Built SLM Ti-6Al-4V subjected to surface mechanical post-treatment. *Procedia CIRP* **2019**, *81*, 1225–1229. [CrossRef]
116. Yakout, M.; Elbestawi, M.A.; Veldhuis, S.C. A study of the relationship between thermal expansion and residual stresses in selective laser melting of Ti-6Al-4V. *J. Manuf. Process.* **2020**, *52*, 181–192. [CrossRef]
117. Vastola, G.; Zhang, G.; Pei, Q.X.; Zhang, Y.W. Controlling of residual stress in additive manufacturing of Ti6Al4V by finite element modeling. *Addit. Manuf.* **2016**, *12*, 231–239. [CrossRef]
118. Ning, J.; Praniewicz, M.; Wang, W.; Dobbs, J.R.; Liang, S.Y. Analytical modeling of part distortion in metal additive manufacturing. *Int. J. Adv. Manuf. Technol.* **2020**, *107*, 49–57. [CrossRef]

metals

MDPI

Article

Effects of Thermal Cycling on Wire and Arc Additive Manufacturing of Al-5356 Components

Markus Köhler *, Jonas Hensel and Klaus Dilger

Institute of Joining and Welding, Technische Universität Braunschweig, Langer Kamp 8, 38106 Braunschweig, Germany; j.hensel@tu-braunschweig.de (J.H.); k.dilger@tu-braunschweig.de (K.D.)
* Correspondence: markus.koehler@tu-braunschweig.de; Tel.: +49-531-391-955-32

Received: 23 June 2020; Accepted: 12 July 2020; Published: 15 July 2020

check for updates

Abstract: Wire and arc additive manufacturing (WAAM) provides a promising alternative to conventional machining for the production of large structures with complex geometry, as well as individualized low quantity components, using cost-efficient production resources. Due to the layer-by-layer build-up approach, process conditions, such as energy input, deposition patterns and heat conduction during the additive manufacturing process result in a unique thermal history of the structure, affecting the build-up properties. This experimental study aims to describe the effects of thermal cycling on the geometrical and material properties of wire arc additive manufactured Al-5356 aluminum alloy. Under consideration, that Al-5356 is a non-heat treatable alloy, a significant effect on geometrical formation is expected. Linear wall samples were manufactured using pulsed cold metal transfer (CMT-P) under variation of wire-feed rate, travel speed and interpass temperatures. The samples were analyzed in terms of geometry; microstructural composition; hardness and residual stress. Furthermore, the mechanical properties were determined in different building directions.

Keywords: direct energy deposition; WAAM; cold metal transfer; 5356-aluminum; temperature distribution; mechanical properties

1. Introduction

In recent years, additive manufacturing (AM) of metallic components gained growing interest in a wide range of industrial sectors such as civil engineering, turbine construction as well as automotive and aerospace industry [1–4]. Thereby, the economic efficiency compared to conventional manufacturing processes is mainly determined by high geometrical complexity and/or small series production [5,6]. In order to meet the requirements of different industrial sectors, a variety of AM processes, following different approaches of material deposition, have been developed. These processes can mainly be classified into powder-bed, powder and wire-feed processes [7]. Due to higher deposition rates and significantly higher maximum part size, wire-based processes such as wire and arc additive manufacturing (WAAM) are particularly suitable for near net-shape production of large components [8]. Furthermore, wire filler metals are more cost-efficient and available in a wider range of materials compared to powder materials [9,10]. On the other hand, the geometrical accuracy and the attainable component complexity are lower compared to powder-based processes [7].

During WAAM processing, the component is generated by a layer-wise deposition of molten filler metal. Melting of the filler material is achieved using electric arc or plasma as a heat source, thus combining well known arc welding processes such as gas metal arc welding (GMAW), gas tungsten arc welding (GTAW) or plasma arc welding (PAW) with wire as the feedstock [11–13]. Due to a simultaneously energy input and wire feed, the wire-feed orientation during GMAW is perpendicular to the forming substrate, offering advantages in automation and productivity [14].

Basic prerequisite for the application of AM technologies is that the resulting component properties meet the requirements of the operation purpose. This applies in particular to mechanical properties such as hardness, strength and ductility as well as to specific requirements such as corrosion resistance. Due to the layer-by-layer build-up approach, both the WAAM process itself and the geometrical and mechanical properties of the resulting components are directly related to the welding parameters and the deposited material. Compared to conventionally manufactured components, the primary material of AM components merely consists of weld metal. Accordingly, there is an increasing number of studies on material-specific processability and component properties. Regarding the resulting geometry, studies in [15,16] have shown an increase in surface waviness with increased energy input per length unit, reporting the main dependencies of geometry formation on heat input and heat accumulation during the AM process. Other than the effects on the resulting geometry, the welding process always results in a heat treatment of the surrounding material during deposition, causing microstructural changes in the current layer and the adjacent material. Depending on the deposited material, this can lead a non-uniform distribution of the mechanical properties [17]. Regarding this, various studies on WAAM of low-alloy and high-grade steels reported an increase in hardness and grain refinement towards the top layers of the deposited structure [11,18,19]. Aiming to reduce the influence of thermal cycling during WAAM and to actively affect the temperature–time regime during deposition as well as the resulting grain structure, recent studies have explored the implementation of active cooling systems [20–22]. However, most of the current research regarding the influence of temperature regimes on the resulting properties during additive manufacturing is focused on steel and titanium alloys. Research on WAAM using various aluminum alloys is mainly focused on microstructural evolution, pore formation as well as the effects of different arc modes on microstructure characteristics and mechanical properties [14,23–26].

Summarizing, the specific objective of this study is to investigate the heat accumulation and the influence of different temperature-time regimes on the resulting component properties during wire and arc additive manufacturing of Al-5356, providing a contribution on geometry formation and mechanical properties.

2. Materials and Methods

The experimental setup is presented in Figure 1a. A Fronius CMT Advanced 4000 R welding system (Fronius International GmbH, Wels, Austria) was used in CMT-Pulse mode with current characteristics according to the deposited material. The welding torch was connected to a KUKA KR22 6-axis robotic system (KUKA Deutschland GmbH, Augsburg, Germany). The presented studies have been conducted using thin-walled specimen with a length of 300 mm and a build-up height of approximately 140 mm. The process sequence is given in Figure 1b. Depending on the underlying process parameters, 100–120 single pass layers were conducted during sample preparation.

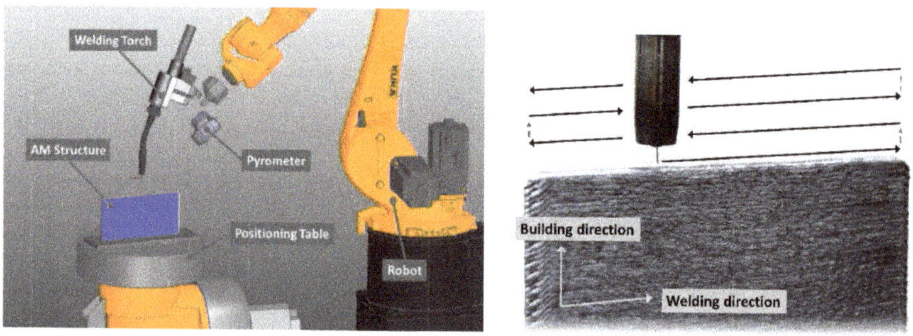

(**a**) (**b**)

Figure 1. Experimental setup (**a**) and process sequence for sample preparation (**b**).

The experimental trials were carried out on Al-6082 base plates with dimensions of 400 × 125 × 15 mm^3. The base plate material was chosen as a typical aluminum alloy ensuring good weldability and thermal conductivity. As the presented research is focused on the material and geometrical properties of the AM component, the base plate material does not affect the findings. Prior to welding, organic residues on the base plate were removed with acetone. The deposited wire was an Al-5356 solid wire electrode with a diameter of 1.0 mm. The chemical composition of the used materials is given in Table 1.

Table 1. Nominal composition of welding wire and substrate material.

Alloy	Chemical Composition (wt%)					
	Al	Mn	Mg	Si	Fe	Cu
Welding Wire: Al-5356	bal	0.15	5.0	0.05	0.15	-
Substrate: Al-6082	bal	0.4–1.0	0.60–1.2	0.7–1.3	<0.5	<0.1

Referring to previous findings, welding parameters such as wire feed and welding speed show a great influence on the geometrical characteristics of the components processed by WAAM. In order to further describe the influence of the process parameters on geometrical properties, thin-walled samples were manufactured with two different settings of process parameters (Series A, Series B). The process parameters of Series A and Series B were chosen to result in similar energy input per unit length to ensure a comparable heat input during welding. Aiming to obtain different temperature profiles during the build-up sequence, thin-walled samples were prepared using different intermediate times (t_i) between layers. Table 2 provides the welding parameters used for processing the samples. For all specimens, the preheating temperature of the base plate was set at approximately 100 °C. During build-up, a contact tube to work piece distance of 10 ± 3 mm was maintained. Pure Argon (Ar ≥ 99.996%) was used as shielding gas with a flow rate of 14 L/min.

Table 2. Welding parameters for processing of samples.

Parameter	Symbol	Unit	Series A	Series B
Wire feed	v_W	m/min	8	10
Welding speed	v_S	cm/min	60	76
Voltage	I	V	15.5–16.5	16.5–17.5
Current	U	A	91–92	111–112
Time between layers	ti	s	30/60/120	30/60/120

In order to describe the occurring temperatures during welding, type-K thermocouples were positioned at the base plate at a distance of 9.5 ± 0.8 mm to the fusion line. Further, an infrared pyrometer (Optris CTlaser 3MH, Optris GmbH, Berlin, Germany) with a measuring range from 100 to 600 °C was used to determine the interpass temperature in the middle of the respective topmost layer during build-up.

Prior to mechanical testing, the specimens were visually examined and measured regarding the resulting height and width. To assess changes in geometry during build-up, a width measurement was conducted at ten measurement points in building direction. Further, residual stress was analyzed by means of X-ray diffraction in the Series A specimen using a cobalt radiation source. Several measurements were taken in welding direction on the top layer and against the building direction. To produce a homogeneous surface for the measurement, electrolytic polishing was performed in the measuring area before testing. Stresses were determined from diffraction patterns by means of the $\sin^2\Psi$ method using the (hkl) crystalline plane {331} of aluminum at eight Ψ-angles between 144° and 154°. The elastic constant was $0.5S_2$ = 18.93 mm^2/N.

Samples for metallographic cross-sections, hardness measurement and tensile testing were sectioned from the AM structures as depicted in Figure 2. The start and end sections of the thin-wall were discarded. Metallographic samples were polished and etched using 2% hydrofluoric acid for macroscopic examination and Kroll etchant for microsections. Hardness measurements (HV1) were conducted equidistant on the cross-sectional samples using hardness testing according to Vickers (TUKON1202, ITW Test & Measurement GmbH, Esslingen, Germany). In addition, the micro hardness (HV0.005) was measured at different areas of the microsections depending on the microstructure composition. Samples for tensile testing were prepared and tested according to EN ISO 6892-1 standards (Figure 2b) at an extension rate of 3.2 mm/min.

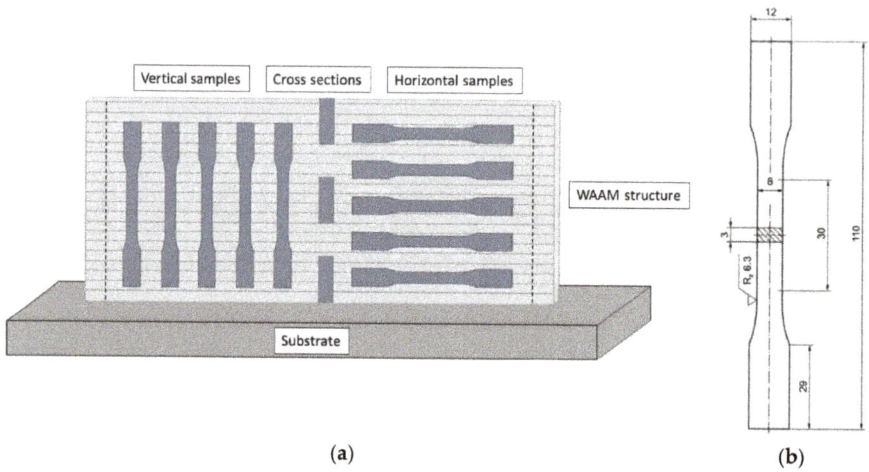

Figure 2. Schematic representation of sample extraction (**a**) and dimensions of tensile samples according to DIN 50125 (**b**).

3. Results

3.1. Temperature Formation During WAAM

In order to describe temperature related effects on structural and mechanical properties of additive manufactured structures, temperature measurements were carried out on both the base plate and on the top layer during build-up. This allowed the determination of the temperature–time regimes during WAAM build-up with different intermediate times between layers. Due to similar energy input per unit length, both regarded parameter settings (Series A, B) resulted in comparable temperature profiles. Differences in the resulting geometry based on varying parameter settings, as described in Section 3.2, only showed a minor influence on the heat distribution due to a similar heat input. For the purpose of improved clarity, only temperature profiles of Series B are discussed below. However, the results can be transferred to Series A likewise.

The resulting temperature-time regime at the base plate is shown in Figure 3. Particular differences in temperature distribution can be detected during the first 10 to 15 build-up layers. For intermediate times of 120 s and 60 s, the graphs show a slight decrease in the average temperature due to heat dissipation at the base plate and the welding table. During build-up with an intermediate time of 30 s, no significant decrease in the base plate temperature could be observed. Further progression of the temperature profiles shows an increase in base plate temperature, reaching a peak at approximately half of the build-up height. During further build-up the base plate temperature decreases. The visible discontinuities in the curves for intermediate times of 120 s and 30 s near the top layers are caused by process interruptions. Due to their location outside of the tensile samples gauge length, the interruptions have no major effect on further results.

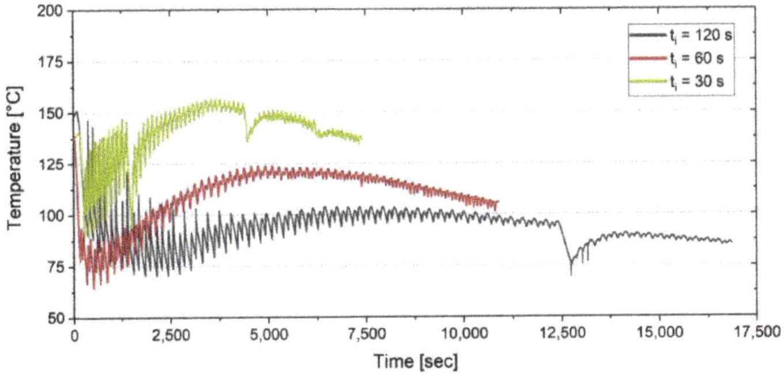

Figure 3. Temperature distribution on the baseplate by example of Series B specimen.

The resulting temperature profiles on the base plate can be explained by basic mechanisms of heat flow. The heat flow in the thin-walled structure is described by terms of heat conduction, convection and thermal radiation. Neglecting the heat flow by thermal radiation, the heat flow at the base plate is mainly determined by heat conduction and convection. The heat conduction in the thin-walled structure is given by (1) with thermal conductivity λ, the cross-section area A and the wall height h:

$$\dot{Q}_h = -\lambda \cdot A \cdot \frac{dT}{dh} \tag{1}$$

Further, the heat flow to the environment due to convection is given by (2) with the heat transfer coefficient η, the surface area of the wall S and the temperature difference between wall and room temperature:

$$\dot{Q}_c = \eta \cdot S \cdot (T_W - T_0) \tag{2}$$

Relating to the periodically growing wall structure, the wall height increases over time, leading to a constant decrease of the heat flow $\dot{Q}_h$. By contrast, the heat flow due to convection increases over time as the wall surface increases, leading to a higher heat transfer to the environment. In combination, the temperature on the base plate decreases with increasing wall height, after reaching a maximum at a build-up height of about 70 mm.

The following Figure 4 shows the development of interpass temperature for the examined intermediate times between layers during additive manufacturing. The data indicates a correlation between interpass and base plate temperature during the first 30 layers. Further build-up leads to increasing interpass temperatures, reaching saturation depending on the intermediate time between the layers. Thereby, the determined interpass temperature substantially exceeds the base plate temperature for intermediate times of 60 s and 30 s.

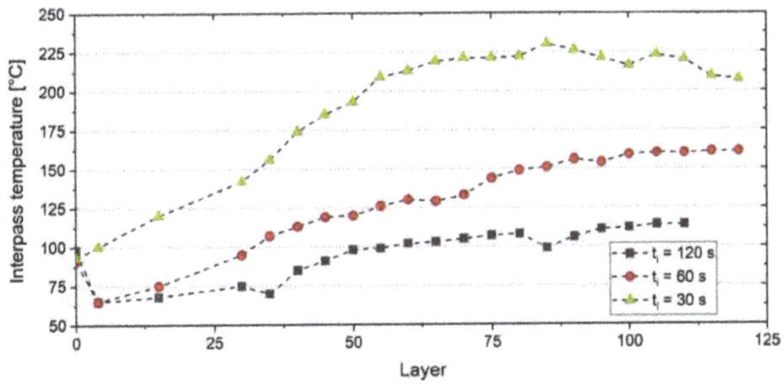

Figure 4. Interpass temperature depending on number of layers and intermediate time between layers by example of Series B specimen.

The results on temperature distribution identify that the measuring point is of key importance in order to describe the heat distribution and correlations between energy input and occurring temperatures. Based on the examined thin-walled structure, a direct influence of the continuously changing geometry during additive manufacturing on the temperature distribution within the component could be shown.

3.2. *Effects on Resulting Geometry*

During additive manufacturing of complex components with high requirements on geometrical and material accuracy, information about the effects of process parameters on the resulting geometry is of primary importance. Aiming to describe the influence of the underlying welding parameters at different temperature–time regimes on the resulting geometry, the average layer height as well as the wall width were determined at the final samples.

The material accumulation during WAAM is presented in Figure 5 by means of cross-section micrographs taken from the center of the respective thin-walled structures. Overall, the micrographs show a defined deposition pattern throughout the considered range of the building parameters. What stands out is a periodical lateral shift of the layers at the Series A (t_i = 30 s) sample. These shifts only show minor influence on the overall waviness of the wall surface. By visual inspection, a more even surface could be obtained using an intermediate time between layers of $t_i \leq 60$ s resulting in interpass temperatures $T_i > 125$ °C. Further, a fine dispersed porosity could be detected in the Series A (t_i = 120 s) sample.

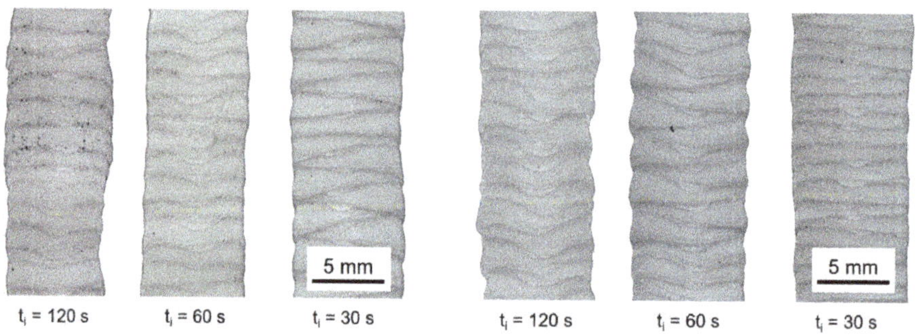

Figure 5. Cross-section micrographs of wire and arc additive manufacturing (WAAM) samples with varying process strategies.

Previous findings reported in [27,28] have shown the influence of wire feed and welding speed on the height and width of the resulting geometry. It was reported that the wall width increases at higher wire-feed rates and decreases at higher welding speeds, whereas the wire feed shows a dominant influence. Further, the layer height increases at higher wire feed and decreases at higher welding speed with dominant influence of the welding speed.

Figure 6 compares the resulting layer height (Figure 6a) and the wall width (Figure 6b) as a function of the intermediate time t_i between each layer for both welding strategies considered. For both welding strategies, the average layer height shows a steady increase with increasing waiting time between layers. This indicates that higher interpass temperatures lead to a reduction in layer height. On average, the layer height of Series A is about 0.15 mm higher compared to Series B, confirming the dominant influence of the welding speed on the average layer height reported in [27]. Although the difference in height is low in regard to single layers, the differences add up during build-up and may cause process instability or errors. The wall width showed higher values for Series B specimen due to the influence of a higher wire feed used compared to Series A specimen. Further, the wall width was found to increase in the building direction, indicating an influence of the interpass temperature.

Since the interpass temperatures increase in building direction (Section 3.1.), it can be concluded that higher interpass temperatures lead to an increase in wall width. This also leads to comparatively high deviations due to average determination on basis of ten measured values in building direction.

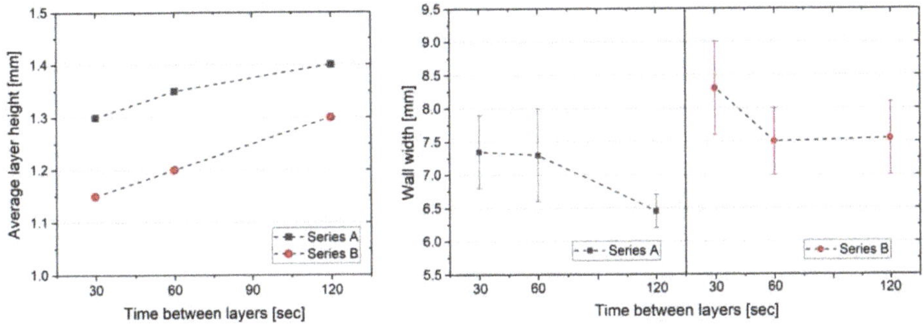

(**a**) (**b**)

Figure 6. Effect of building strategy on average layer height (**a**) and wall width (**b**).

The investigation of the resulting geometry indicated a different geometrical formation despite similar energy input per length unit for the welding strategies considered and thus confirming a different emphasis of the main welding parameters influence on geometrical formation during WAAM. Accordingly, a verification of the geometry after any adaption of the process parameters is suggested.

3.3. *Residual Stress and Mechanical Properties*

The longitudinal residual stress was analyzed on samples of Series A in the as-welded condition manufactured. Figure 7a shows the stress distribution determined on the top layer of the specimens. The residual stress in welding direction varies along the measuring path from approximately −75 MPa to +150 MPa. The individual scatter at discrete measuring locations from the $\sin^2$-calculation is in the order of ±30 to ±50 MPa due to coarse microstructure. However, a clear influence of the intermediate time between layers is not visible.

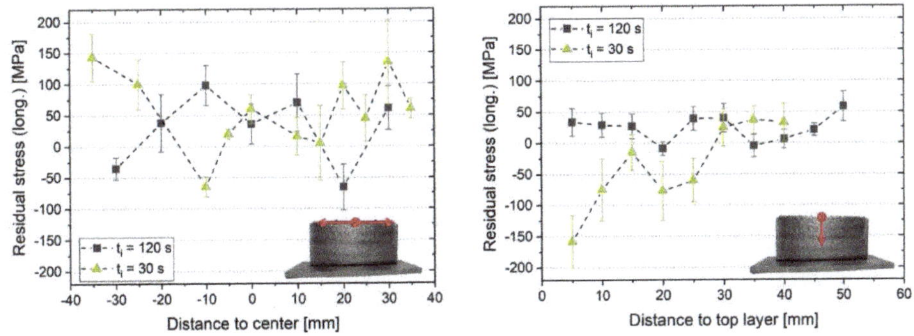

(**a**) (**b**)

Figure 7. Residual stress in welding (**a**) and building (**b**) direction (Series A).

Figure 7b depicts the residual stress distribution at the side of the specimens, measured from top layer downwards. The abscissa indicates the distance of the measuring point from the top layer with positive values directing to the base plate. In addition, shown here are the residual stresses in the welding direction. The residual stress magnitudes are more homogeneous than on the top layer and the scatter is smaller. The residual stress ranges from approximately 0 MPa to +50 MPa at an intermediate time of 120 s. A tendency of decreasing residual stress can be observed near the top layers at an intermediate time of 30 s.

Overall, the residual stress in the investigated WAAM aluminum structure appears to be of low magnitude. This can be explained by the heat conduction properties of the alloy in combination with its low yield strength. This leads to homogeneous temperature distributions during relatively fast cooling times. Further, the metal is relatively soft causing yielding of residual stress. These effects are also known from conventional fusion welding of aluminum alloys.

The hardness test according to Vickers was conducted using cross-section samples from the bottom, middle and top part from the thin-walled samples (see Figure 2). The hardness distributions for different intermediate time between the layers in the as-build condition are given in Figure 8 by example of Series B specimen. Hardness values ranged from 66 to 72 HV1 showing no significant dependency on the build-up height and therefore stating independency of hardness values on temperature distribution. This result can be explained by the fact that the investigated aluminum Al-5356 is a non-heat treatable alloy.

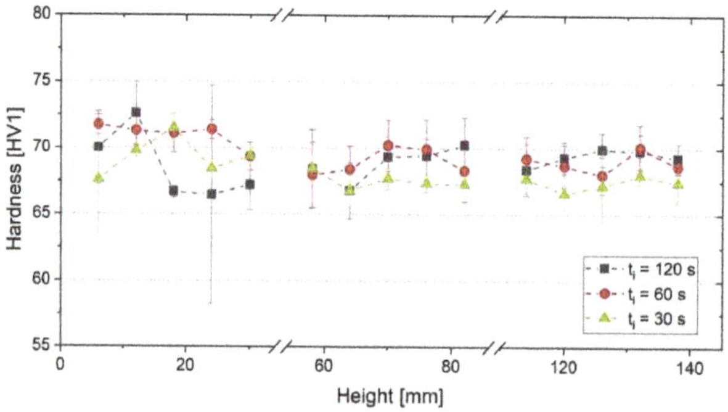

Figure 8. Hardness distribution depending on build-up height and intermediate time between layers by example of Series B.

Aiming to describe the tensile properties of additive manufactured Al-5356 depending on the investigated process settings, yield strength ($R_{p0.2}$), tensile strength and elongation were determined horizontally and vertically to the welding direction. The results on tensile properties can be compared in Figure 9.

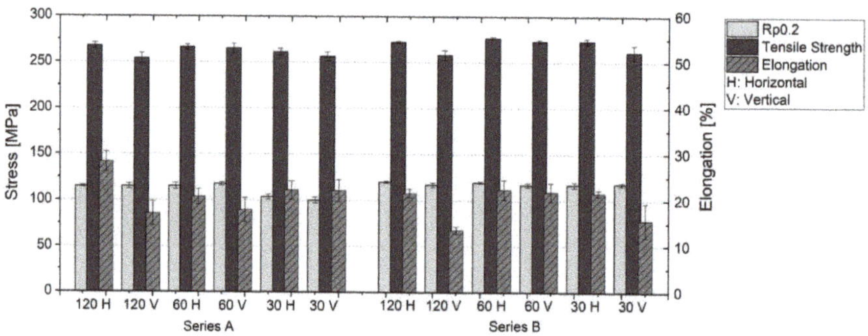

Figure 9. Tensile properties of WAAM processed Al-5356.

Depending on the welding parameters and test direction, the average values of yield strength reach 115–120 MPa, ultimate tensile strength 254–277 MPa and elongation 14–28%. Therefore, all batches achieved both strength and ductility properties of wrought material (EN AW-5019 (F): $R_m \geq 250$ MPa; $R_{p0,2} \geq 110$ MPa; $A \geq 14\%$). What stands out is that tensile strength and elongation were generally found to be lower for batches in vertical direction compared to horizontal samples (see Figure 9). Especially at long intermediate times between layers, i.e., lower interpass temperatures, the elongation showed up to 40% lower values in vertical direction. The higher deviation in the values for vertical specimens from Series B welded with $t_i = 30$ s can be attributed to fracture locations near the end of the gage length and therefore measurement of elongation could contain errors. The result of an anisotropic material behavior may be explained by an inhomogeneous microstructure at the fusion line boundaries, resulting in less favorable conditions for the vertical load direction. On basis of the microscopic examination shown in Figure 10, three distinct areas of different metallurgical properties in the layer structure could be determined. Near the fusion line boundary, regions with a fine grain microstructure and comparatively high micro hardness are formed (Figure 10a). By contrast, regions below the interlayer boundary are characterized by coarse grain microstructure with an increased formation of segregations (Al_8Mg_5) at the grain boundaries and lower hardness (Figure 10c). The grain growth may be explained by the thermal influence of the following layer. An interlayer boundary zone with a further increased occurrence of segregations can be identified (Figure 10b). In these regions, primary stages and segregations merged at the grain boundaries as a result of comparatively long heat exposure. Due to the difference in hardness, grain size and the formation of segregations, these areas can act as metallurgical notches, thus affecting the mechanical properties in vertical loading direction. The most consistent results regarding the anisotropic material behavior could be found for Series B specimen with intermediate time between layers of 30 s. Although interpass temperature was increasing during the built-up sequence, no distinct effect of sample allocation on the tensile properties could be observed.

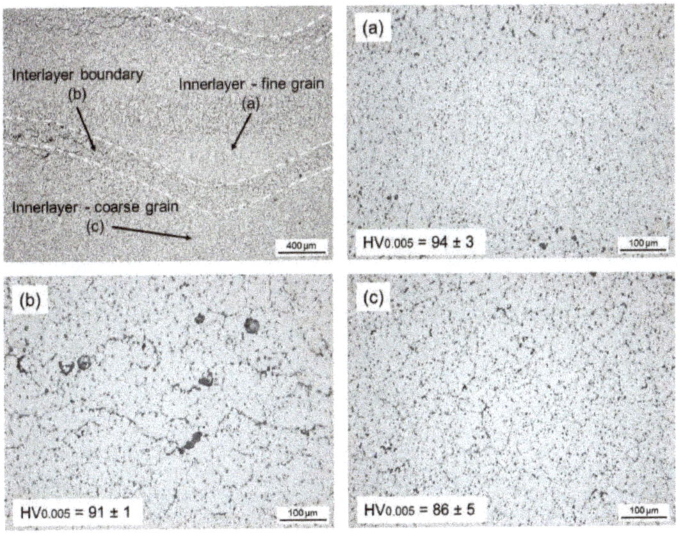

Figure 10. Microstructure and micro hardness of innerlayer fine grain (**a**), interlayer boundary (**b**) and innerlayer coarse grain (**c**) regions for Series A specimen with $t_i = 120$ s.

4. Conclusions

Within the underlying study, the effects of different temperature-time regimes during wire and arc additive manufacturing of Al-5356 on both the geometrical and mechanical properties have been systematically investigated. Linear thin-walled samples were built and analyzed with the following conclusions:

- Temperature profiles during WAAM processing indicated that the temperature distribution is directly related to the geometry of the additive manufactured component. Due to a continuously changing geometry, a non-steady temperature distribution during WAAM can be assumed. However, steady geometrical build-up results can be expected by obtaining uniform temperature fields.
- Although different process strategies, using the same energy input per length unit, lead to comparable temperature distributions and cooling rates, they showed differences in the resulting geometry of the component. An increase in wire feed and welding speed resulted in lower layer height and increased wall width. Furthermore, higher interpass temperatures led to an increase in wall width and lower layer height.
- Residual stresses in the top layer and on the surface of the wall specimen appeared of low magnitude due to small temperature gradients combined with a comparably low yield strength of the regarded material.
- Material properties were evenly distributed over the build-up geometry with values comparable to AlMg5 wrought material. No significant dependency on the temperature-time regime during processing could be detected. However, the mechanical properties, especially the elongation at fracture, showed dependencies on the loading in relation to the build-up direction.

Author Contributions: Conceptualization, M.K. and J.H.; methodology, M.K. and J.H.; validation, J.H.; formal analysis, M.K.; investigation, M.K.; resources, K.D.; data curation, J.H.; writing—original draft preparation, M.K.; writing—review and editing, J.H.; visualization, M.K.; supervision, K.D.; project administration, J.H. and K.D. All authors have read and agreed to the published version of the manuscript.

Funding: This research received no external funding.

Acknowledgments: We acknowledge support by the German Research Foundation and the Open Access Publication Funds of the Technische Universität Braunschweig.

Conflicts of Interest: The authors declare no conflict of interest.

References

1. Buchanan, C.; Gardner, L. Metal 3D printing in construction: A review of methods, research, applications, opportunities and challenges. *Eng. Struct.* **2019**, *180*, 332–348. [CrossRef]
2. Frazier, W.E. Metal Additive Manufacturing: A Review. *J. Mater. Eng. Perform.* **2014**, *23*, 1917–1928. [CrossRef]
3. Grunwald, R.; Mayer, M.; Schörghuber, M. WAAM-Technologie und aktuelle Anwendungen: DVS-Berichte Congress 2018. *DVS Media GmbH* **2018**, *344*, 97–103.
4. Brandl, E.; Baufeld, B.; Leyens, C.; Gault, R. Additive manufactured Ti-6Al-4V using welding wire: Comparison of laser and arc beam deposition and evaluation with respect to aerospace material specifications. *Phys. Procedia* **2010**, *5*, 595–606. [CrossRef]
5. Busachi, A.; Erkoyuncu, J.; Colegrove, P.; Martina, F.; Watts, C.; Drake, R. A review of Additive Manufacturing technology and Cost Estimation techniques for the defense sector. *CIRP J. Manuf. Sci. Technol.* **2017**, *19*, 117–128. [CrossRef]
6. Cunningham, C.; Wikshåland, S.; Xu, F.; Kemakolam, N.; Shokrani, A.; Dhokia, V.; Newman, S. Cost Modelling and Sensitivity Analysis of Wire and Arc Additive Manufacturing. *Procedia Manuf.* **2017**, *11*, 650–657. [CrossRef]
7. Ding, D.; Pan, Z.; Cuiuri, D.; Li, H. Wire-feed additive manufacturing of metal components: Technologies, developments and future interests. *Int. J. Adv. Manuf. Technol.* **2015**, *81*, 465–481. [CrossRef]
8. Thomas-Seale, L.E.J.; Kirkman-Brown, J.C.; Attallah, M.M.; Espino, D.M.; Shepherd, D.E.T. The barriers to the progression of additive manufacture: Perspectives from UK industry. *Int. J. Prod. Econ.* **2018**, *198*, 104–118. [CrossRef]
9. Baufeld, B.; Widdison, R.; Dutilleul, T.; Bridger, K. Electron Beam Additive Manufacturing at the Nuclear AMRC. *Elektron. Elektrotech.* **2016**, *51*, 25.

10. Taminger, K.M.; Hafley, R.A. Electron Beam Freeform Fabrication for Cost Effective Near-Net Shape Manufacturing. In Proceedings of the Meeting on Cost Effective Manufacture via Net Shape Processing, Amsterdam, The Netherlands, 15–17 May 2006.
11. Liberini, M.; Astarita, A.; Campatelli, G.; Scippa, A.; Montevecchi, F.; Venturini, G.; Durante, M.; Boccarusso, L.; Minutolo, F.M.C.; Squillace, A. Selection of Optimal Process Parameters for Wire Arc Additive Manufacturing. *Procedia CIRP* **2017**, *62*, 470–474. [CrossRef]
12. Hoefer, K.; Haelsig, A.; Mayr, P. Arc-based additive manufacturing of steel components—Comparison of wire-and powder-based variants. *Weld World* **2018**, *62*, 243–247. [CrossRef]
13. Reisgen, U.; Sharma, R.; Oster, L.; Zanders, E. Plasma-Mehrdraht-Schweißen zum Herstellen gradierter Strukturen. In Proceedings of the DVS CONGRESS, Friedrichshafen, Germany, 17–18 September 2018; DVS 344. pp. 109–114.
14. Fang, X.; Zhang, L.; Li, H.; Li, C.; Huang, K.; Lu, B. Microstructure Evolution and Mechanical Behavior of 2219 Aluminum Alloys Additively Fabricated by the Cold Metal Transfer Process. *Materials* **2018**, *11*, 812. [CrossRef]
15. Kazanas, P.; Deherkar, P.; Almeida, P.; Lockett, H.; Williams, S. Fabrication of geometrical features using wire and arc additive manufacture. Proceedings of the Institution of Mechanical Engineers. *Part B J. Eng. Manuf.* **2012**, *226*, 1042–1051. [CrossRef]
16. Venturini, G.; Montevecchi, F.; Scippa, A.; Campatelli, G. Optimization of WAAM Deposition Patterns for T-crossing Features. *Procedia CIRP* **2016**, *55*, 95–100. [CrossRef]
17. Workowski, M.; Nitschke-Pagel, T.; Dilger, K. Load induced inhomogeneous plastic deformations in welded aluminium joints. *Weld. World* **2014**, *58*, 529–538. [CrossRef]
18. Ge, J.; Lin, J.; Chen, Y.; Lei, Y.; Fu, H. Characterization of wire arc additive manufacturing 2Cr13 part: Process stability, microstructural evolution, and tensile properties. *J. Alloys Compd.* **2018**, *748*, 911–921. [CrossRef]
19. Suryakumar, S.; Karunakaran, K.P.; Chandrasekhar, U.; Somashekara, M.A. A study of the mechanical properties of objects built through weld-deposition. *Proc. Inst. Mech. Eng. Part B J. Eng. Manuf.* **2013**, *227*, 1138–1147. [CrossRef]
20. Henckell, P.; Günther, K.; Ali, Y.; Bergmann, J.P.; Scholz, J.; Forêt, P. The Influence of Gas Cooling in Context of Wire Arc Additive Manufacturing—A Novel Strategy of Affecting Grain Structure and Size. In *The Minerals, Metals & Materials Series, TMS 2017 146th Annual Meeting & Exhibition Supplemental Proceedings*; BSpringer International Publishing: Cham, Switzerland, 2017; pp. 147–156.
21. Li, F.; Chen, S.; Shi, J.; Zhao, Y.; Tian, H. Thermoelectric Cooling-Aided Bead Geometry Regulation in Wire and Arc-Based Additive Manufacturing of Thin-Walled Structures. *Appl. Sci.* **2018**, *8*, 207. [CrossRef]
22. Shi, J.; Li, F.; Chen, S.; Zhao, Y.; Tian, H. Effect of in-process active cooling on forming quality and efficiency of tandem GMAW–based additive manufacturing. *Int. J. Adv. Manuf. Technol.* **2019**, *101*, 1349–1356. [CrossRef]
23. Gu, J.; Ding, J.; Williams, S.; Gu, H.; Ma, P.; Zhai, Y. The effect of inter-layer cold working and post-deposition heat treatment on porosity in additively manufactured aluminium alloys. *J. Mater. Process. Technol.* **2016**, *230*, 26–34. [CrossRef]
24. Cong, B.; Ding, J.; Williams, S. Effect of arc mode in cold metal transfer process on porosity of additively manufactured Al-6.3%Cu alloy. *Int. J. Adv. Manuf. Technol.* **2015**, *76*, 1593–1606. [CrossRef]
25. Fang, X.; Zhang, L.; Chen, G.; Dang, X.; Huang, K.; Wang, L.; Lu, B. Correlations between Microstructure Characteristics and Mechanical Properties in 5183 Aluminium Alloy Fabricated by Wire-Arc Additive Manufacturing with Different Arc Modes. *Materials* **2018**, *11*, 2075. [CrossRef] [PubMed]
26. Derekar, K.; Lawrence, J.; Melton, G.; Addison, A.; Zhang, X.; Xu, L. Influence of Interpass Temperature on Wire Arc Additive Manufacturing (WAAM) of Aluminium Alloy Components. *MATEC Web Conf.* **2019**, *269*, 5001. [CrossRef]
27. Köhler, M.; Tóth, T.; Hensel, J.; Dilger, K. Prozesscharakteristiken bei der lichtbogenbasierten generativen Fertigung metallischer Komponenten. *DVS Ber. 39 Assist. Fügetechnik* **2019**, *356*, 96–102.
28. Köhler, M.; Fiebig, S.; Hensel, J.; Dilger, K. Wire and Arc Additive Manufacturing of Aluminum Components. *Metals* **2019**, *9*, 608. [CrossRef]

metals

MDPI

Article

Development of a High Strength Magnesium Alloy for Wire Arc Additive Manufacturing

Stefan Gneiger [1], Johannes A. Österreicher [1,*], Aurel R. Arnoldt [1], Alois Birgmann [1] and Martin Fehlbier [2]

[1] LKR Light Metals Technologies, Austrian Institute of Technology, A-5282 Ranshofen, Austria; stefan.gneiger@ait.ac.at (S.G.); aurel.arnoldt@ait.ac.at (A.R.A.); alois.birgmann@ait.ac.at (A.B.)

[2] Chair of Foundry Technology, University of Kassel, Kurt-Wolters-Strasse 3, D-34109 Kassel, Germany; fehlbier@uni-kassel.de

* Correspondence: johannes.oesterreicher@ait.ac.at; Tel.: +43-(0)664-825-1151

Received: 23 April 2020; Accepted: 8 June 2020; Published: 10 June 2020

check for updates

Abstract: Due to their high specific strength, magnesium alloys are promising materials for further lightweighting in mobility applications. In contrast to casting and forming processes, additive manufacturing methods allow high degrees of geometrical freedom and can generate significant weight reductions due to load-specific part design. In wire arc additive manufacturing processes, large parts can be produced with high material utilization. Process-inherent high melt temperatures and solidification rates allow for the use of magnesium alloys which are otherwise complicated to process; this enables the use of unconventional alloying systems. Here, we report the development of a Mg-Al-Zn-Ca-rare earth alloy for wire arc additive manufacturing (WAAM). Compared to parts made of commercially available filler wire, the newly developed alloy achieves a higher strength (approx. +9 MPa yield strength, +25 MPa ultimate tensile strength) in WAAM.

Keywords: wire arc additive manufacturing; WAAM; microstructure; magnesium; mechanical properties; scanning electron microscopy; electron backscattered diffraction method

1. Introduction

Magnesium is the lightest construction metal available and used mainly in the form of cast components. It can be processed in a wide variety of casting processes including gravity casting, low- and high pressure die casting, and thixomolding [1]. Compared to casting, additive layer manufacturing (ALM) allows for higher part complexity at the cost of reduced productivity. This enables further potential for weight savings in part design [2,3]. For Mg, two directions of ALM processes have been proven to be feasible: powder-based and wire-based methods. While powder-based additive manufacturing of Mg is already in use to produce high quality parts for medical applications [4], widespread use is prevented by the complex production process and slow building rates. Mg powder tends to oxidize fast, so the use of a chamber with controlled gas shielding is mandatory, limiting the maximum part size [5]. In selective laser melting (SLM), the vaporizing metal reduces the visibility in the chamber and, therefore, the laser efficiency, which further reduces processability [6,7].

Very good mechanical properties can be achieved using Mg additive manufacturing processes. Typically, SLM Mg parts show similar or even superior strength compared to cast and wrought parts due to the small volume melt pools where rapid solidification rates and short segregation paths are achieved [8,9].

Tandon et al. reported the mechanical values of direct energy deposited (DED) as well as powder bed fusioned MAP + 43 alloy powders in the "HIPed" (hot isostatic pressed) and subsequently T6 treated condition [10]. For the DED material, they achieved a yield strength (YS) of 170 MPa, an ultimate tensile strength (UTS) of 245 MPa, and an elongation of 6.4%. In contrast, the hot isostatic pressing led to an essentially pore-free material, which showed superior strength and ductility in T6 condition: 194 MPa YS, 312 MPa UTS, and elongation of up to 14%.

Wire-based processes (e.g., wire arc additive manufacturing—WAAM) have the advantage of easier and safer handling of filler metal compared to powder-based processes because commercial, state-of-the-art wire arc welding equipment can be used. Due to the existence of a local gas shielding provided by the welding gun, working inside a chamber is not necessary. Thus, there are practically no limitations in part size. Mg wire arc additive manufacturing offers the ability to produce large components with a high material efficiency without the need to handle powders or large quantities of molten magnesium.

Despite their advantages, wire-based processes for magnesium are relatively uncommon to date, not least due to insufficient filler wire availability. Only a limited number of filler wire grades are known, namely AZ31B, AZ61A, AZ91D, EZ33A, AZ101A, QE22A, AM50A, and AZ92A [11–14] and not all of them are commercially available at the moment. These grades are generally used as all-purpose wires for magnesium welding, where the filler wire is selected to best fit the base material composition. As these alloys are not specifically designed for welding or additive layer manufacturing purposes, their performance is limited.

To date, the number of published results of Mg WAAM is very small. To the best of our knowledge, no results for WAAM of Mg alloys were published yet using other than the before-mentioned standard alloys. Guo et al. reported WAAM of AZ31 wires and achieved 104 MPa YS, 263 MPa UTS, and elongation of 23% by adjusting the pulse frequency during processing [15]. Takagi et al. reported in [16] the wire arc additive manufacturing of bulk samples using AZ31 wires and varying processing conditions. They report a very low porosity of 2.5×10^{-4}% in the central part of the object with the biggest defect being 7.74×10^{-4} mm^3. The mechanical properties achieved are 95 MPa YS, 239 MPa UTS, and an elongation of 21% in the WAAM part, which is comparable to the values reported by Guo et al. in [15]. Holguin et al. and Han et al. reported WAAM of AZ91 alloy, showed microstructures and XRD-analysis but did not report mechanical properties [17,18]. Graf et al. used numerical approaches for predicting temperature fields of WAAM processes using AZ31 filler wire [19].

Additive manufacturing of magnesium parts is especially promising for applications where minimum weight and high specific strength are crucial, like aeronautics and space applications. However, magnesium parts must possess good flame resistance to achieve a certification. Tests carried out by the Federal Aviation Administration (FAA) showed that typical Mg-Al alloys are not able to fulfill these requirements while alloys including high amounts of rare earth elements (e.g., WE43 using around 4 wt.% of Y and 3 wt.% of rare earths or Elektron 21 using around 2.6 wt.% Nd and 1.3 wt.% Gd) [20,21] or minor amounts of Ca (AZ80 + 0.3 wt.% Ca and 0.35 wt.% Y) [22] are able to successfully pass the tests. We have previously shown that Mg-Al-Zn alloys using Ca as flame retardant can be used for welding magnesium sheets [23].

High specific strength values can be achieved using the WAAM process, as the fast solidification speeds lead to solid solution strengthening which also allows for subsequent precipitation hardening. Therefore, Mg alloys with high Al contents and additional strengthening phase forming elements are suitable candidates for generating high strength WAAM parts. A promising strategy is the addition of rare earth elements which have been shown to increase the strength of Mg-Al alloys due to the formation of intermetallic phases of the type $Mg_{12}RE$, Al_2RE, Al_3RE and $Al_{11}RE_3$ [24,25]. This approach was followed in this study.

The novel alloy reported here—AEX11—is based on the casting alloys AZ91 (Mg-Al9-Zn1) and AZ61A (Mg-Al6-Zn1). Compared to these alloys, it has an increased Al content as well as minor additions of Ca and rare earth elements (Ce-rich mischmetal). The Al-content was chosen based on CALPHAD-calculations to match the amount of $Mg_{17}Al_{12}$ compared to AZ91 after the formation of Al-RE- and Al-Ca-phases (Figure 5). This allows the use heat treatment times and temperatures analogous to the ones used for AZ91. Ca was chosen to increase the flame resistance of the alloy and rare earth elements to increase the strength. Cerium-rich mischmetal (65 wt.% Ce, 35 wt.% La) was used due to its low price compared to single rare earth elements. Mechanical properties were evaluated from thin walled structures made from AZ61A and from plates (walls) made from AEX11, both produced via WAAM.

2. Materials and Methods

We produced AEX11 wire via casting and subsequent direct extrusion at LKR in Ranshofen, Austria, as described below. The nominal chemical compositions are given in Table 1 along with the compositions of AZ61A and AZ91D for comparison.

Table 1. Nominal compositions of the alloys AZ61A (acc. to ASTM B107), AZ91D (acc. to ASTM B94), and AEX11.

Alloy	Al (wt.%)	Zn (wt.%)	Ce-Mischmetal (wt.%)	Ca (wt.%)	Mn (wt.%)	Mg (wt.%)
AZ61A	5.8–7.2	0.40–1.5	-	-	0.15–0.5	Rem.
AZ91D	8.5–9.5	0.45–0.9	-	-	≥0.17	Rem.
AEX11	10.1–11.5	0.3–0.6	2.5–3.5	0.2–0.44	0.15–0.5	Rem.

AEX11 was prepared in a resistance heated furnace Nabertherm Liquitherm® (Nabertherm GmbH, Lilienthal, Germany) T20/H in a mild steel crucible with a melt capacity of 30 kg. The melt was covered with Ar + 1% SF_6 shielding gas during preparation and casting. Casting temperature was 720 °C.

The alloy was gravity cast into billets with a diameter of 65 mm and a height of 240 mm in a tempered (150 °C) mild steel mold. The billets were turned to a diameter of 58.5 mm and a length of 180 mm and homogenized at 420 °C for 8 h with subsequent cooling using forced air resulting in a cooling rate of approximately 25 K/min. Afterwards, the billets were extruded in two single steps to the desired filler wire diameter (Φ58.5 mm → Φ19.5 mm → Φ1.2 mm) using a 1.5 MN direct extrusion press NEHP 1500.01 manufactured by Müller Engineering, Todtenweis, Germany. A more detailed explanation of the wire manufacturing process can be found elsewhere [17].

Directly before welding, the wires were cleaned using isopropanol and a fiber mat. Wire arc additive manufacturing of AEX11 was carried out using the gas-metal arc-welding cold metal transfer (CMT) process (Fronius, Pettenbach, Austria). The power source was a Fronius CMT TPS with the following parameters: welding curve KL 1904, weld speed 12 mm/s, wire feed rate 3.5 m/min (base layer) and 2.2 m/min (building layers), current 51 A, pulse frequency 80–120 Hz, layer thickness 2.5 mm each, and a wall thickness of 5 mm. The distance between workpiece and tip was 13 mm. Ar with 30% He was used as shielding gas. The nozzle diameter was 14 mm and the gas flow was 14 L/min. Figure 1 shows the AM welding facility at LKR.

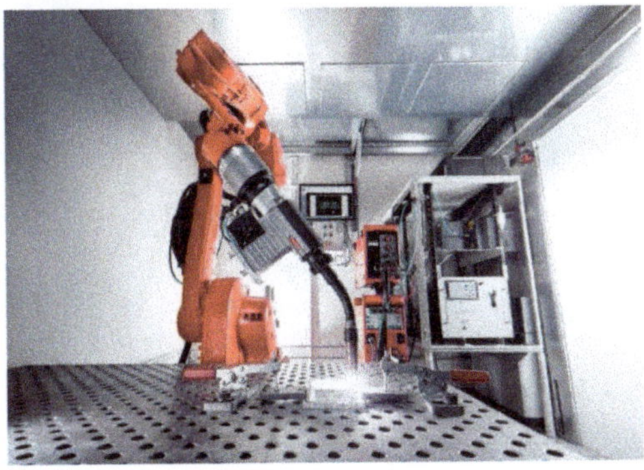

Figure 1. Additive manufacturing lab at LKR, Ranshofen.

Thin-walled structures with a height of 80 mm, a length of 130 mm and a width of 30 mm were manufactured continuously from AZ61 wires on a AZ31 base plate (Mg-3.0Al-0.6Zn). Walls with dimensions of 140 mm × 120 mm × 5 mm were manufactured discontinuously (layer for layer) using AEX11 wires for microstructural and mechanical investigations. In the plates, the welding direction was alternating and changed every single layer. The walls of AEX11 manufactured via WAAM are intended to represent a section of the structures manufactured out of AZ61A. The AZ61 samples were obtained from AZ61A filler wires sourced from DRATEC Drahttechnik GmbH, Krefeld, Germany. The resulting structures can be seen in Figure 2.

(a) (b)

Figure 2. Additive manufactured plate (**a**) out of AEX11 with schematic representation of sample extraction. The red area indicates the position of samples for microstructural and EBSD analysis. Additive manufactured sample structure out of AZ61A (**b**). The power source was a Fronius CMT TPS with the following parameters: weld speed 12 mm/s, wire feed rate 3.5 m/min (base layer) and 2.2 m/min (building layers), current 51 A, layer thickness 2.5 mm each, and a wall thickness of 5 mm.

After milling off the wavy surface, tensile test specimens were taken out of the WAAM parts in welding direction (WD) and transversal direction (TD). The positions of specimen extraction are shown schematically in Figure 3 (all dimensions in mm), and the geometry of the specimens is shown in Figure 4. In the as fabricated state, five samples per direction were taken and, in the T6 condition, four samples per direction were taken. Only one sample per direction was taken for analyzing the solution heat treatment state. Tensile tests were carried out at room temperature using a DIL 805 A/D deformation dilatometer (Bähr Thermoanalyse GmbH, Hüllhorst, Germany) at a deformation rate of $0.006\ s^{-1}$.

For solution heat treatment, some of the tensile specimens were placed in a preheated furnace (415 °C) for 6 h and then heated up to 430 °C with holding for another 6 h, followed by water quenching. Afterwards, the solution heat treated samples were placed in a preheated furnace (260 °C) for 30 min with subsequent water quenching.

Thermodynamic calculations were done using Thermo-Calc software (Version 2017a) and database TCMG5 (both Thermo-Calc Software AB, Solna, Sweden).

The samples for microstructural investigations were extracted from the WAAM parts (Figure 3) perpendicular to all three directions (WD welding direction, ND normal direction and TD transversal direction) from at least 30 mm distance to the base plate. The samples were mechanically polished to $\frac{1}{4}$ µm and then finished using colloidal silica suspension. Additionally, selected samples were etched using picric acid to visualize the grain boundaries and allows for investigating the grain structure.

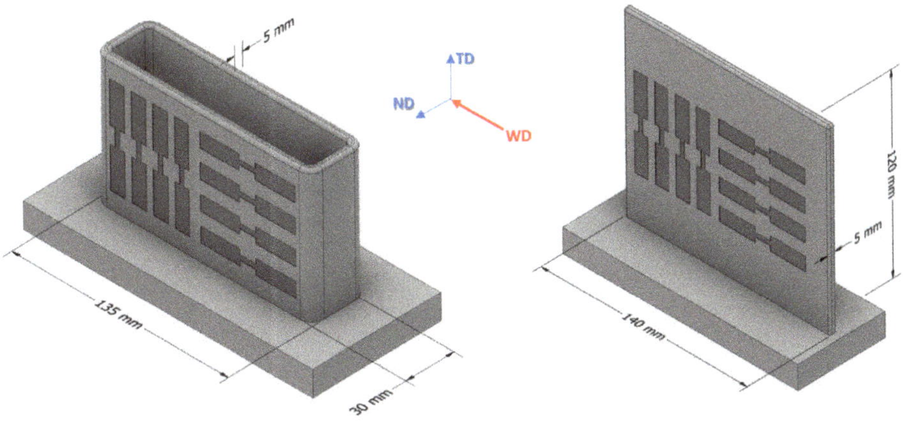

(a) (b)

Figure 3. Schematic position of specimen extraction for AZ61A (**a**) and AEX11 (**b**). Specimens were extracted in welding direction (WD) and transversal direction (TD). The walls of AEX11 manufactured via WAAM represent a section of the structures manufactured out of AZ61A.

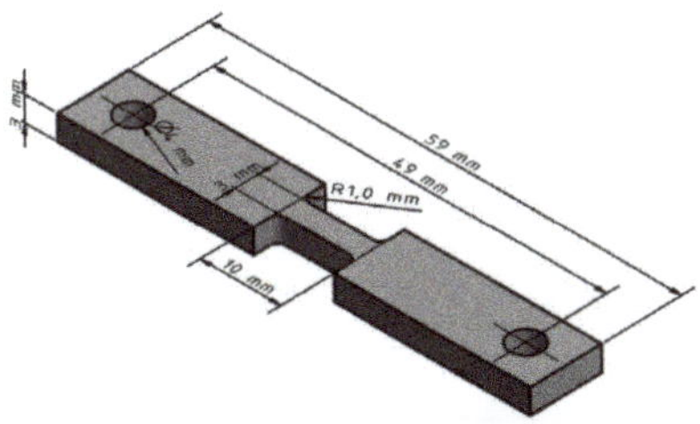

Figure 4. Dimensions of tensile test specimen.

Microstructural investigations in all directions were carried out using optical light microscopy and scanning electron microscopy (SEM). A Tescan MIRA3 field emission gun (FEG) scanning electron microscope (Brno, Czech Republic) operated at 15 kV and a 4-quadrant solid-state backscattered electron (BSE) detector were used for imaging and energy dispersive X-ray spectroscopy (EDS) with an EDAX (Mahwah, NJ, USA) Octane Elect silicon drift detector system. Electron backscattered diffraction (EBSD) was performed at 20 kV and a working distance of ~7 mm on a Zeiss Sigma 500 VP FEG scanning electron microscope (AZ61A) (Carl Zeiss AG, Oberkochen, Germany) and at ~15 mm working distance on the aforementioned Tescan Mira 3 (AEX11). On both microscopes, an EDAX Hikari Plus EBSD camera (AMETEK, Berwyn, PA, USA) was used. EBSD data analysis was carried out with EDAX OIM 8 (AMETEK, Berwyn, PA, USA). The samples were taken about 30 mm away from the base plate. For cleanup, we used of grain dilatation with a grain tolerance angle of 5° and a minimum grain size of 2 px (AZ61A) or 20 px (AEX11). For grain size analysis, a misorientation angle of 15° was used to determine grain boundaries.

3. Results and Discussion

Figure 5 shows the results of the equilibrium thermodynamic calculations carried out for AZ61A, AZ91D, and AEX11 alloys. According to the calculations, the amount of the intermetallic phase $Mg_{17}Al_{12}$ is predicted to be significantly higher in the alloy AEX11 than in AZ61A, whilst, for AZ91D, it is approximately equal as in AEX11. Additionally, the alloy AEX11 shows small amounts of intermetallic phases $Al_{11}RE_3$ and Al_2Ca. While the $Mg_{17}Al_{12}$ in the divorced eutectic can be completely dissolved into the matrix by solution heat treatment, the Al-RE and Al_2Ca phases cannot.

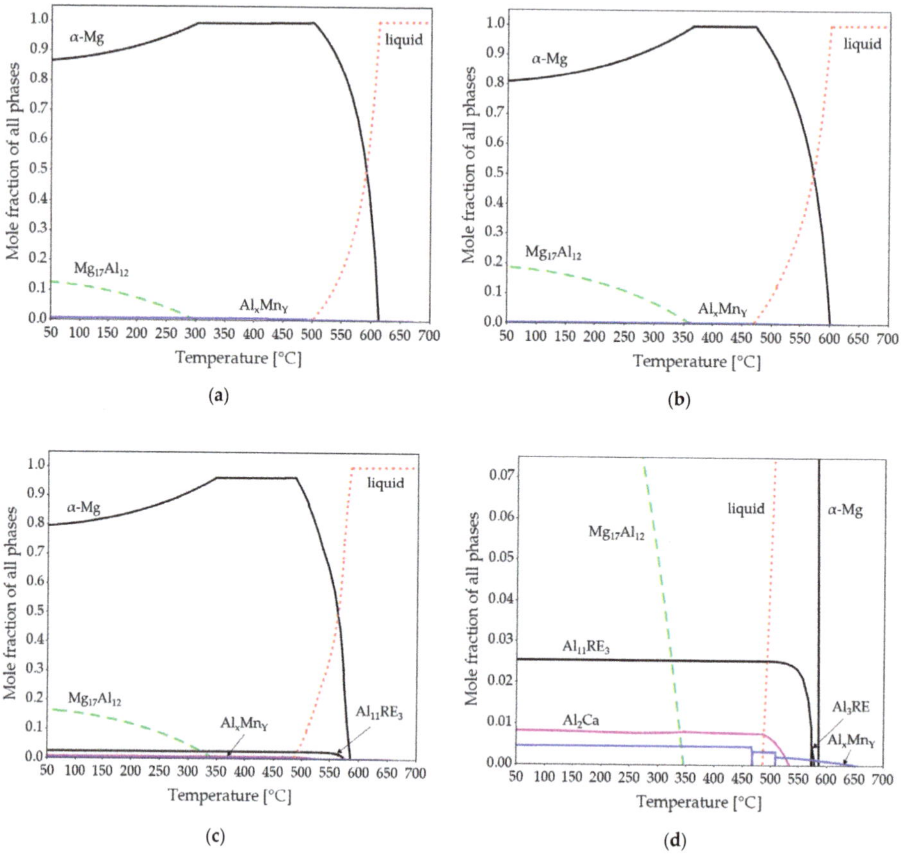

Figure 5. Equilibrium thermodynamic calculations of AZ61A (**a**), AZ91D (**b**) and AEX11 (**c**,**d**). (**d**) represents a section of (**c**).

As the manganese content is about equal in all three alloys, the resulting Al_xMn_y phases are almost identical with regard to their stoichiometry depending on the relative Al and Mn amounts present. The calculated total amount of intermetallic phases is highest for AEX11 because it contains the highest amount of alloying elements. The calculated solidification ranges are 116 °C for AZ61A, 124 °C for AEX11, and 136 °C for AZ91D.

Figure 6 shows SEM of the microstructure of AEX11 fabricated via WAAM in the "as fabricated" and the "T6 heat treated" states as well as the microstructure of AZ61A in the "as fabricated" state. EDS measurements identify the present intermetallic phases as Al-RE, Al-Ca, and Mg/Al-rich phases. With reference to the thermodynamic calculations and literature [24,26,27], these are presumed to be $Al_{11}RE_3$, Al_2Ca, and β-$Mg_{17}Al_{12}$. By indexing of EBSD patterns, the Al-RE phases were identified with high confidence as orthorhombic $Al_{11}RE_3$ phase with large d-spacings (a = 4.4 Å, b = 10.1 Å, c = 13.0 Å).

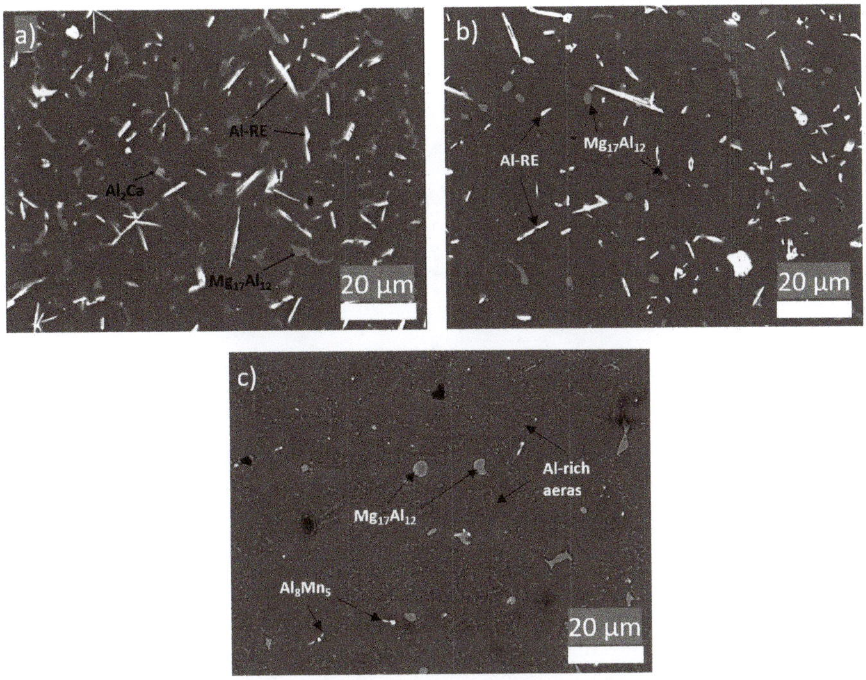

Figure 6. Microstructure of parts manufactured via WAAM. (**a**) AEX11 "as fabricated" state; (**b**) AEX11 after T6 heat treatment; (**c**) AZ61A "as fabricated"; (backscattered SEM).

Due to the higher cooling rates of WAAM compared to classical casting processes (with the exception of high pressure die casting), typical grain sizes and phase fractions of coarsely precipitated (divorced) $Mg_{17}Al_{12}$ are lower, resulting in generally higher mechanical properties [28–30]. The $Al_{11}RE_3$ intermetallic phases found in the AEX11 are relatively large and needle-like; therefore, their contribution to strengthening is not clear. In [31], it is stated that the addition of RE increases the strength of Mg-Al. Nevertheless, it is also shown that, with an increasing amount of RE, the size of Al-RE phases increases, leading to a decrease of mechanical properties. Additionally, a positive contribution to creep resistance is expected due to the high thermal stability of the phases [32,33]. Figure 6b shows that the T6 heat treatment resulted in spheroidization of $Mg_{17}Al_{12}$ phases, while no change can be detected in the Al-RE phases. This is also consistent with the thermodynamic calculation (Figure 5c,d), which shows that the phase dissolution temperature of Al-RE is significantly higher than the heat treatment temperature, while the $Mg_{17}Al_{12}$ is dissolved and precipitated in the T6 heat treatment.

While in wrought material the grains are deformed and often partially recrystallized, WAAM material typically possesses columnar grains in the welding direction and a low grain size compared to conventional casting methods resulting from the low amount of molten volume and high solidification rates. The resulting grain structure is mainly dependent on the component geometry and the energy input during processing, which is dependent on the wire feed, the wire thickness, power settings, and polarity. Additionally, the grain sizes are strongly dependent on the cooling rate. Therefore, the grain size can be minimized using optimized processing parameters [34]. Guo et al. [15] have achieved grain sizes of 20–39 µm while the AZ61A shown in this work has a grain size of 31 ± 12 µm and the grain size of AEX11 is roughly 55 ± 19 µm. The grain size was determined via EBSD (Figure 7). No formation of columnar grains was observed and the grains were formed approximately equiaxially.

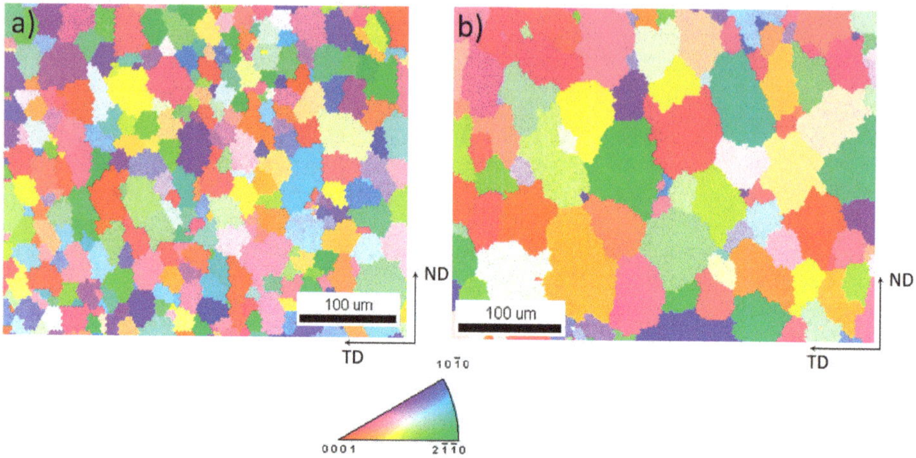

Figure 7. Detail of inverse pole figure [001] map of (**a**) AZ61A and (**b**) AEX11; (cross section in welding direction; TD: transversal direction; ND: normal direction).

Figure 8 gives a comparison of the microstructure of AEX11 (Figure 8a,b) and AZ61A (Figure 8c,d) in the "as fabricated" state, manufactured via WAAM. In agreement with SEM, potential Al-Mn (dark grey) and $Mg_{17}Al_{12}$ (light grey) can be found. In AZ61A, potential Al-Mn is the dominating phase while potential $Mg_{17}Al_{12}$ is only visible in very small quantities, coming from the high cooling speed during processing. Al has a high solubility limit in Mg and the formation of $Mg_{17}Al_{12}$ can be suppressed when the solidification speed is fast enough. This is in agreement with the thermodynamic calculations which show that the phase is only stable at temperatures below 300 °C for AZ61A and 350 °C for AEX11. Due to the higher amount of Al in AEX11, the amount of already precipitated potential $Mg_{17}Al_{12}$ is substantially higher. Additionally, needle shaped Al-RE phases are present in AEX11. According to the thermodynamic calculations, their temperature of formation is above the solidus temperature. Thus, achieving a complete solution of these phases via heat treatment is not possible.

A number of small defects (pores, cavities) were found in the samples. The pores have a size of <20 µm and are distributed evenly across the cross section of the samples. It is believed that the pores are mainly a result of impurities in the filler wires and defects in the surface region of underlying layers [35]. As the process was carried out only using local gas shielding instead of a gas chamber, insufficient shielding might have led to porosity [36]. The cavities occur preferably in transitions between layers (Figure 9). Possibly, a reduction of porosity could be achieved by further adaptions of processing parameters and proper preparation (cleaning) of base material and filler wires in future research. A better indication of the quantity and size of the voids to take targeted measures could be provided by a non-destructive in-line analysis method such as that presented by Javadi et al. [37].

Tensile test results of WAAM magnesium parts are given in Figure 10 and Table 2. In the "as fabricated" state, AEX11 has an elongation at break of around 6% while AZ61A achieved values of around 16%. The amount of $Mg_{17}Al_{12}$ in AEX11 is about equal to the amount of what one would expect to find in AZ91D; therefore, a T6 heat treatment can be performed with approximately the same efficiency. A relatively high annealing temperature was chosen to keep the treatment time low. Nevertheless, reducing this temperature may lead to even higher strength at the cost of increased treatment time.

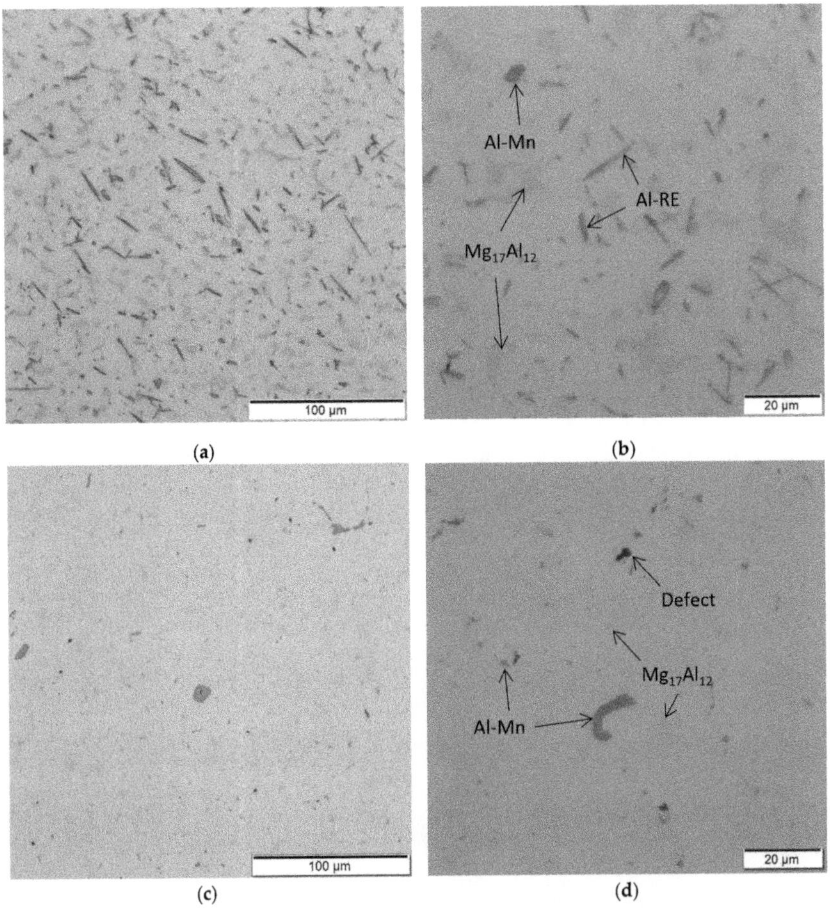

Figure 8. Microstructures of AEX11 (**a**,**b**) and AZ61A (**c**,**d**) in the "as fabricated" state, manufactured via WAAM.

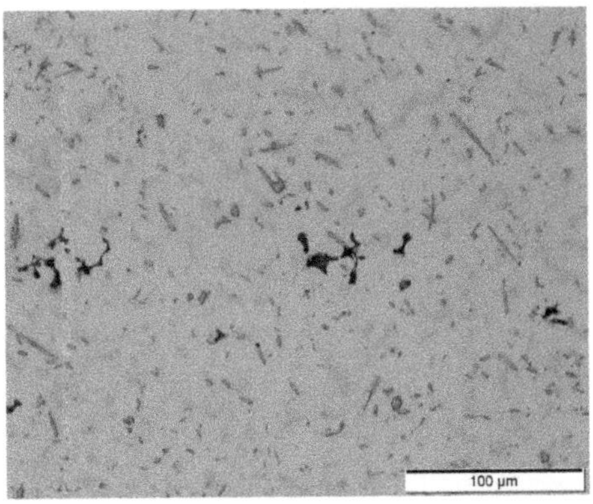

Figure 9. Cross section of AEX11 sample manufactured via WAAM. Voids occur in the transitions between single layers.

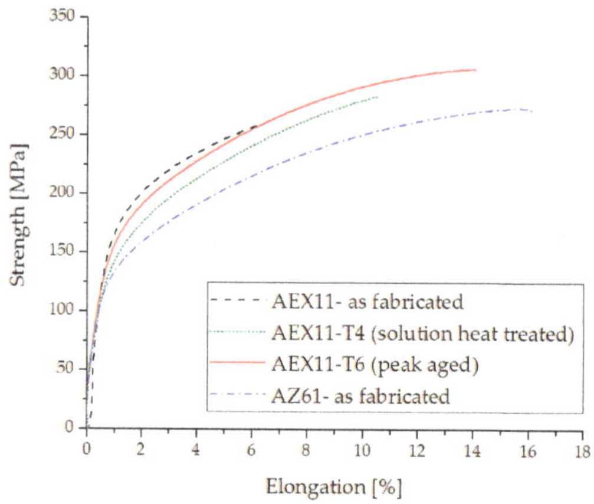

Figure 10. Typical tensile test results of additive manufactured structures out of AZ61 and AEX11 filler wires.

Table 2. Mechanical properties (tensile) of AZ61A and AEX11 in detail.

Alloy	State	Direction	UTS (MPa)	SD (MPa)	Elongation (%)	SD (%)
AZ61A	as fabricated	WD	275	1.0	16.8	0.4
		TD	272	4.5	17.2	1.3
AEX11	as fabricated	WD	243	17.3	4.5	2.3
		TD	233	12.3	5.2	0.7
	T6	WD	299	8.0	12.6	1.5
		TD	298	6.3	11.0	0.8

After dissolution of $Mg_{17}Al_{12}$, high elongation was achieved while at the same time the strength was decreased. Annealing at 260 °C was carried out to exploit the precipitation strengthening effect of $Mg_{17}Al_{12}$. We achieved a maximum UTS of 307 MPa at an elongation of 14% using the alloy AEX11, designed specifically for WAAM processing in the T6 state. With a density of ~1.84 g/cm^3, this results in a specific strength of ~166.8 Nms^{-1}, which is comparable to the aluminum alloy Al-2024-T6 (171.5 Nms^{-1}) but inferior to Al-7075-T6 forgings (196.4 Nms^{-1}) [38]. The UTS is 15% higher than that of parts made from our reference material, commercial AZ61A wires in the as fabricated state, using classical welding parameters without frequency modulation. The strength of the baseline material is comparable to previous reports [15].

The elongation of the AEX11 samples increased after every single heat treatment step. This can be attributed to the spheroidization of brittle phases (mainly $Mg_{17}Al_{12}$). In AZ61A, the amount of precipitated β-$Mg_{17}Al_{12}$ is inherently lower, leading to a good deformability already in the as fabricated state. It is known that β-$Mg_{17}Al_{12}$ has a detrimental effect on the ductility of Mg-Al alloys [11,39]. In the WAAM AZ61A samples, a large proportion of Al was found to be already in solid solution due to the rapid solidification as only a small amount of Al-rich phases was found in the microstructural analysis. Additionally, according to [40,41], hardness increase by heat treatment is not very effective for AZ61A. Therefore, no further heat treatment was applied.

The mechanical properties of the WAAM manufactured parts shown in this work can be compared to the ones achieved in wrought processes such as forging [42] and extrusion [43]. Forged AZ61 parts reach an ultimate tensile strength of 295 MPa and an elongation of 12% [44], while, for extruded AZ61, UTS values of 310 MPa at elongations of 16% are stated [45].

Figure 11 shows the fracture surface of a tensile test specimen of AEX11 in the as fabricated state. The matrix deforms around the intermetallic phases before detaching. All samples showed ductile fracture behavior because no continuous network of intermetallic phases is formed on the grain boundaries, which could act as crack propagation sites.

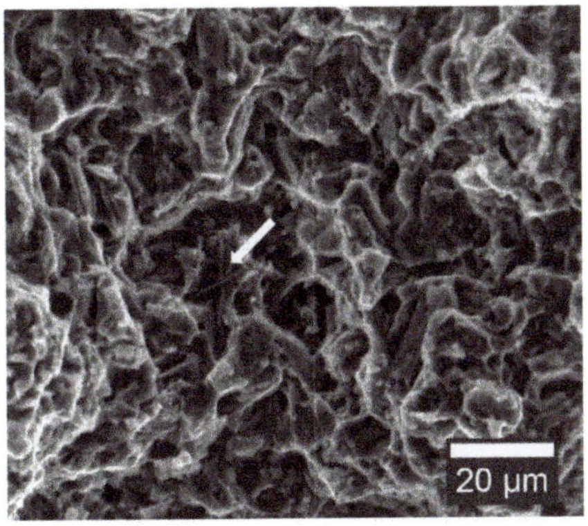

Figure 11. Fracture surface of AEX11 tensile test specimen in the as fabricated state (secondary electron SEM). Al-RE phase indicated with an arrow.

4. Conclusions

WAAM is a suitable production process for low quantity Mg parts and prototypes. Especially for large dimensional parts, this processing route can be faster and easier to implement than traditional routes like milling and casting.

In this research, we showed that commercial AZ61 filler wire as well as custom made wire from AEX11 can be used for the fabrication of WAAM parts. We demonstrated that new alloys with a high proportion of soluble elements can be used to take advantage of the high cooling rate of the process and, thus, achieve superior mechanical properties.

While WAAM can be used to produce components with attractive properties, there are still some challenges along the process chain to be solved. For wire manufacturing, impurities must be reduced to a minimum as these may lead to porosity in the parts. For WAAM itself, temperature control and optimization of the welding path are essential to avoid warpage.

Author Contributions: This paper is the work of several authors, the individual contributions are subsequently listed: conceptualization, S.G., J.A.Ö., and M.F.; methodology, S.G. and J.A.Ö.; software, S.G.; validation, J.A.Ö.; formal analysis, J.A.Ö. and M.F.; investigation, S.G., J.A.Ö., A.R.A., and A.B.; data curation, S.G.; writing—original draft preparation, S.G. and J.A.Ö.; writing—review and editing, S.G., J.A.Ö., A.R.A., A.B., and M.F.; visualization, S.G.; supervision, J.A.Ö. and M.F.; project administration, S.G. All authors have read and agreed to the published version of the manuscript.

Funding: This research was funded by the European Regional Development Fund (EFRE) in the framework of the EU-program "IWB Investition in Wachstum und Beschäftigung Österreich 2014–2020" and the federal state Upper Austria.

Acknowledgments: The authors thank Martin Schnall for his efforts in the search for suitable processing parameters and his support in the production of AM structures, Iris Baumgartner for the preparation of metallographic samples, and René de Kloe (EDAX/AMETEK) for providing crystallographic files for $Al_{11}RE_3$.

Conflicts of Interest: The authors declare no conflict of interest. The founding sponsors had no role in the design of the study; in the collection, analyzes, or interpretation of data; in the writing of the manuscript, or in the decision to publish the results.

References

1. Mordike, B.L.; Ebert, T. Magnesium: Properties—Applications—Potential. *Mater. Sci. Eng. A* **2001**, *302*, 37–45. [CrossRef]

2. Emmelmann, C.; Sander, P.; Kranz, J.; Wycisk, E. Laser Additive Manufacturing and Bionics: Redefining Lightweight Design. *Phys. Procedia* **2011**, *12*, 364–368. [CrossRef]
3. Herzog, D.; Seyda, V.; Wycisk, E.; Emmelmann, C. Additive manufacturing of metals. *Acta Mater.* **2016**, *117*, 371–392. [CrossRef]
4. Windhagen, H.; Radtke, K.; Weizbauer, A.; Diekmann, J.; Noll, Y.; Kreimeyer, U.; Schavan, R.; Stukenborg-Colsman, C.; Waizy, H. Biodegradable magnesium-based screw clinically equivalent to titanium screw in hallux valgus surgery: Short term results of the first prospective, randomized, controlled clinical pilot study. *Biomed. Eng. OnLine* **2013**, *12*, 62. [CrossRef]
5. Gieseke, M.; Kiesow, T.; Noelke, C.; Kaierle, S.; Maier, H.J.; Matena, J.; Kampmann, A.; Nolte, I.; Gellrich, N.-C.; Haferkamp, H. Challenges of Processing Magnesium and Magnesium Alloys by Selective Laser Melting. In Proceedings of the World PM 2016 Congress and Exhibition. European Powder Metallurgy Association (EPMA), Hamburg, Germany, 9–13 October 2016.
6. Jauer, L.; Meiners, W.; Vervoort, S.; Gayer, C.; Zumdick, N.A.; Zander, D. Selective Laser Melting of Magnesium Alloys. In Proceedings of the World Powder Metallurgy Conference 2016, Hamburg, Germany, 9–13 October 2016.
7. Gieseke, M.; Noelke, C.; Kaierle, S.; Wesling, V.; Haferkamp, H. Selective Laser Melting of Magnesium and Magnesium Alloys. In *Magnesium Technology 2013*; Hort, N., Mathaudhu, S.N., Neelameggham, N.R., Alderman, M., Eds.; Springer International Publishing: Berlin/Heidelberg, Germany, 2016; pp. 65–68. ISBN 978-3-319-48150-0.
8. Manakari, V.; Parande, G.; Gupta, M. Selective Laser Melting of Magnesium and Magnesium Alloy Powders: A Review. *Metals* **2017**, *7*, 2. [CrossRef]
9. Frazier, W.E. Metal Additive Manufacturing: A Review. *J. Mater. Eng. Perform.* **2014**, *23*, 1917–1928. [CrossRef]
10. Tandon, R.; Palmer, T.; Gieseke, M.; Noelke, C.; Kaierle, S. Additive manufacturing of magnesium alloy powders: Investigations into process development using elektron®MAP+43 via laser powder bed fusion and directed energy deposition. In Proceedings of the World PM 2016 Congress and Exhibition; European Powder Metallurgy Association (EPMA), Hamburg, Germany, 9–13 October 2016.
11. Avedesian, M.M.; Baker, H. *ASM Specialty Handbook: Magnesium and Magnesium Alloys*; ASM International: Russel Township, OH, USA, 1999; ISBN 978-0-87170-657-7.
12. Products–Mag Specialties. Available online: https://www.magspecialtiesinc.com/products/ (accessed on 10 July 2019).
13. Magnesium Alloys|Products|Washington Alloy Company. Available online: https://weldingwire.com/products/Magnesium-Alloys (accessed on 10 July 2019).
14. Magnesium-DRATEC. Available online: http://www.dratec.de/magnesium.html (accessed on 10 July 2019).
15. Guo, J.; Zhou, Y.; Liu, C.; Wu, Q.; Chen, X.; Lu, J. Wire Arc Additive Manufacturing of AZ31 Magnesium Alloy: Grain Refinement by Adjusting Pulse Frequency. *Materials* **2016**, *9*, 823. [CrossRef]
16. Takagi, H.; Sasahara, H.; Abe, T.; Sannomiya, H.; Nishiyama, S.; Ohta, S.; Nakamura, K. Material-property evaluation of magnesium alloys fabricated using wire-and-arc-based additive manufacturing. *Addit. Manuf.* **2018**, *24*, 498–507. [CrossRef]
17. Han, S.; Zielewski, M.; Martinez Holguin, D.; Michel Parra, M.; Kim, N. Optimization of AZ91D Process and Corrosion Resistance Using Wire Arc Additive Manufacturing. *Appl. Sci.* **2018**, *8*, 1306. [CrossRef]
18. Holguin, D.A.M.; Han, S.; Kim, N.P. Magnesium Alloy 3D Printing by Wire and Arc Additive Manufacturing (WAAM). *MRS Adv.* **2018**, *3*, 2959–2964. [CrossRef]
19. Graf, M.; Hälsig, A.; Höfer, K.; Awiszus, B.; Mayr, P. Thermo-Mechanical Modelling of Wire-Arc Additive Manufacturing (WAAM) of Semi-Finished Products. *Metals* **2018**, *8*, 1009. [CrossRef]
20. Marker, T. *Evaluating the Flammability of Various Magnesium Alloys During Laboratory- and Full-Scale Aircraft Fire Test*; US Department of Transportation, Federal Aviation Administration: Atlantic City, NJ, USA, 2013.
21. Marker, T.R. *Development of a Laboratory-Scale Flammability Test for Magnesium Alloys Used in Aircraft Seat Construction*; Federal Aviation Administration: Washington, DC, USA, 2014.
22. Gneiger, S.; Frank, S.; Betz, A. Untersuchung des Elementeinflusses von Ca und Y auf Magnesiumlegierungen. In Proceedings of the 9th Ranshofener Leichtmetalltage, Bad Ischl, Austria, 9–10 November 2016.

23. Gneiger, S.; Gradinger, R.; Betz, A. Manufacturing route for producing tailor-made magnesium wires for welding and ALM purposes. In Proceedings of the 2017 World Magnesium Conference, Singapore, 21–23 May 2017.
24. Bakke, P.; Westengen, H. The Role of Rare Earth Elements in Structure and Property Control of Magnesium Die Casting Alloys. In *Essential Readings in Magnesium Technology*; Springer: Berlin/Heidelberg, Germany, 2016; pp. 313–318. ISBN 978-3-319-48588-1.
25. Rokhlin, L.L. *Magnesium Alloys Containing Rare Earth Metals: Structure and Properties*; CRC Press: Boca Raton, FL, USA, 2003; ISBN 978-0-429-17922-8.
26. Li, P.; Tang, B.; Kandalova, E.G. Microstructure and properties of AZ91D alloy with Ca additions. *Mater. Lett.* **2005**, *59*, 671–675. [CrossRef]
27. Nami, B.; Shabestari, S.G.; Razavi, H.; Mirdamadi, S.; Miresmaeili, S.M. Effect of Ca, RE elements and semi-solid processing on the microstructure and creep properties of AZ91 alloy. *Mater. Sci. Eng. A* **2011**, *528*, 1261–1267. [CrossRef]
28. Dini, H.; Andersson, N.-E.; Jarfors, A. Effect of Mg17Al12 Fraction on Mechanical Properties of Mg-9%Al-1%Zn Cast Alloy. *Metals* **2016**, *6*, 251. [CrossRef]
29. Cai, H.; Guo, F.; Su, J. Combined effects of cerium and cooling rate on microstructure and mechanical properties of AZ91 magnesium alloy. *Mater. Res. Express* **2018**, *5*, 016503. [CrossRef]
30. Zhu, T.; Chen, Z.W.; Gao, W. Effect of cooling conditions during casting on fraction of β-Mg17Al12 in Mg–9Al–1Zn cast alloy. *J. Alloys Compd.* **2010**, *501*, 291–296. [CrossRef]
31. Cui, X.; Liu, H.; Meng, J.; Zhang, D. Microstructure and mechanical properties of die-cast AZ91D magnesium alloy by Pr additions. *Trans. Nonferrous Met. Soc. China* **2010**, *20*, s435–s438. [CrossRef]
32. Easton, M.; Zhu, S.; Gibson, M.; Abbott, T.; Ang, H.Q.; Chen, X.; Birbilis, N.; Savage, G. Performance Evaluation of High-Pressure Die-Cast Magnesium Alloys. In *Magnesium Technology 2017*; Solanki, K.N., Orlov, D., Singh, A., Neelameggham, N.R., Eds.; Springer International Publishing: Berlin/Heidelberg, Germany, 2017; pp. 123–129. ISBN 978-3-319-52391-0.
33. Zhu, S.; Easton, M.A.; Abbott, T.B.; Nie, J.-F.; Dargusch, M.S.; Hort, N.; Gibson, M.A. Evaluation of Magnesium Die-Casting Alloys for Elevated Temperature Applications: Microstructure, Tensile Properties, and Creep Resistance. *Metall. Mater. Trans. A* **2015**, *46*, 3543–3554. [CrossRef]
34. Klein, T.; Schnall, M. Control of macro-/microstructure and mechanical properties of a wire-arc additive manufactured aluminum alloy. *Int. J. Adv. Manuf. Technol.* **2020**. [CrossRef]
35. Monteiro, W.A. *New Features on Magnesium Alloys*; InTech: Rijeka, Croatia, 2012; ISBN 978-953-51-0668-5.
36. Liu, L.; Song, G.; Liang, G.; Wang, J. Pore formation during hybrid laser-tungsten inert gas arc welding of magnesium alloy AZ31B—Mechanism and remedy. *Mater. Sci. Eng. A* **2005**, *390*, 76–80. [CrossRef]
37. Javadi, Y.; MacLeod, C.N.; Pierce, S.G.; Gachagan, A.; Lines, D.; Mineo, C.; Ding, J.; Williams, S.; Vasilev, M.; Mohseni, E.; et al. Ultrasonic phased array inspection of a Wire + Arc Additive Manufactured (WAAM) sample with intentionally embedded defects. *Addit. Manuf.* **2019**, *29*, 100806. [CrossRef]
38. Kaufman, J.G. *Properties of Aluminum Alloys: Tensile, Creep, and Fatigue Data at High and Low Temperatures*; ASM International: Russel Township, OH, USA, 1999; ISBN 978-0-87170-632-4.
39. Kammer, C. *Magnesium-Taschenbuch*; Aluminium-Verlag: Düsseldorf, Germany, 2000; ISBN 978-3-87017-264-0.
40. Jo, S.M.; Go, Y.; You, B.S.; Kim, Y.M. Precipitation Behavior of Mg–Al–Sn–Zn–(Na) Alloy. In *Magnesium Technology 2017*; Solanki, K.N., Orlov, D., Singh, A., Neelameggham, N.R., Eds.; Springer International Publishing: Berlin/Heidelberg, Germany, 2017; pp. 203–207.
41. Manivannan, S.; Gopalakrishnan, S.K.; Saravana Murthy, L.N.; Babu, K.; Sundarrajan, S. Microstructure and Mechanical Properties of Cast Mg-6Al- 1Zn-xRE Alloy with Additions of Yttrium at Different Ageing Temperature. *Eur. J. Sci. Res.* **2014**, *122*, 381–391.
42. Papenberg, N.P.; Gneiger, S.; Weißensteiner, I.; Uggowitzer, P.J.; Pogatscher, S. Mg-Alloys for Forging Applications—A Review. *Materials* **2020**, *13*, 985. [CrossRef] [PubMed]
43. Gneiger, S.; Papenberg, N.; Frank, S.; Gradinger, R. Investigations on Microstructure and Mechanical Properties of Non-flammable Mg–Al–Zn–Ca–Y Alloys. In *Proceedings of the Magnesium Technology 2018*; Orlov, D., Joshi, V., Solanki, K.N., Neelameggham, N.R., Eds.; Springer International Publishing: Berlin/Heidelberg, Germany, 2018; pp. 105–113.

44. ASM Handbook Volume 2: Properties and Selection: Nonferrous Alloys and Special-Purpose Materials-ASM International. Available online: https://www.asminternational.org/materials-resources/results/-/journal_content/56/10192/06182G/PUBLICATION (accessed on 24 September 2019).
45. Zeng, Z.; Stanford, N.; Davies, C.H.J.; Nie, J.-F.; Birbilis, N. Magnesium extrusion alloys: A review of developments and prospects. *Int. Mater. Rev.* **2019**, *64*, 27–62. [CrossRef]

metals

MDPI

Article

Thermal Frequency Drift of 3D Printed Microwave Components

Gregory Peter Le Sage

U.S. Government, Washington, DC 20001, USA; lesagegp@gmail.com

Received: 14 March 2020; Accepted: 26 April 2020; Published: 29 April 2020

check for updates

Abstract: Fabrication of microwave slot array antennas and waveguide bandpass and notch filters using 3D printing has significant advantages in terms of speed and cost even for parts with high mechanical complexity. One disadvantage of Stereolithography (SLA) 3D printed, copper plated microwave components is that some SLA resins have a high Coefficient of Thermal Expansion (CTE), quoted in micrometers per meter per degree or 10^{-6} per degree. Compared to typically used metals such as aluminum (CTE $24 \times 10^{-6} \cdot K^{-1}$) and copper (CTE $17 \times 10^{-6} \cdot K^{-1}$), SLA resin can have CTE above $100 \times 10^{-6} \cdot K^{-1}$. Resonant structures experience significant frequency drift with temperature changes on the order of 10–50 °C. The issue of 3D printed microwave structures changing frequency characteristics significantly with temperature shift has not been addressed or reviewed in current literature. We measured and simulated the effect of temperature change on a slot array, cavity notch filters, and post loaded waveguide bandpass filters. We tested several types of SLA resin, different plating techniques, and also Direct Metal Laser Sintering (DMLS) and Binder Infusion metal 3D printing. Performance as a function of temperature is presented for these alternatives.

Keywords: cavity resonators; filters; microwave; plating; stereolithography; thermal expansion; three-dimensional printing

---

1. Introduction

Microwave slot array antennas and bandpass and notch filters can be produced quickly and inexpensively using 3D printing [1–3]. The high CTE of SLA resins causes frequency drift with temperature to be notably worse than solid metal structures [4]. Somos Taurus SLA resin [5] has CTE of $105.3 \times 10^{-6} \cdot K^{-1}$ over the temperature range 0–50 °C. Choosing lower CTE resin, strengthening metal coatings using multiple layers of nickel and copper, and switching to 3D metal printing with DMLS or binder infusion improves frequency dependence on temperature. In this paper, the frequency dependence with temperature is presented for multiple SLA resins, multiple metal coatings, and 3D metal printed structures. Theoretical fits to expected temperature variation are presented to confirm the predictability of performance. Armed with these predictions and results, requirements for temperature stability can be applied to make decisions regarding the materials and processes to use for 3D printed fabrication of microwave devices.

Another notable advantage of 3D printed parts is that they are much lighter than metal parts. Invar is a 36% nickel alloy with iron. It is used when temperature stability is important. It is ductile and weldable. Its CTE over 20–100 °C is $<1.3 \times 10^{-6} \cdot K^{-1}$, with $\sigma = 0.202 \times 10^6$ S/m (must be plated for use in microwave circuits). Its density is 8.1 to 8.2 $g \cdot cm^{-3}$. Accura Bluestone [6] has solid density of 1.78 $g \cdot cm^{-3}$ at 25 °C, so invar is 4.6 times heavier than Accura Bluestone. Using metals with low CTE [7] is a conventional way to attack the temperature variation problem. There are also various methods of compensating temperature variations in microwave structures [8–19].

It is possible to achieve metal CTE characteristics with 3D printing using DMLS. The AlSi10Mg alloy used for DMLS 3D printing has a CTE of $20.5 \times 10^{-6} \cdot K^{-1}$ [20] and $\sigma = 11.3 \times 10^6$ S/m [21], which is better CTE than pure aluminum. There are a range of CTE values from $53 \times 10^{-6} \cdot K^{-1}$ to $145 \times 10^{-6} \cdot K^{-1}$ available for a desktop SLA 3D printer [22]. Electroplating can also use layers of copper and nickel to strengthen the metal coating and resist the expansion of plastic. Relatively broad band structures such as slot arrays and bandpass filters are not as badly affected by frequency drift as narrow band notch filters.

2. Materials and Methods

We measured the effect of temperature change on a slot array, cavity notch filters with single and multiple cavities, and post loaded waveguide bandpass filters. We tested several types of SLA resin, different plating techniques, and also DMLS and Binder Infusion metal 3D printed structures.

Analytical predictions of frequency shifts with temperature and calculation of effective CTE are estimates based on bulk scaling of resonators. Analytical scaling and HFSS bulk scaling results are consistent. Mechanical complications of the expansion of the structure causing it to expand in a more complicated way than bulk scaling cannot be simulated analytically or with HFSS. Metal coatings decrease the effect of plastic expansion. Quoted commercial CTE values for 3D printed resin are assumed to be accurate, and their quoted values change with temperature ranges. There is no quoted CTE for the binder infusion metal. The DMLS metal CTE is based on a publication. Measurements present the actual frequency drift performance of each device. Calculations and predictions offer a way to know the impact of material and construction choices.

The oven used for temperature testing was a Test Equity model 107, with temperature range −42 to +130 °C. The network analyzer used in the measurements was a Keysight N5242A. Commercial waveguide to coaxial transitions were used to test both filters and a slot array antenna.

An 18 GHz slot array was cycled in temperature from −10 to +60 °C six times in 24 h intervals. Only S_{11} was measured. After six cycles, there was visible warping of the structure. The S_{11} minimum changed permanently after each cycle, so there was a hysteresis effect with a 70 °C temperature excursion. Our theory is that the plastic expanding during heating pushes the metal plating, and when the plastic shrinks during cooling, the metal does not return exactly to its original position.

As shown in Figure 1, the minimum S_{11} was −20.047 dB at 18.07 GHz on the first cycle, −33.16 dB at 18.12 GHz after the second cycle, −32.11 dB at 18.13 GHz after the third cycle, −28.94 dB at 18.14 GHz after the fourth cycle, −22.19 dB at 18.15 GHz after the fifth cycle, and −27.36 dB at 18.14 GHz after the sixth cycle.

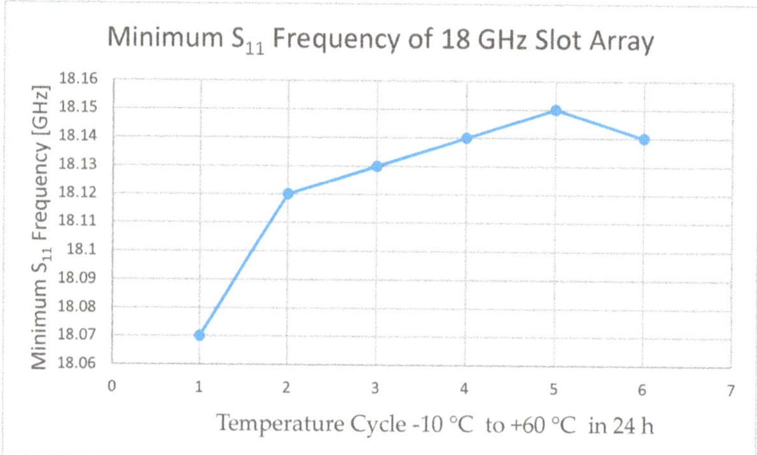

Figure 1. Minimum S_{11} frequency of slot array temperature cycled −10 °C to +60 °C.

The maximum frequency excursion was 80 MHz, which is a 0.44% permanent change. The −10 dB bandwidth did not change from 100 MHz or 0.55%. A 70 °C temperature shift, a minimum of −10 °C and a maximum of +60 °C cause permanent damage to a 3D printed, copper plated microwave slot array. An obvious alternative that would avoid hysteresis and damage is metal 3D printing. Metal 3D printing with DMLS becomes expensive for part dimensions greater than six inches.

Three slot arrays were simulated with CTE of $50 \times 10^{-6} \cdot K^{-1}$ to examine the effects of temperature change on gain and S_{11} parameters. A 315 × 323 mm array designed to operate at 18.4 GHz showed an S_{11} minimum at 18.375 GHz, with S_{11} = −30.52 dB with no temperature offset. The −10 dB S_{11} width was from 18.3 to 18.44 GHz (140 MHz bandwidth). With a 20-degree temperature offset, the minimum S_{11} shifted to 18.356 GHz (Δf = −19 MHz), with S_{11} = −30.78 dB. The −10 dB S_{11} width was from 18.28 to 18.43 GHz. With a 50-degree temperature offset, the minimum S_{11} shifted to 18.330 GHz (Δf = −45 MHz), with S_{11} = −35.5613 dB. The −10 dB S_{11} width was from 18.25 to 18.40 GHz. The temperature scaling is shown in Figure 2. This array used a four-point corporate network, so it had a wide bandwidth (145 MHz or 7.9%). The realized gain at the design frequency of 18.4 GHz was 34.16 dB with no temperature offset, and reduced to 33.92 dB with a 20-degree temperature offset and to 33.53 dB with a 50-degree temperature offset (−0.62 dB). This is due to the shift in optimum S_{11}.

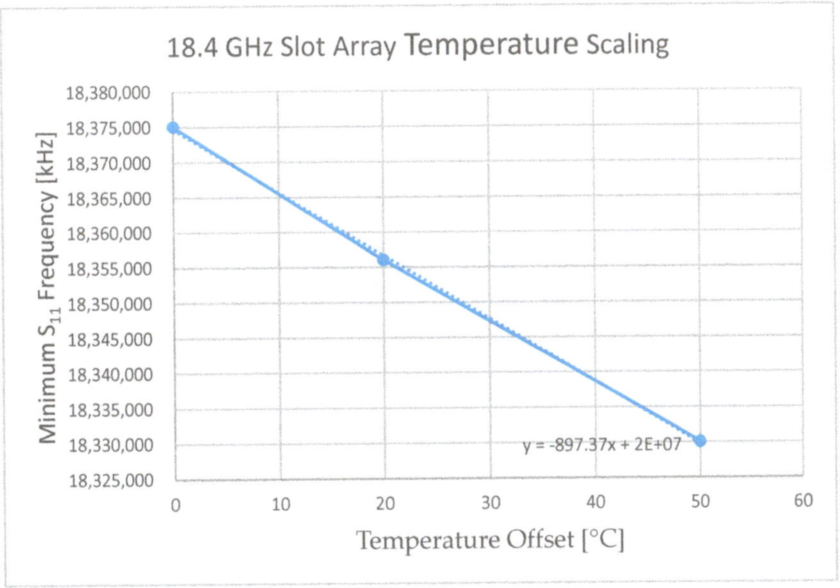

Figure 2. 18.4 GHz slot array temperature scaling.

A 301 × 319 mm array designed to operate at 10.73 GHz had a minimum S_{11} at 10.7305 GHz, with S_{11} = −20.26 dB with no temperature offset as shown in Figure 3. The −10 dB S_{11} width was from 10.69 to 10.76 GHz (70 MHz bandwidth). With a 20-degree temperature offset, the minimum S_{11} shifted to 10.721 GHz (Δf = −9.5 MHz), with S_{11} = −20.1103 dB. The −10 dB width was from 10.6842 to 10.7530 GHz. With a 50-degree temperature offset, the minimum S_{11} shifted to 10.705 GHz (Δf = −25.5 MHz), with S_{11} = −20.87 dB. The −10 dB S_{11} width was from 10.67 to 10.74 GHz. This array was coaxial fed in the center of a centered feeding waveguide. The realized gain at the design frequency of 10.73 GHz reduced from 28.42 dB with no temperature offset to 28.45 dB with a 20-degree temperature offset and 28.13 dB with a 50-degree temperature offset (−0.29 dB).

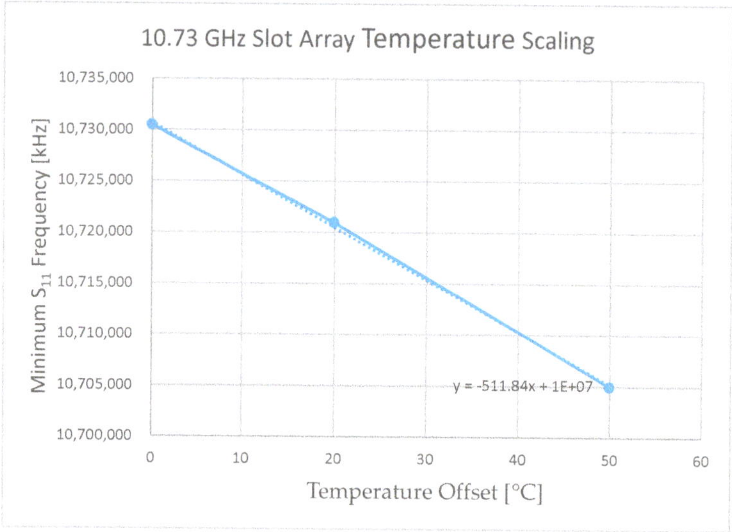

Figure 3. 10.73 GHz slot array temperature scaling.

A 448 × 420 mm array designed to operate at 8.18 GHz showed an S_{11} minimum at 8.171 GHz, with S_{11} = −30.11 dB with no temperature offset as shown in Figure 4. The −10 dB S_{11} width was from 8.13 to 8.22 GHz (90 MHz bandwidth). With a 20-degree temperature offset, the minimum S_{11} shifted to 8.161 GHz (Δf = −10 MHz), with S_{11} = −32.42 dB. The −10 dB S_{11} width was from 8.12 to 8.21 GHz. With a 50-degree temperature offset, the minimum S_{11} shifted to 8.147 GHz (Δf = −24 MHz), with S_{11} = −35.4 dB. The −10 dB S_{11} width was from 8.11 to 8.19 GHz. The array used a center feeding waveguide fed from the end, which is why it has smaller bandwidth. The realized gain at the design frequency of 8.18 GHz was 29.84 dB with no temperature offset, reduced to 29.81 dB with a 20 °C offset, and reduced to 29.66 dB with a 50-degree temperature offset (−0.18 dB).

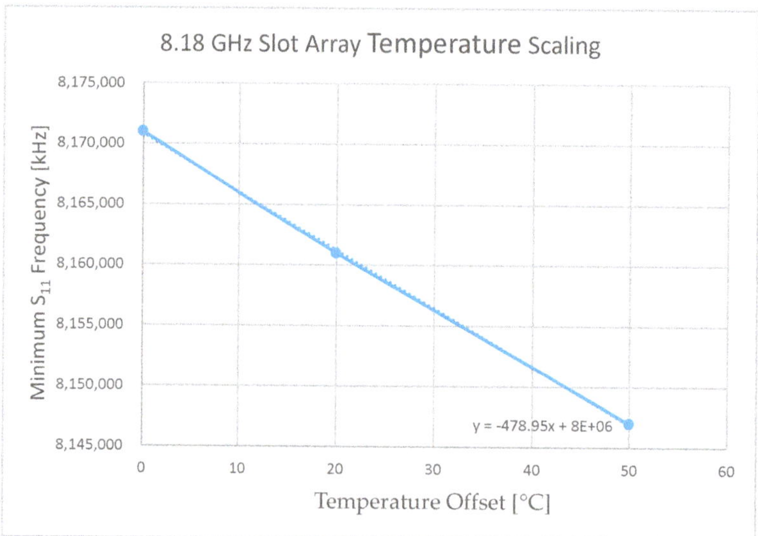

Figure 4. 8.18 GHz slot array temperature scaling.

Given that all three arrays lose less than one dB of realized gain with a 50 °C temperature increase, slot array performance is apparently not significantly affected by operating at high temperatures. The effect of temperature variation is much less pronounced than it is for narrow band notch filters and to a lesser degree bandpass filters. On the other hand, a high enough temperature will cause permanent changes to the array, and with a high enough temperature, visible damage.

A microwave notch filter can be created by coupling a single or multiple resonant cavities to a waveguide [23]. The cavity resonance causes a dip in S_{21}, notching the frequency out of the passband. Not only do the resonant cavity dimensions grow with increasing temperature, but the coupling apertures will also increase in size, which also decreases the resonant frequency assuming that the cavity was originally under-coupled.

The resonant frequency of a TE_{101} rectangular cavity is given by

$$f_{101} = \frac{c}{2\pi\sqrt{\mu_r \varepsilon_r}} \sqrt{\left(\frac{\pi}{a}\right)^2 + \left(\frac{\pi}{d}\right)^2} \quad (1)$$

For a resonant frequency of 10 GHz based on WR-90 waveguide (a = 0.9 inches = 22.86 mm), the guide wavelength is given by

$$\lambda_g = \frac{\lambda_0}{\sqrt{1 - \left(\frac{\lambda_0}{\lambda_c}\right)^2}} \quad (2)$$

where λ_c is the cutoff wavelength given by two times the waveguide width 'a.' At 10 GHz, λ_0 = 29.979 mm and λ_g = 39.707 mm.

The length of a TE_{101} cavity is exactly $\lambda_g/2$ so d = 19.854 mm. As temperature increases, cavity dimensions change according to the following formulas:

$$a = 22.86\text{mm} \times (1 + CTE \times \Delta T) \quad (3)$$

$$d = 19.854\text{mm} \times (1 + CTE \times \Delta T) \quad (4)$$

The resonant frequency as a function of temperature and CTE is shown in Figure 5 below.

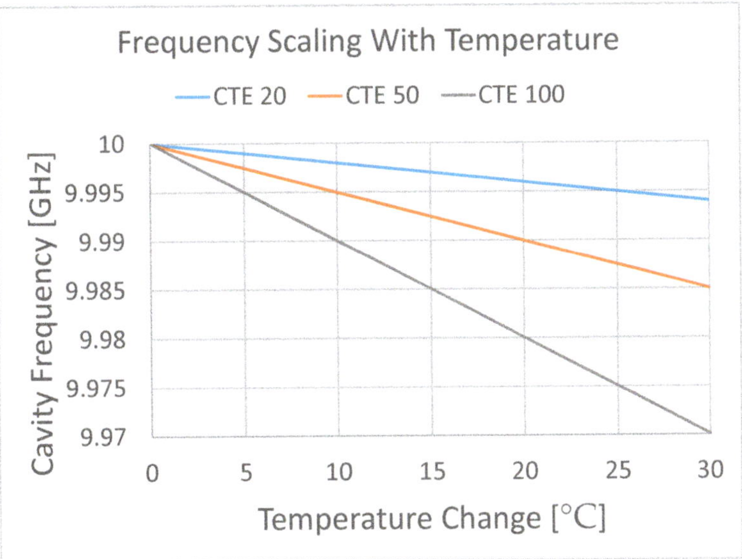

Figure 5. Frequency scaling of 10 GHz TE_{101} Resonant Cavity – Temperature versus CTE.

For a notch filter centered at 10 GHz, the slope for CTE 20 × 10^{-6}·K^{-1} is −200 kHz·K^{-1}. For CTE of 50 × 10^{-6}·K^{-1}, the slope is −499 kHz·K^{-1}. For CTE of 100 × 10^{-6}·K^{-1}, the slope is −997 kHz·K^{-1}. Equation (1) combined with Equations (3) and (4) yields the approximate result that the slope of resonant frequency change with temperature is –CTE multiplied by the unmodified resonant frequency. This matches the fit slopes for CTE 20, 50, and 100 × 10^{-6}·K^{-1}.

For a TE_{101} notch filter centered at 15.7 GHz, using the same analytical scaling, CTE of 20 × 10^{-6}·K^{-1} produces a slope of −300 kHz·K^{-1}, a CTE of 50 × 10^{-6}·K^{-1} produces a slope of −800 kHz·K^{-1}, and a CTE of 100 × 10^{-6}·K^{-1} produces a slope of −1.6 MHz·K^{-1}. For a DMLS 3D printed cavity filter made with AlSi10Mg, with a notch at 15.644 GHz, with CTE published as 20.5 × 10^{-6}·K^{-1}, we measured −16 MHz shift for a temperature range from 23 to 55 °C, corresponding to −500 kHz·K^{-1}. Measurement of S_{21} at 20, 23, 35, 40, 45, and 50 °C was followed by plotting the S_{21} null versus temperature and applying a linear fit to the data. An analytical fit to this frequency change corresponds to a CTE of 32 × 10^{-6}·K^{-1}. An image of the single cavity DMLS metal 3D printed notch filter is shown in Figure 6.

Figure 6. DMLS metal 3D printed single cavity notch filter.

A single cavity filter SLA 3D printed with Formlabs Tough resin at 15.9 GHz using WR-62 waveguide, and plated with 12.7 μm copper, 25.4 μm nickel, and 12.7 μm copper measured 15.933 GHz with S_{21} = −16.04 dB at 20 °C, 15.91 GHz with S_{21} = −16.26 dB at 35 °C, 15.9036 GHz with S_{21} = −15.48 dB at 40 °C, and 15.913 GHz with S_{21} = −28.97 dB at 45 °C. The linear slope fit to these data is −856.23 kHz per °C. The analytical fit to CTE for this resonant cavity is 53.95 × 10^{-6}·K^{-1}. The published CTE of Formlabs Tough resin [24] in the cured state is 119 × 10^{-6}·K^{-1}, so the stronger metal coating did have an effect. The heated plastic expands, but a stronger metal coating more strongly resists this expansion. In the limit where the metal coating was very thick, it is obvious that the plastic would be completely prevented from expanding. Returning to 20 °C, the S_{21} null was −29.89 dB (−13.85 dB change) at 15.939 GHz (Δf = +6 MHz). There was some hysteresis effect.

Images of SLA 3D printed, copper plated single cavity notch filters using WR-42 and WR-62 waveguide are shown in Figure 7.

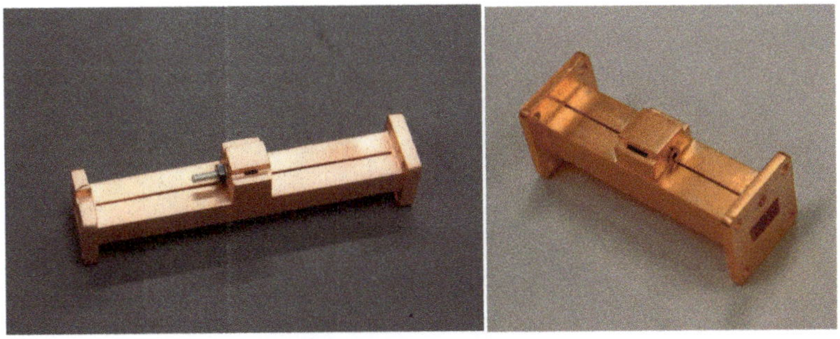

Figure 7. SLA 3D printed single cavity notch filter using WR-42 and WR-62 waveguide.

A single cavity filter SLA 3D printed with Taurus resin at 17.1 GHz using WR-62 waveguide, and plated with copper measured an S_{21} minimum at 17.068 GHz with S_{21} = –9 dB at 20 °C, 17.037 GHz with S_{21} −8.15 dB at 35 °C, 17.027 GHz with S_{21} = −8.31 dB at 40 °C, 17.015 GHz with S_{21} = −6.52 dB at 45 °C, and 17.001 GHz with S_{21} = −7.61 dB at 50 °C. As shown in Figure 8, the linear slope fit to these data is −2197.9 kHz per °C. The analytical fit to CTE for this resonant cavity is $129 \times 10^{-6} \cdot K^{-1}$. The published CTE of Somos Taurus is $105.3 \times 10^{-6} \cdot K^{-1}$. Returning to 20 °C, the S_{21} null was –7.61 dB (+ 1.39 dB change) at 17.055 GHz (Δf = −13 MHz). There was some hysteresis effect.

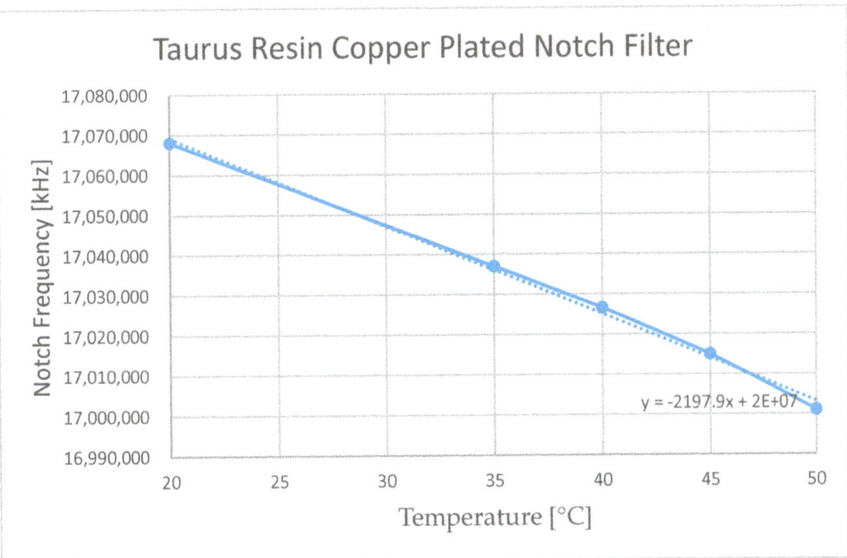

Figure 8. Somos Taurus SLA 3D printed copper plated notch filter.

A three-cavity filter at 17.2 GHz was fabricated using binder infusion metal 3D printing. In this process, a binder is printed on tungsten powder and infused with bronze [25]. The percentage of tungsten is 50–55%. For a TE_{101} notch filter centered at 17.2 GHz, using the same analytical scaling, CTE of $20 \times 10^{-6} \cdot K^{-1}$ produces a slope of -343 kHz$\cdot K^{-1}$, a CTE of $50 \times 10^{-6} \cdot K^{-1}$ produces a slope of −859 kHz K^{-1}, and a CTE of $100 \times 10^{-6} \cdot K^{-1}$ produces a slope of −1.71 MHz$\cdot K^{-1}$. The binder infusion three cavity notch filter measured a frequency shift of 3 MHz over a temperature range from 23 to 50 °C, corresponding to −111.11 kHz$\cdot K^{-1}$. Compared to analytical scaling, this corresponds to an effective CTE of $6.5 \times 10^{-6} \cdot K^{-1}$. Since bronze has CTE of $17 \times 10^{-6} \cdot K^{-1}$ and tungsten has CTE of $4.5 \times 10^{-6} \cdot K^{-1}$, assuming 50% of each one should expect overall CTE of $10.75 \times 10^{-6} \cdot K^{-1}$. An image of the binder infusion metal 3D printed notch filter is shown in Figure 9.

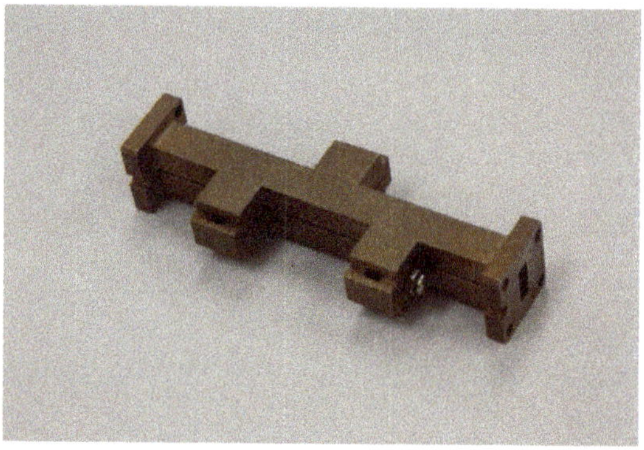

Figure 9. Binder infusion metal 3D printed cavity notch filter.

One remedy for the temperature drift problem is to adjust the coupling aperture of a single cavity filter to be near critical coupling. This increases the resonant bandwidth. If the notch width is 150 MHz and the interfering signal has a bandwidth of 50 MHz or less, significant temperature drift can be tolerated while still attenuating the target signal. One problem with this technique is that it does not allow filtering of an interfering signal that is closer in frequency than the notch bandwidth.

The bandpass filter design that was 3D printed is a post loaded waveguide. The inductive posts form coupled resonant cavities. This filter was designed based on a technique described in [24]. The filter has openings along the center of the waveguide broad wall, following the technique described in [1] to enable internal electroplating of a 3D printed structure.

The SLA 3D printed, copper electroplated bandpass filter measured a passband shift of -549 kHz·K^{-1}. The DMLS 3D printed bandpass filter measured a passband shift of -243.4 kHz·K^{-1}. This was determined by measuring the $S_{21} = -15$ dB point of each bandpass S_{21} characteristic plot as a function of frequency. These were measured at 20, 23, 35, 40, 45, and 50 °C with frequencies 12.6892 GHz (20 °C), 12.6883 GHz (23 °C), 12.6850 GHz (35 °C), 12.6842 GHz (40 °C), 12.6833 GHz (45 °C), and 12.6817 GHz (50 °C) as shown in Figure 10. The value of f for $S_{21} = -15$ dB at each temperature was plotted and a linear fit was used to characterize the overall temperature variation. The high temp SLA, copper plated bandpass filter measured a passband shift of -365.89 kHz·K^{-1}. The measured frequencies were 12.6742 GHz (20 °C), 12.6733 GHz (23 °C), 12.6708 GHz (35 °C), 12.6658 GHz (40 °C), 12.6650 GHz (45 °C), and 12.6642 GHz (50 °C). Images of the SLA and DMLS 3D printed, post-loaded, waveguide microwave bandpass filters are shown in Figure 11.

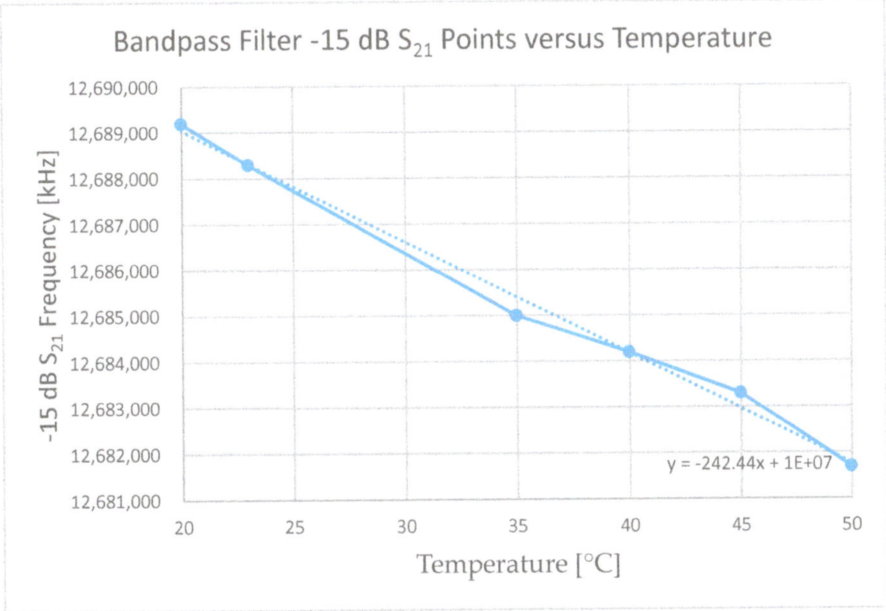

Figure 10. Bandpass Filter -15 dB S_{21} Points versus Temperature.

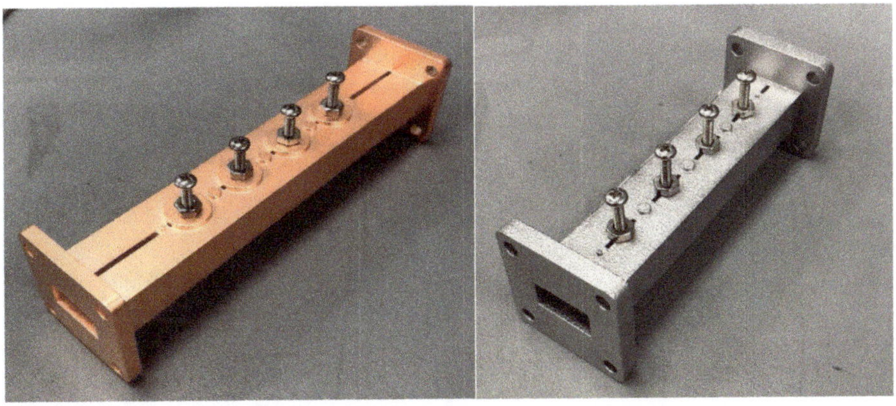

Figure 11. SLA 3D printed, copper plated and DMLS microwave bandpass filter.

Simulation with HFSS using bulk scaling of the structure corresponding to CTE of $20 \times 10^{-6} \cdot K^{-1}$ showed a frequency shift of -300 kHz$\cdot K^{-1}$, very close to the measured value for DMLS with actual reported CTE of $20 \times 10^{-6} \cdot K^{-1}$. The HFSS simulation result is shown in Figure 12 for $\Delta T = 0$ to 50 °C.

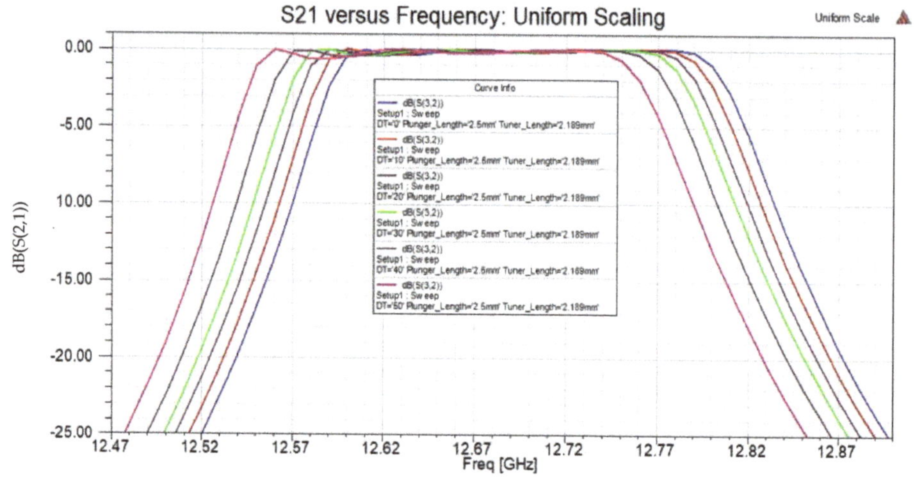

Figure 12. Bandpass filter scaling in HFSS ΔT = 0–50 °C.

3. Conclusions

Resins used for stereolithography 3D printing typically have high CTE. Microwave structures manufactured with 3D printing using SLA and copper plated can have strong frequency dependence on temperature. In particular, multi-cavity filters with narrow bandwidth are strongly affected by dependence of resonant frequency on temperature. This can preclude their usefulness if the frequency drift approaches the bandwidth of the device. Electroplating with layers of copper and nickel to strengthen the metal coating resists the expansion of plastic and reduces temperature dependence. Using DMLS metal printing approaches metal structure performance, so that metal 3D printed structures can approach temperature characteristics of solid metal structures. Single cavity notch filters printed with lower CTE resin can be useful if a wide enough notch bandwidth is used to accommodate frequency drift. One must choose the right 3D printing materials and processes to achieve the temperature stability required by the application.

Funding: This research received no external funding.

Conflicts of Interest: Page: 9The authors declare no conflict of interest.

References

1. Le Sage, G.P. 3D printed waveguide slot array antennas. *IEEE Access* **2016**, *4*, 1258–1265. [CrossRef]
2. Le Sage, G.P. Dielectric steering of a 3-D printed microwave slot array. *IEEE Antennas Wirel. Propag. Lett.* **2018**, *17*, 2141–2144. [CrossRef]
3. Hindle, P. First flight-worthy metal printed RF filter developed by airbus and 3D system. *Microw. J.* **2017**.
4. Tamayo-Dominguez, A.; Fernandez-Gonzales, J.; Sierra-Perez, M. Groove Gap Waveguide in 3-D Printed Technology for Low Loss, Weigh, and Cost Distribution Networks. *IEEE Trans. Microw. Theory Tech.* **2017**, *65*, 4138–4147. [CrossRef]
5. Somos Taurus Resin Specification. Available online: https://www.dsm.com/solutions/additive-manufacturing/en_US/resource-center/user-guide/somos-taurus.html (accessed on 1 March 2020).
6. Accura Bluestone Specification. Available online: https://www.buildparts.com/materials/accurabluestone (accessed on 1 March 2020).

7. Martín-Iglesias, P.; Raadik, T.; Teberio, F.; Percaz, J.M.; Martín-Iglesias, S.; Pambaguian, L.; Arregui, I.; Arnedo, I.; Lopetegui, T.; Laso, M.A.G. Multiphysic Analysis of High Power Microwave Filter Using High Performance Aluminium Alloy. In Proceedings of the 2019 IEEE MTT-S International Microwave Workshop Series on Advanced Materials and Processes for RF and THz Applications (IMWS-AMP), Bochum, Germany, 16–18 July 2019; pp. 58–60.
8. Djerafi, T.; Wu, K.; Deslandes, D. A Temperature-Compensation Technique for Substrate Integrated Waveguide Cavities and Filters. *IEEE Trans. Microw. Theory Tech.* **2012**, *60*, 2448–2455. [CrossRef]
9. Keats, B.F. Bimetal Temperature Compensation for Waveguide Microwave Filters. Ph.D. Thesis, University of Waterloo, Waterloo, ON, Canada, 2007. Available online: https://uwspace.uwaterloo.ca/bitstream/handle/10012/3529/BFKPhDThesis.pdf;jsessionid=5258AA80CC2EEDBB50BBB511BC90A52E?sequence=1 (accessed on 1 March 2020).
10. Martin, T.; Ghiotto, A.; Vuong, T.; Lotz, F. Self-Temperature-Compensated Air-Filled Substrate-Integrated Waveguide Cavities and Filters. *IEEE Trans. Microw. Theory Tech.* **2018**, *66*, 3611–3621. [CrossRef]
11. Lisi, M. Temperature compensation of microwave resonators and filters for space applications. *Microw. J.* **2015**, *58*, 100–108.
12. Lisi, M. A Review of Temperature Compensation Techniques for Microwave Resonators and Filters. In Proceedings of the Micro and Millimetre Wave Technology and Techniques Workshop 2014, Estec, Noordwijk, The Netherlands, 25 November 2014.
13. Temperature Compensated Cavity Filters. EverythingRF. Available online: https://www.everythingrf.com/product-datasheet/521-334-temperature-compensated-cavity-filters (accessed on 1 March 2020).
14. Hsieh, L.; Dai, S. *Temperature Compensated Bandpass Filters in LTCC*; SAND2013-2744C; Sandia National Lab.: Albuquerque, NM, USA, 2013.
15. Temperature Compensated Microwave Resonator. US Patent 4677403, 30 June 1987.
16. Temperature Compensated Resonator and Method. US Patent 6131256A, 17 October 2000.
17. Filter with Temperature Compensated Tuning Screw. US Patent CA2206942C, 17 January 1999.
18. Temperature Compensated Microwave Filter. US Patent US5867077A, 2 February 1999.
19. Temperature Compensated High Power Bandpass Filter. US Patent US6529104B1, 4 March 2003.
20. Gumbleton, R.; Cuenca, J.A.; Klemencic, G.M.; Jones, N.; Porch, A. Evaluating the coefficient of thermal expansion of additive manufactured AlSi10Mg using microwave techniques. *Addit. Manuf.* **2019**, *30*, 100841. [CrossRef]
21. Silbernagel, C.; Ashcroft, I.; Dickens, P.; Galea, M. Electrical resistivity of additively manufactured AlSi10Mg for use in electric motors. *Addit. Manuf.* **2018**, *21*, 395–403. [CrossRef]
22. Formlabs Resin Specifications. Available online: https://formlabs.com/3d-printers/form-3/tech-specs/#data-sheets (accessed on 1 March 2020).
23. Matthaei, G.; Young, L.; Jones, E.M.T. *Microwave Filters, Impedance-Matching Networks, and Coupling Structures*; Artech House: London, UK, 1980.
24. Available online: https://formlabs-media.formlabs.com/datasheets/Tough_Technical.pdf (accessed on 1 March 2020).
25. Available online: http://exone.com/Admin/getmedia/79471d50-8b99-41e5-945d-4f151409ef42/X1_MaterialData_Tungsten-Bronze_Tekna_062519.pdf (accessed on 1 March 2020).

metals MDPI

Article

Mechanical Properties of High Strength Aluminum Alloy EN AW-7075 Additively Manufactured by Directed Energy Deposition

Anika Langebeck [1,*], Annika Bohlen [1], Rüdiger Rentsch [2] and Frank Vollertsen [1,3]

1 BIAS—Bremer Institut für angewandte Strahltechnik GmbH, Klagenfurter Str. 5, 28359 Bremen, Germany; bohlen@bias.de (A.B.); info-mbs@bias.de (F.V.)
2 Leibniz Institute for Material-Oriented Technologies, Badgasteiner Straße 3, 28359 Bremen, Germany; rentsch@lfm.uni-bremen.de
3 Faculty Production Technology, University of Bremen, Bibliothekstr. 1, 28359 Bremen, Germany
* Correspondence: langebeck@bias.de; Tel.: +49-421-218-58035

Received: 31 March 2020; Accepted: 23 April 2020; Published: 29 April 2020

Abstract: A manifold variety of additive manufacturing techniques has a significant positive impact on many industry sectors. Large components are often manufactured via directed energy deposition (DED) instead of using powder bed fusion processes (PBF). The advantages of the DED process are a high build-up rate with values up to 300 cm^3/h and a nearly limitless build-up volume. In combination with the lightweight material aluminum it is possible to manufacture large lightweight components with geometries adapted to customer requirements in small batches. This contributes the pursuit of higher efficiency of machines through lightweight materials as well as lightweight design. A low-defect additive manufacturing of high strength aluminum EN AW-7075 powder via DED is an important challenge. The laser power has a significant influence on the remaining porosity. By increasing the laser power from 2 kW to 4 kW the porosity in single welding tracks can be lowered from 2.1% to only (0.09 ± 0.07)% (n = 3). However, when manufacturing larger specimens; the remaining porosity is higher than in single tracks; which can be attributed to the oxide skin on the preceding welding tracks. Further investigations regarding the mechanical properties were carried out. In tensile tests an ultimate tensile strength of (222 ± 17) MPa (n = 6) was measured. The DED processed EN AW-7075 shows comparable mechanical properties to PBF processed EN AW-7075.

Keywords: directed energy deposition; EN AW-7075; porosity; ultimate tensile strength

1. Introduction

Metal additive manufacturing for producing components and component parts is becoming more and more relevant, especially in customized single and small series production [1]. Two relevant process groups are powder bed fusion (PBF) and directed energy deposition (DED) (ISO/ASTM 52900:2018-06). The latter is commercialized by several organizations and known under different acronyms such as LENS (Laser Engineering Net Shaping) [2]. These DED systems are used besides additive manufacturing purposes for repairing high-value components [3]. A common classification of DED processes is to divide them into powder-feed processes and wire-feed processes [4]. For PBF processes the powder material is supplied in a powder bed. For powder-feed DED, powder nozzles are used to deliver the powder directly into the process zone. In contrast to PBF processes, DED is used in case of larger structures with lower resolution [4] due to high build-up rates which can reach values up to 1.8 kg/h for the powder-feed DED and a nearly limitless build-up volume [5]. Besides a laser as energy source for DED an electron beam or an arc can be used. A five times higher build-up rate can be achieved in DED with wire feed [6]. However, by using a laser as energy source the heat input is significantly lower compared to an arc and therefore the distortion is smaller [7].

When aluminum alloys are used in additive processes, hydrogen related porosity is a disadvantage that must be considered. Hydrogen related porosity is formed during solidification of the melt pool, when the solubility of the aluminum for hydrogen decreases significantly [8]. An important source for hydrogen besides the oxide layer on the substrate's surface is the humidity of the ambient air near to the process zone [9]. There are two approaches to reduce hydrogen related porosity. First, by improving the shielding gas coverage to protect the process zone from hydrogen. Secondly, by adapting the process parameters for a better degassing of hydrogen during the solidification of the melt pool.

A low porosity is necessary to achieve high mechanical characteristics such as ultimate tensile strength and Young's modulus. For additively manufactured tensile specimens made of aluminum EN AW-7075 laser powder bed fusion (LPBF) is often used [10]. Only few scientists report on produced tensile specimens using DED [11]. When additively manufactured tensile specimens are tested, it is shown that the orientation of the specimen's single tracks to the tensile load direction has a significant influence on the results for the mechanical characteristics. The ultimate tensile strength is highest when building direction is parallel to the load direction [11]. EN AW-7xxx alloys can be precipitation hardened in a subsequent heat treatment that results in higher hardness and therefore in higher ultimate tensile strength [12]. A T6 heat treatment is often used to achieve an increase in hardness through precipitation hardening [13]. Therefore, solution heat treatment followed by quenching is carried out to achieve a supersaturated solid solution. Precipitation of magnesium and zinc take place during subsequent artificial aging and an increase in hardness is reached. For DED manufactured specimens of an EN AW-7xxx alloy with high zinc content of 11.9% and high magnesium content of 2.7% the measured Vickers hardness increased after a T6 heat treatment from (133 ± 6) HV0.05 to (219 ± 4) HV0.05 [13]. When processing EN AW-7xxx alloys in DED processes a loss in volatile elements such as zinc and magnesium is observed [14]. This alters the initial alloy composition and can affect the (mechanical) properties.

In this study low-defect specimens of EN AW-7075 will be manufactured via powder-feed DED with an adapted shielding gas shroud and the influence of the laser power on the porosity will be determined. The low-defect specimens will be used to examine the influence of artificial aging on hardness and mechanical characteristics of the low-defect specimens will be evaluated as well.

2. Materials and Methods

The specimens were manufactured by powder-feed DED. Therefore, a lamp pumped Nd:YAG rod laser (Trumpf GmbH + Co. KG, Ditzingen, Germany) was used. The Trumpf HL 4006 D generates a wavelength of 1064 nm with a maximum power of 4 kW. The laser beam is guided via an optical fiber with a core diameter of 600 μm to the processing head. The processing head consists out of an optical unit by Trumpf (type BEO D70). Collimation lens as well as focus lens have a focal length of 200 mm. By setting the laser focus with a top-hat profile 24.5 mm above the substrate surface a calculated spot size of 4.5 mm with a gaussian intensity distribution on the substrate's surface was reached. With increasing the laser power to the maximum of 4 kW at a spot size of 4.5 mm a maximum laser intensity of 251 W/cm^2 can be reached. This ensures that the threshold intensity for forming a keyhole is not exceeded during the DED process. The aluminum powder was supplied by a rotating disk powder feeder made by Plasma-Technik AG (Twin10C, Wohlen, The Switzerland) to a coaxial three-jet nozzle by Ixun Lasertechnik GmbH (Aachen, Germany) with a working distance for the powder focus of 12 mm. Argon was used as carrier and shielding gas. Substrates for the specimens and the processing head were mounted on a 3-axis CNC by Föhrenbach GmbH (Löffingen-Unadingen, Germany) and which was driven by a control unit from Power Automation GmbH (type PA8000, Pleidelsheim, Germany).

As substrate material aluminum wrought alloys EN AW-5083 and EN AW-6082 with a thickness of 10 mm each were used. The used aluminum powder EN AW-7075 was supplied by NMD—New Materials Development GmbH (Heemsen, Germany) with a grain size distribution of D_{10} 37 µm, D_{50} 68 µm and D_{90} 122 µm. It was sieved with 50 µm and 125 µm mesh size. See Table 1 for the chemical composition of the powder material.

Table 1. Chemical composition of EN AW-7075 in wt. % supplied by NMD—New Materials Development GmbH (Heemsen, Germany). Measurements in accordance with EN 10204 3.1.

Al	Si	Fe	Mn	Zn	Mg	Cu	Cr
89.9	0.11	0.09	0.01	5.51	2.42	1.60	0.21

In order to improve the shielding gas coverage of the processing zone an additional shielding gas shroud was developed. See Figure 1 for a schematic view of the developed shielding gas shroud. The shroud is made of sinter metal with an internal radius of 19 mm and an external radius of 47 mm and has four ports for the gas supply to provide an extended and homogenous shielding gas coverage for the process zone.

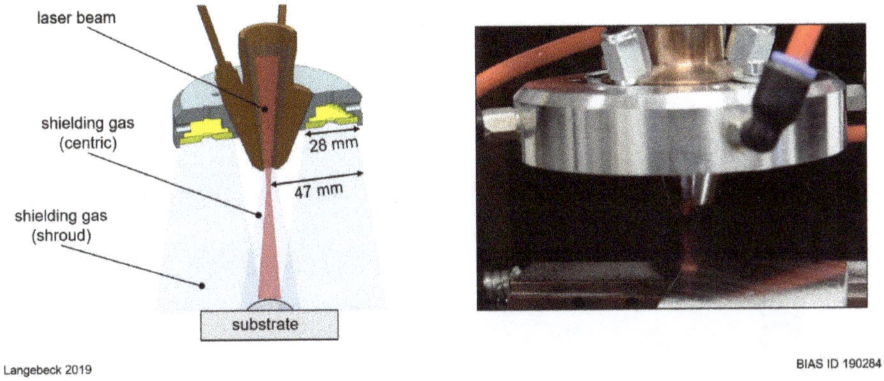

Figure 1. Improved shielding gas concept with an additional shielding gas shroud.

The effect of the new shielding gas concept was evaluated by manufacturing single tracks (see Table 2). The porosity was measured at cross sections. A decrease in porosity from over 7.0% to 5.9% was measured. This corresponds to a reduction by 1/8 for the used parameter set. For all subsequent experimental studies regarding DED with EN AW-7075 the additional shielding gas shroud was used.

The influence of the laser power on pore volume was determined at cross sections in the area of the center of the specimens. The laser power was varied between 2 kW and 4 kW in steps of 0.5 kW. Each parameter combination was examined in randomized triple determination.

For measuring the tensile strength as well as the hardness of the additively manufactured aluminum parts larger specimens consisting of 9 layers in total with 6 single tracks per layer were manufactured. A horizontal overlapping (y-direction) between the single layers of 30% was used. The vertical overlapping (z-direction) was 13%. Between two subsequent layers an offset in y-direction of 2.5 mm was chosen. The building direction (x-direction) was parallel to the load direction chosen for the tensile tests. For the evaluation of the porosity, the microscopic images of non-etched cross-sections were binarized using Otsu's method. See Figure 2 for the position of the cross-sections. The binarized images were used to calculate the porosity by measuring the proportion of black (pores) to white pixels (EN AW-7075). The Vickers hardness measurements were carried out on ground and polished cross sections directly after the DED process and after artificial aging according DIN EN ISO 6507-3. By using wire-cut EDM (electrical discharge machining) tensile specimens were manufactured out of the artificially aged DED parts with an overall length of 100 mm, a width of grip section of 25 mm, a width of 10 mm and a thickness of 4 mm (see Figure 2). The artificial aging includes stretching to 2% elongation, aging for 3 days at room temperature and two step artificial aging for 18 h at 120 °C and for 5 h at 175 °C.

Table 2. Process parameters for single tracks to determine the influence of shielding gas concept on pore volume and the influence of energy input per unit length on pore volume.

Parameters	Single Tracks for Shielding Gas Investigations	Single Tracks for Porosity Investigations
Laser power	2.4 kW	2 kW to 4 kW
Laser spot diameter	2.0 mm	4.5 mm
Shielding gas flow conv.	7.5 L/min (centric)	-
Shielding gas flow with additional shroud	7.5 L/min (centric)	7.5 L/min (centric)
	30 L/min (additional shroud)	30 L/min (additional shroud)
Carrier gas flow	5.5 L/min	4 L/min
Powder feed rate	(9.9 ± 0.2) g/min	(9.3 ± 0.3) g/min
Substrate material	EN AW-5083	EN AW-5083
Welding speed	400 mm/min	400 mm/min

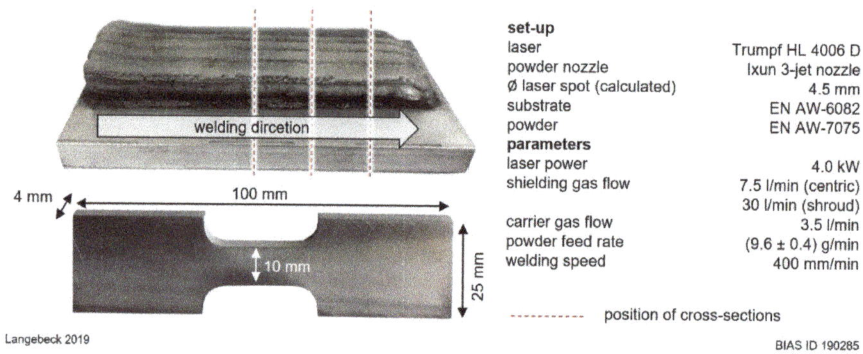

Figure 2. Additively manufactured tensile specimen with building direction parallel to load direction before and after wire-cut electrical discharge machining (EDM).

3. Results

As shown in Figure 3 with increased laser power from 2 kW to 4 kW the measured porosity decreased from 2.1% to just (0.09 ± 0.07)% (n = 3). With a powder feed rate of (9.3 ± 0.3) g/min and a powder usage efficiency of (82.0 ± 3.4)% (n = 14) a build-up rate of (164 ± 10) cm^3/h (n = 14) was reached.

When manufacturing larger parts consisting of several layers with several single tracks the low pore volume of (0.09 ± 0.07)% (n = 3) measured in single tracks could not be reached anymore. However, a pore volume of (2.2 ± 0.9)% (n = 15) was reached. Figure 4 shows an exemplary cross section. The cross section reveals a preferred localization of the pores in the area of the melt pool borders. Additionally to porosity some cracks between the lower layers and the substrate material on the left and right side are visible.

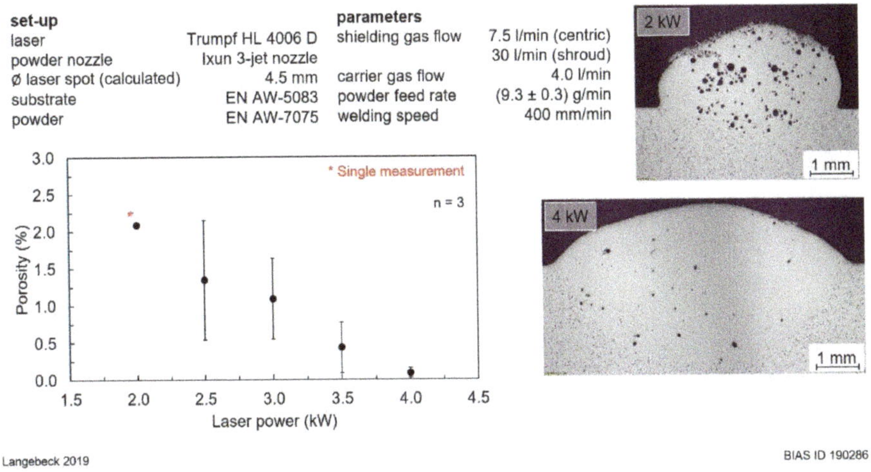

Figure 3. Influence of laser power on porosity.

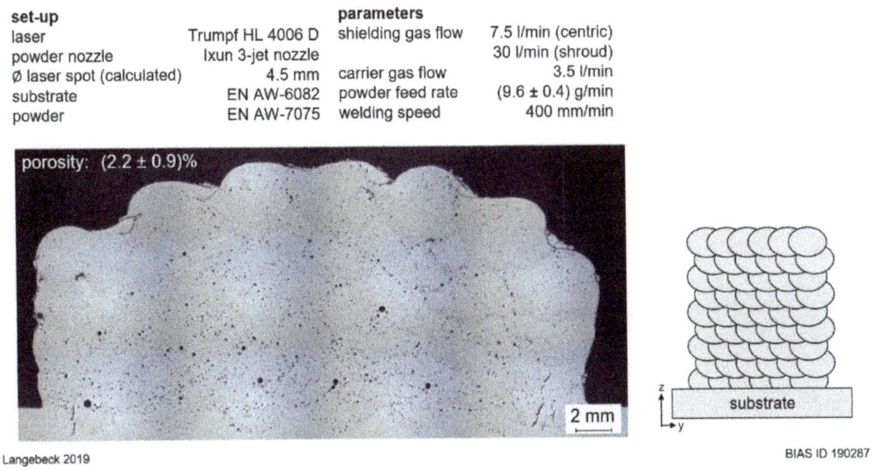

Figure 4. Schematic view of built-up strategy; cross section of large part for hardness measurements as well as tensile tests.

After the DED process Vickers hardness measurements were carried out. A hardness of (85 ± 4) HV0.5 (n = 4) was measured. A slight improvement in hardness up to (93 ± 2) HV0.5 (n = 3) could be reached after artificial aging.

After artificial aging tensile specimens were manufactured and tested. The results are shown in Figure 5. An ultimate tensile strength of (222 ± 17) MPa (n = 6) was calculated.

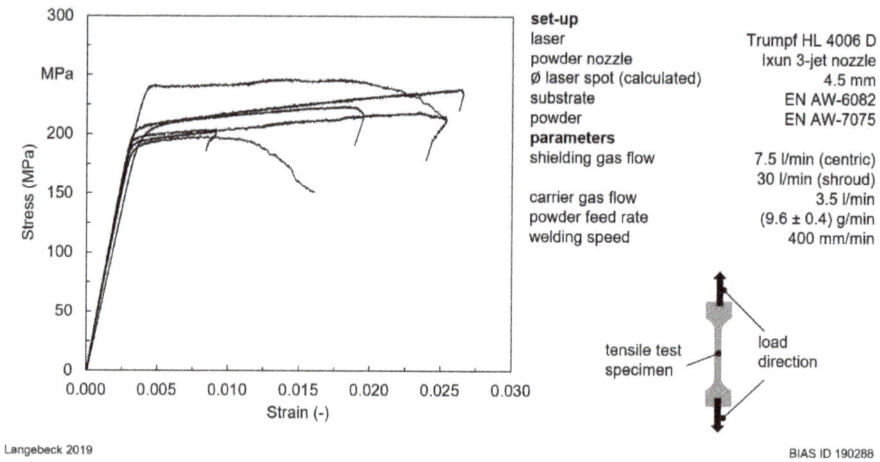

Figure 5. Results of tensile tests for additively manufactured specimens made of EN AW-7075.

4. Discussion

In this work, low-defect specimens were manufactured via DED with an additional shielding gas shroud. An increase in laser power leads to a significant decrease in porosity. This parameter set were then used to evaluate the mechanical characteristics of (tensile) specimens made of EN AW-7075.

The pore volume of DED processed EN AW 7075 decreased from 2.09% to just (0.09 ± 0.07)% (n = 3) in single track experiment by the increase of laser power from 2 kW to 4 kW. The increased laser power led to a more overheated and larger melt pool. Additionally, the thermo-capillary convection in the melt pool increases with higher laser power [15]. This aspect results in lower cooling rates which leave the precipitated gaseous hydrogen and other volatile elements such as zinc and magnesium from the alloy system longer time for degassing out. When larger specimens are manufactured with the same parameter set, which was used for single tracks, the low pore volume of single track experiment could not be achieved. A reason for this can be hydrated oxide layers which form on the substrate surfaces as well as on previous weld beads [16]. The weld beads have a convex macroscopic geometry and have a rough surface due to adhering, partially melted powder particles. This significantly increases the surface area that can oxidize and hydrate. In addition, the increased temperatures during the welding of the individual layers further increase the thickness of the resulting oxide layer [17]. The surface of the weld beads can be identified as a significant source for additional hydrogen that causes porosity. The previous welding tracks are partially remelted during the DED process and the oxide layer with its hydrogen content can cause porosity in the area of the melt pool borders of the new welding track.

EN AW-7075 can be precipitation hardened. Highest hardness can be reached after a T6 heat treatment. The large specimens for hardness measurements were artificially aged without a previous solution heat treatment. The results show a slight increase in hardness from (85 ± 4) HV0.5 (n = 4) to (93 ± 2) HV0.5 (n = 3). This indicates that after the DED process a microstructure is already present which achieves precipitation hardening by artificial aging. For conventionally produced EN AW-7075 T6 a Vickers hardness of (177 ± 7) HV0.5 (n = 3) was measured, which is considerably higher. It is uncertain whether a higher increase in hardness can be achieved by previous solution heat treatment with quenching and subsequent artificial aging reaching a T6 condition. The measured ultimate tensile strength of (222 ± 17) MPa (n = 6) and elongation of about 2% for the DED specimens are in good agreement with the mechanical characteristics of specimens additively manufactured by LPBF (laser powder bed fusion) with an ultimate tensile strength of (206 ± 26) MPa and an elongation smaller than 1% [10]. These comparative figures were determined for both as-built specimens and solution annealed, quenched and artificially aged specimens with no significant deviations [10]. This suggests that a complete T6 heat treatment after the DED process would have had a similar negligible effect on the ultimate tensile strength as for the LPBF specimens. Though when comparing the measured ultimate tensile strength of (222 ± 17) MPa (n = 6) with the tensile strength of conventionally produced EN AW-7075 for T6 temper with 572 MPa, it becomes apparent that the additively manufactured specimen's tensile strength is considerably lower yet it shows good agreement with the tensile strength of soft annealed aluminum with 228 MPa [18]. However, the elongation of only 2% for the additively manufactured specimens is uncharacteristic for the aluminum alloy EN AW-7075 which normally has an elongation of 11% to 17% [18]. This leads to the assumption that the additively manufactured specimens could reach higher ultimate tensile strength if the elongation could be enlarged. The loss in elongation can be attributed to the defects within the specimens such as porosity and hot cracks (see Figure 4) and therefore further improvement for low-defect DED is crucial.

5. Conclusions

Low-defect aluminum specimens can be manufactured via DED by increasing the time for degassing through an increase in laser power. Whereby the surface of the previous welding tracks has a significant impact on porosity, too. After the DED process the EN AW-7075 specimens have a microstructure which allows for a slight increase in hardness from (85 ± 4) HV0.5 (n = 4) to (93 ± 2) HV0.5 (n = 3) through precipitation hardening. The mechanical characteristics do not reach those of conventionally produced EN AW-7075 T6 specimens, which is mostly attributable to the remaining defects. When comparing the measured ultimate tensile strength of (222 ± 17) MPa (n = 6) with that of other additively manufactured EN AW-7075 specimens, good agreement is obtained irrespective of whether the specimens were heat treated or not.

Author Contributions: Conceptualization, A.L., A.B. and R.R.; data curation, A.L.; formal analysis, A.L. and R.R.; investigation, A.L. and R.R.; methodology, A.L. and R.R.; project administration, A.B.; resources, F.V.; supervision, A.B. and F.V.; validation, A.L., A.B., R.R. and F.V.; visualization, A.L.; writing—original draft, A.L.; writing—review & editing, A.B. and F.V. All authors have read and agreed to the published version of the manuscript.

Funding: This research was funded by the Federal Ministry for Economic Affairs and Energy (BMWi) via the German Federation of Industrial Research Associations (AiF) in accordance with the policy to support the Central Innovations of Medium-Sized Enterprises (ZIM) on the basis of a decision by the German Bundestag, grant number ZF4063003SU7.

Acknowledgments: The authors gratefully acknowledge the collaboration with the company IBO GmbH regarding their support of knowledge over the course of the research. The "BIAS ID" numbers are part of the figures and allow the retraceability of the results with respect to the mandatory documentation required by the funding organization.

Conflicts of Interest: The authors declare no conflict of interest.

References

1. Schmidt, M.; Merklein, M.; Bourell, D.; Dimitrov, D.; Hausotte, T.; Wegener, K.; Overmeyer, L.; Vollertsen, F.; Levy, G.N. Laser based additive manufacturing in industry and academia. *CIRP Ann.* **2017**, *66*, 561–583. [CrossRef]
2. Saboori, A.; Gallo, D.; Biamino, S.; Fino, P.; Lombardi, M. An overview of additive manufacturing of titanium components by directed energy deposition: Microstructure and mechanical properties. *Appl. Sci.* **2017**, *7*, 883. [CrossRef]
3. Saboori, A.; Aversa, A.; Marchese, G.; Biamino, S.; Lombardi, M.; Fino, P. Application of directed energy deposition-based additive manufacturing in repair. *Appl. Sci.* **2019**, *9*, 3316. [CrossRef]
4. Jahn, S.; Matthes, S.; Scheller, D.; Huber, M. Process principles for metal additive manufacturing. In Proceedings of the IIW Commissions I, IV, XII, SG 212 Joint Intermediate Meeting, Greifswald, Germany, 7–12 July 2019.
5. Ding, D.; Pan, Z.; Cuiuri, D.; Li, H. Wire-feed additive manufacturing of metal components: Technologies, developments and future interests. *Int. J. Adv. Manuf. Technol.* **2015**, *81*, 465–481. [CrossRef]
6. Wohlers, T. *Wohlers Report 2014: 3D Printing and Additive Manufacturing State of the Industry Annual Worldwide Progress Report*; Wohlers Associates: Fort Collins Col, CO, USA, 2014.

7. Ding, J.; Colegrove, P.; Mehnen, J.; Ganguly, S.; Sequeira Almeida, P.M.; Wang, F.; Williams, S. Thermo-mechanical analysis of Wire and Arc Additive Layer Manufacturing process on large multi-layer parts. *Comput. Mater. Sci.* **2011**, *50*, 3315–3322. [CrossRef]
8. Zhao, H.; White, D.R.; DebRoy, T. Current issues and problems in laser welding of automotive aluminium alloys. *Int. Mater. Rev.* **1999**, *44*, 238–266. [CrossRef]
9. Guo, G.; Zhang, M.; Chen, H.; Chen, J.; Li, P.; Yang, Y.P. Effect of humidity on porosity, microstructure and fatigue strength of A7N01S-T5 aluminum alloy welded joints in high-speed trains. *Mater. Des.* **2015**, *85*, 309–317.
10. Reschetnik, W.; Brüggemann, J.-P.; Aydinöz, M.E.; Grydin, O.; Hoyer, K.-P.; Kullmer, G.; Richard, H.A. Fatigue crack growth behavior and mechanical properties of additively processed EN AW-7075 aluminium alloy. *Procedia Struct. Integr.* **2016**, *2*, 3040–3048. [CrossRef]
11. Khazan, P.; Stroth, M.; Freiße, H.; Köhler, H. Mechanical Properties of Large Three-Dimensional Specimens Build through Direct Powder Deposition. *WGP Congr.* **2014**, *1018*, 525–532. [CrossRef]
12. Brüggemann, J.-P.; Risse, L.; Kullmer, G.; Richard, H.A. Optimization of the fracture mechanical properties of additively manufactured EN AW-7075. *Procedia Struct. Integr.* **2018**, *13*, 311–316. [CrossRef]
13. Wang, P.; Li, H.C.; Prashanth, K.G.; Eckert, J.; Scudino, S. Selective laser melting of Al-Zn-Mg-Cu: Heat treatment, microstructure and mechanical properties. *J. Alloys Compd.* **2017**, *707*, 287–290. [CrossRef]
14. Singh, A.; Ramakrishnan, A.; Dinda, G. Direct laser metal deposition of Al 7050 alloy. *SAE Tech. Paper* **2017**. [CrossRef]
15. Dai, D.; Gu, D. Influence of thermodynamics within molten pool on migration and distribution state of reinforcement during selective laser melting of AlN/AlSi10Mg composites. *Int. J. Mach. Tools Manuf.* **2016**, *100*, 14–24. [CrossRef]
16. Haboudou, A.; Peyre, P.; Vannes, A.B.; Peix, G. Reduction of porosity content generated during Nd:YAG laser welding of A356 and AA5083 aluminium alloys. *Mater. Sci. Eng. A* **2003**, *363*, 40–52. [CrossRef]
17. Jeurgens, L.P.H.; Sloof, W.G.; Tichelaar, F.D.; Mittemeijer, E.J. Growth kinetics and mechanisms of aluminum-oxide films formed by thermal oxidation of aluminum. *J. Appl. Phys.* **2002**, *92*, 1649–1656. [CrossRef]
18. Davis, J.R. *Properties and Selection*, 10th ed.; ASM International: Materials Park, OH, USA, 2000.

metals

MDPI

Article

Fabrication of Single Crystals through a μ-Helix Grain Selection Process during Electron Beam Metal Additive Manufacturing

Martin R. Gotterbarm [1,2,*], Alexander M. Rausch [2] and Carolin Körner [1,2]

[1] Joint Institute of Advanced Materials and Processes, University of Erlangen-Nürnberg, Dr.-Mack-Str. 81, 90762 Fürth, Germany; carolin.koerner@fau.de

[2] Chair of Materials Science and Engineering for Metals, University of Erlangen-Nürnberg, Martensstr. 5, 91058 Erlangen, Germany; alexander.m.rausch@fau.de

* Correspondence: martin.gotterbarm@fau.de; Tel.: +49-(0)911-65078-65121

Received: 13 January 2020; Accepted: 26 February 2020; Published: 28 February 2020

check for updates

Abstract: Selective Electron Beam Melting (SEBM) is a powder bed-based additive manufacturing process for metals. As the electron beam can be moved inertia-free by electromagnetic lenses, the solidification conditions can be deliberately adjusted within the process. This enables control over the local solidification conditions. SEBM typically leads to columnar grain structures. Based on numerical simulation, we demonstrated how technical single crystals develop in IN718 by forcing the temperature gradient along a μ-Helix. The slope of the μ-Helix, i.e., the deviation of the thermal gradient from the build direction, determined the effectiveness of grain selection right up to single crystals.

Keywords: additive manufacturing; powder bed fusion; single crystal; grain selection

1. Introduction

Additive Manufacturing (AM) is an integral part of Industry 4.0, the fourth industrial revolution. AM is a revolutionary way of manufacturing, and is thus subject to intense research and development efforts. With respect to powder bed fusion AM of metals, Selective Electron Beam Melting (SEBM) and the more common Selective Laser Melting (SLM) were investigated. The use of an electron beam as a heat source offers unique possibilities for the manufacturing process. Namely, operating temperatures of more than 1000 °C become feasible as the electron beam can not only be used in a focused way for selective melting. The defocused beam can also be used as a source of global heat input. This makes SEBM suitable for processing of the most challenging materials like nickel- and cobalt-based superalloys or intermetallics, such as titanium aluminides [1]. As the grain structure governs the mechanical properties, it is highly desirable to control or manipulate the developing grain structure within any manufacturing process. SEBM typically yields a columnar grain structure, resulting in a highly anisotropic mechanical behavior [2–4]. The elongated grains originate from epitaxial growth and steep temperature gradients oriented mostly parallel to the build direction (BD) [3,5–7]. However, the grain structure can be manipulated by altering the SEBM processing strategy, and thus influencing the solidification conditions. The possibility of tailoring the grain structure towards a more equiaxed one has been demonstrated in various publications [8–10]. For extremely demanding applications like turbine blades, single crystalline parts are state-of-the-art. Until now, the successful generation of single crystals via SEBM has only been reported by Ramsperger et al. [11] and Chauvet et al. [12], and the underlying mechanisms are not yet fully understood. This study aims to provide a deeper understanding of the underlying grain selection processes leading to single crystals. The nickel-based superalloy IN718 was used as the model alloy to fabricate technical single crystals via SEBM. The solidification conditions were subsequently analyzed using numerical simulations of the process.

2. Materials and Methods

In the present study, trapezoidal prisms, as depicted in Figure 1b, were fabricated from IN718 powder on an Arcam A2 machine (Arcam AB, Mölndal, Sweden), which was modified to meet the specifications of an Arcam A1. As a scan strategy, cross snake hatching (CS, see Figure 1a) without contour melting was used for all samples. The SEBM processing parameters are listed in Table 1 and originate from prior processing window experiments. Two different variations of CS were employed. On the one hand, a standard hatching (SH) approach with a moderate line offset of 100 μm, and a scan speed of 3 m/s was used. On the other hand, the multi passing (MPCol) strategy introduced by Helmer et al. [8] was employed, where small line offsets and high scan speeds are used. The μ-Helix parameter setting is a special variant of multi passing, which enables controlling the solidification paths to resemble a μm-sized helix. Three trapezoidal prisms with varying aperture angles in the wedge section were manufactured with each parameter setting. A total of nine samples were built. Note, that the samples were rotated by 45° with respect to the scan vectors S_X and S_Y, in order to achieve a constant scan length of 15 mm in the middle of the sample. Additionally, cuboid samples with the dimensions 15 × 15 × 120 mm^3 (XYZ) were manufactured with the μ-Helix parameter setting for comparison. In this case, the edges of the cuboids were aligned with the scan vectors. As a start plate, a polycrystalline IN718 disk with a thickness of 16 mm, and a diameter of 136 mm was used. The samples were connected to the start plate via cylindrical support structures. The build temperature was held constant at 950 °C, as measured by a thermocouple attached to the start plate. A build chamber pressure of 2×10^{-3} mbar under Helium atmosphere was applied during the build. The raw material IN718 powder, with a particle size distribution of 45–105 μm, was used, provided by TLS Technik GmbH & Co. Spezialpulver, Bitterfeld, Germany. The powder particles showed spherical morphology with few satellites and moderate residual porosity. The chemical composition is given in Table 2. For analysis, all manufactured samples were metallographically prepared as longitudinal microsections. They were subsequently etched with a HCl + H_2O_2 etching solution to reveal the grain structure. No heat treatments were applied, in order to document the as-built state.

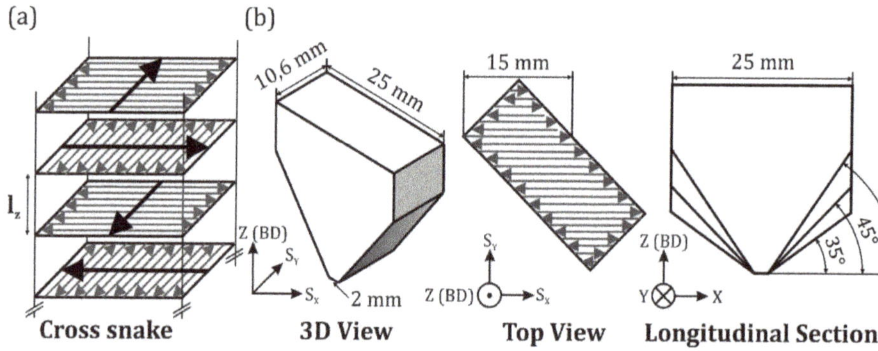

Figure 1. (**a**) Schematic of the cross snake scan strategy used; (**b**) Schematic of the trapezoidal prism geometry used in the present study. The build direction (BD) Z and the scan vectors S_X and S_Y are indicated. Note the rotation of the sample coordinates X and Y in the longitudinal sections by 45° with respect to S_X and S_Y.

Table 1. Selective Electron Beam Melting (SEBM) processing parameters for trapezoid samples.

Parameter Setting	Scan Strategy	Scan Speed [m/s]	Line Offset [μm]	Beam Power [W]	Build Temperature [°C]	Layer Thickness [μm]
Standard hatching (SH)	Cross snake	3	100	660	950	50
Multi passing (MPCol)	Cross snake	13	30	858	950	50
Multi passing (μ-Helix)	Cross snake	7	30	546	950	50

Table 2. Chemical composition (wt.%) of the IN718 powder used.

Fe	Ni	Cr	Nb	Mo	Ti	Al	Co	C
Bal.	53.6	18.1	5.38	3.07	0.97	0.6	0.22	0.024

In order to gain further insights into the grain orientation, electron backscatter diffraction (EBSD) measurements were performed on four samples. Multiple mappings from the center of the sample, close to the top surface, were collected with a step size of 5 μm, and subsequently stitched. The measurements and post processing were performed on a Helios NanoLab DualBeam 600 FIB/SEM (FEI Company, Hillsboro, OR, USA) equipped with a NordlysNano EBSD detector using the AZtecHKL and the HKL CHANNEL5 software (Oxford Instruments, Abingdon, UK), as well as the MATLAB toolbox MTEX. In order to allow for a quantitative comparison of the crystallographic texture, the texture index J_{ODF} was computed by MTEX according to Equation (1) [13]:

$$J_{ODF} = \int |f(g)|^2 dg, \tag{1}$$

where f is the orientation distribution function (ODF) that is calculated from the EBSD data, and g represents an orientation. Accordingly, $J_{ODF} = 1$ for a completely uniform distribution of orientations and it is infinitely large for a single orientation.

Numerical simulations were used to analyze the solidification conditions during SEBM as a function of the processing parameters. The simulations were conducted using the 2D version of our in-house software tool, SAMPLE^{2D} (Simulation of Additive Manufacturing on the Powder scale using a Laser or Electron beam), which is described elsewhere in detail [14]. It is based on a Lattice Boltzmann framework that solves full hydrodynamics in the melt pool, and thermodynamics in the entire domain. The setup used in the current study is depicted in Figure 2a, and the used material parameters [15–17] are listed in Table 3. As the simulation domain is two dimensional (15 mm width), only rotations of the hatching pattern by 180° are possible. A cell size of 5×5 μm^2 and a time step of 2×10^{-7} s were used. The scan vectors, with a length of 15 mm, run perpendicular to the simulation plane. In order to assess the solidification conditions during SEBM without the stochastic influence of powder, a block of solid material with a height of 50 μm was added, instead of placing individual powder particles in each layer. For each parameter setting listed in Table 1, three layers were simulated. While a large quantity of outputs can be sourced from SAMPLE^{2D}, the main focus of this study was the thermal gradient G and the solidification interface velocity v_s. Both values were taken at the solidification front. This was implemented by tracking the isotherm of the average temperature between solidus and liquidus. The orientation of G plays a key role as it dictates the growth direction of the dendrites during solidification [18]. As depicted in Figure 2b, the maximum deviation of v_s from the BD, labeled φ_{int}, is always present at the interface between two layers. Consequently, φ_{int} was tracked directly at the interface between the first layer and the respective next higher layer.

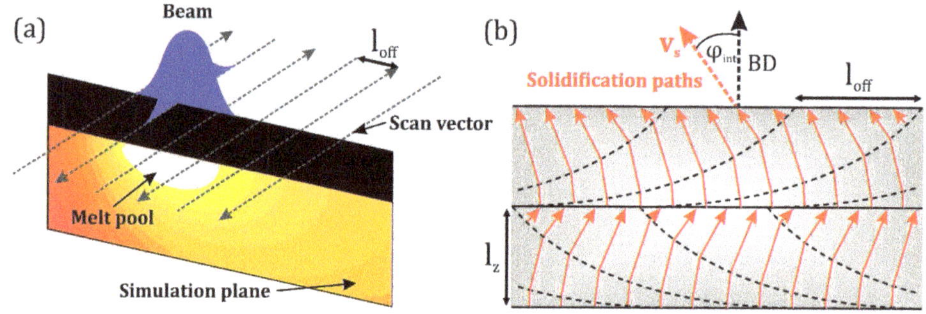

Figure 2. (**a**) Schematic of the simulation domain. The temperature field and the melt pool shape are shown. For convenience, the whole domain (width 15 mm) is not illustrated. The beam's hatch pattern is schematically shown. (**b**) Schematic illustration of the solidification directions within overlapping melt pools. The maximum angle between thermal gradient G (proportional to the solidification velocity v_S) and the BD, φ_{int}, is reached at the interface between two layers, while φ_{int} is about 0° at the bottom of the next layer.

Table 3. Material parameters for IN718 used in SAMPLE^{2D} [15–17]. The thermal conductivity in the solid and the liquid are linearly approximated with the temperature (T) in K.

Parameter	Value	Unit
Density	7580	Kgm^{-3}
Dynamic Viscosity	4.0×10^{-3}	Pas
Atomic Mass	26.73	u
Surface Tension	1.73	Nm^{-1}
Solidus Temperature	1528	K
Liquidus Temperature	1610	K
Heat Capacity (solid)	652	$JKg^{-1}K^{-1}$
Heat Capacity (liquid)	778	$JKg^{-1}K^{1}$
Thermal Conductivity (solid)	1.66×10^{-2} T + 5.80	$Wm^{-1}K^{-1}$
Thermal Conductivity (liquid)	1.35×10^{-2} T + 5.15	$Wm^{-1}K^{-1}$
Latent Heat of Fusion	$2.27 \cdot 10^{5}$	JK^{-1}

3. Results and Discussion

3.1. Grain Structure Characterization

Figure 3 shows overview images of macro etched longitudinal sections of the three different microstructures examined in this study. Deviations from the CAD geometry are mainly due to specimen preparation. The standard hatching strategy (SH) (Figure 3a) leads to a columnar grain structure, with grains growing epitaxially over numerous layers. A certain grain coarsening over the build height is evident. The mean grain width increases from about 100 µm at the very bottom of the sample up to around 500 µm near the top surface. In the vicinity of the side surfaces, the grain structure is dominated by inward growing fine columnar grains, which are triggered by nucleation from the surrounding powder bed. The section is about 1 mm wide, and is a well-known phenomenon in powder bed fusion additive manufacturing [7]. As can be seen from the detailed picture in Figure 3d, the grain boundaries are, overall, clearly aligned parallel to the BD, yet somewhat serrated. The corresponding results from the large area EBSD mappings in Figure 6 provide detailed information about the crystallographic orientation of the grains. The IPF-Z map, where the coloring is chosen according to the BD, is almost solely colored in red, which indicates a very strong <100> fiber texture parallel to the BD. This is quantified by a high texture index of $J_{ODF} = 26.07$, whereas a value of 1 represents no texture at all. Minor agglomerations of stray grains can also be observed, but these are quickly overgrown by <100> oriented grains over the course of a few layers. The IPF-X and Y maps show a different picture. Here, green is the predominant color, indicating orientations around <110>. This can be seen even more clearly as distinct peaks in the pole figures, e.g., on the outer rim of the {100} pole figure. It can thus be noted that in addition to the strong anisotropy in the orientation of the primary dendrite arms, there is also anisotropy in the orientation of the secondary dendrite arms. Taking the 45° rotation of the samples with respect to the scan vectors into account, the preferred orientation of the secondary <100> directions coincides with the scan vectors. This grain structure can be seen as the reference grain structure for SEBM-IN718, manufactured with standard cross snake hatching scan strategies, and is comparable to what is reported in the literature [1,5,19,20].

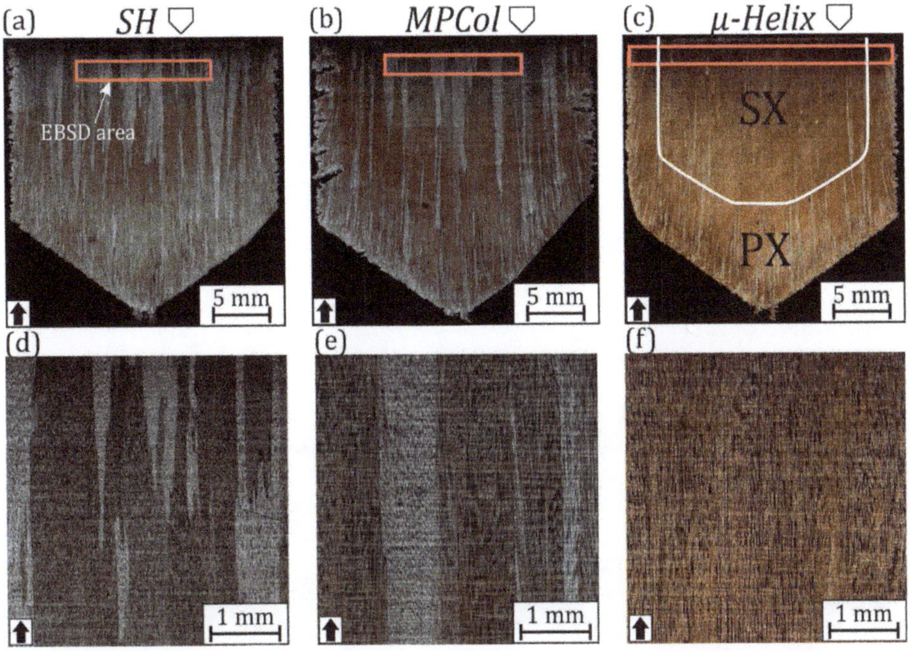

Figure 3. Macro-etched longitudinal sections (**a**–**c**) and corresponding detailed pictures (**d**–**f**) from trapezoidal prisms manufactured with different SEBM processing parameters and aperture angles of 35°. The build direction is indicated by the black arrows. The μ-Helix parameter setting allows the fabrication of technical single crystals (SXs). For convenience, the SX region (white), as well as the areas for the electron backscatter diffraction (EBSD) measurements are highlighted (red). SH—standard hatching; MPCol—multi passing.

A small line offset combined with a very high scan speed (MPCol) leads to columnar grains, as depicted in Figure 3b. However, the grain selection along the BD is slightly more pronounced here. The grain width increases faster within the first 10 mm of the build compared to SH. From the detailed picture in Figure 3e, it is evident that the grain boundaries are significantly less serrated and generally tend to run more parallel to the BD. Qualitatively, the grain length appears to be larger as well. With an expansion of about 2.5 mm, the region affected by nucleation from the powder bed at the side surfaces is notably enlarged in comparison to SH. This region also exhibits significant notches, which drastically deteriorate the surface quality. This phenomenon can be explained as follows: During hatching, the beam generally passes any given spot on the surface multiple times, due to the fact, that the beam diameter (300–400 μm) is significantly larger than the line offset. Using multi passing, the necessary energy input per area to produce dense materials is higher compared to standard hatching, although the line energies are significantly lower. The first hatch lines within any layer are generally needed to heat up the material and initially form a stable melt pool. The low line energies consequently lead to a prolonged region of reduced melt depth at the beginning of a layer, and may thus cause delamination issues at the edges since no contour melting is applied. In terms of crystallographic texture, the grain structure exhibits similar results in the EBSD measurements compared to the standard hatching sample. The <100> fiber texture is even more pronounced, as indicated by a texture index of 36.93. However, the anisotropy of the secondary dendrite arms is not as distinct, which becomes evident from the more diffuse appearance at the rim of the {100} pole figure.

As can be seen in Figure 3c, a small line offset paired with moderate scan speeds (μ-Helix) allows the realization of technical single crystals (SXs). It is already evident from the macro etched state, that the SX region in the center of the sample, which is about 21 mm wide, is surrounded by a polycrystalline (PX) shell of about 2 mm width, similar to the other samples examined in this study. There are some isolated stray grains present within the SX region, which are overgrown in subsequent layers (see Figure 3f). The orientation of the grains within the first layers is rather randomly distributed due to heterogeneous nucleation from the powder bed [21]. A grain selection region of approximately 9 mm in build height is present at the bottom, until the SX orientation prevails. It is interesting to note, that the shape of the SX-region is similar to the shape of the sample, as the wedged bottom section leads to nucleation from the powder, as long as the cross section of the part is still increasing with the build height. Once the final cross section is established, the SX grain selection process will again take place over the course of the next approximately 9 mm build height. This behavior is also seen in the fabrication of SEBM-CMSX-4 technical single crystal samples [11].

In order to analyze the impact of the part geometry on the grain selection process, trapezoid samples with different aperture angles in the wedge section were manufactured. The corresponding macro-etched longitudinal sections are depicted in Figure 4. Please note, that the notch-like feature seen in Figure 4b is due to an error during sample cutting. In all cases, the grain selection in the sample center takes place within the first 9 mm of build height, regardless of the part geometry. For comparison purposes, large cuboid samples were manufactured with the μ-Helix parameter setting at a later stage, with an improved build temperature control during the process. A macro etched longitudinal section of the lower part of such a cuboid sample is depicted in Figure 5. Again, an SX region is evident in the center of the sample surrounded by a PX shell. The SX region is stable along the whole build height of 120 mm. The section demonstrates the possibility of obtaining large, high quality SX samples with the μ-Helix parameter setting, and the importance of tight temperature control during SEBM build jobs.

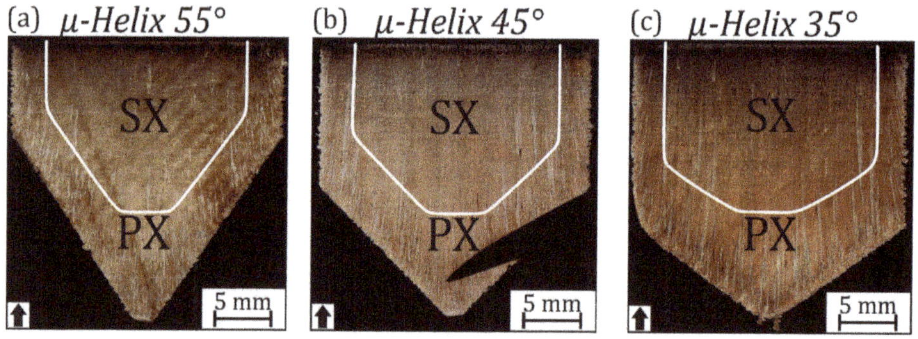

Figure 4. Macro-etched longitudinal sections from trapezoidal prisms manufactured with the μ-Helix parameter setting. The build direction is indicated by the black arrows. The aperture angle of (**a**) 55°, (**b**) 45° and (**c**) 35° in the wedge section influences the overall shape of the SX region (white), but the SX selection height of about 9 mm remains constant.

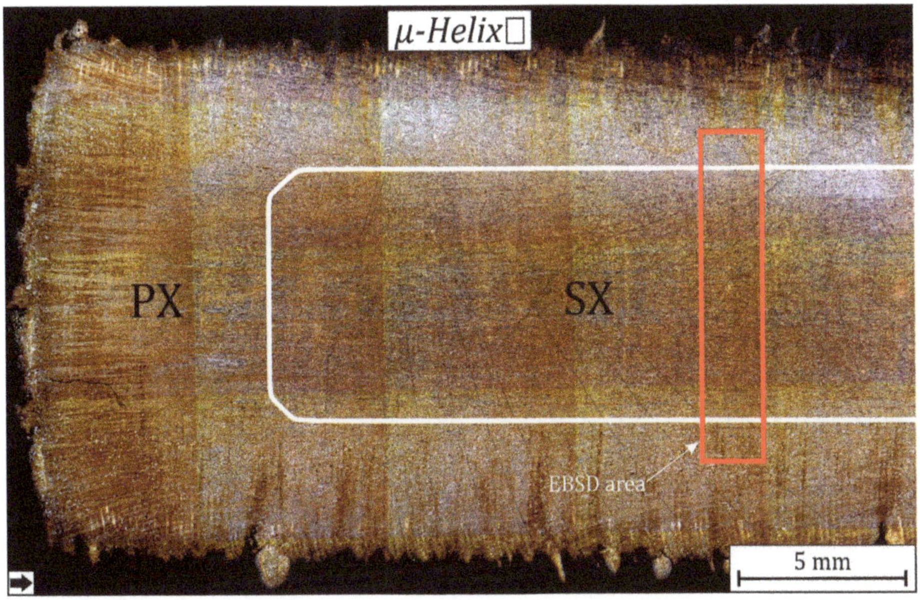

Figure 5. Macro-etched longitudinal section of a cuboid sample manufactured with the μ-Helix parameter setting. The build direction is indicated by the black arrows. A high quality SX region is present in the center of the sample. For convenience, the SX region (white), as well as the areas for the EBSD measurements are highlighted (red).

While there is no generally accepted definition, technical single crystals typically allow for low angle grain boundaries, with a misorientation angle of up to 10°, to be present within the material. In order to verify the compliance of the presented samples with this requirement, large area EBSD mapping was performed over the whole sample width. The results are shown in Figure 6c and confirm that a technical single crystal was indeed built, albeit flawed. The only grain boundaries exhibiting a misorientation angle of > 10° are the isolated stray grains, and the PX shell, which were both already discussed. Consequently, the texture index of 120.29 within the SX region is extremely high. It should be noted however, that the present stray grains may significantly affect the material's properties. The results of the EBSD analysis of the SX cuboid sample are depicted in Figure 7 (ODF pole figures) and Figure 8 (IPF maps). This SX exhibits a higher quality by means of the absence of stray grains. Consequently, the texture index within the SX region of 145.17 is even higher compared to the trapezoidal prism SX. The resulting pole figures clearly show that there is only one distinct crystal orientation present in the SX regions. The primary orientation of the dendrites is once again <100> parallel to the BD, analogous to the columnar grain structures. The key difference lies within the uniform orientation of all secondary dendrite arms, which are rotated by 45° with respect to the scan vectors (note the orientation of the samples in the build chamber for the trapezoidal prisms, see Figure 1).

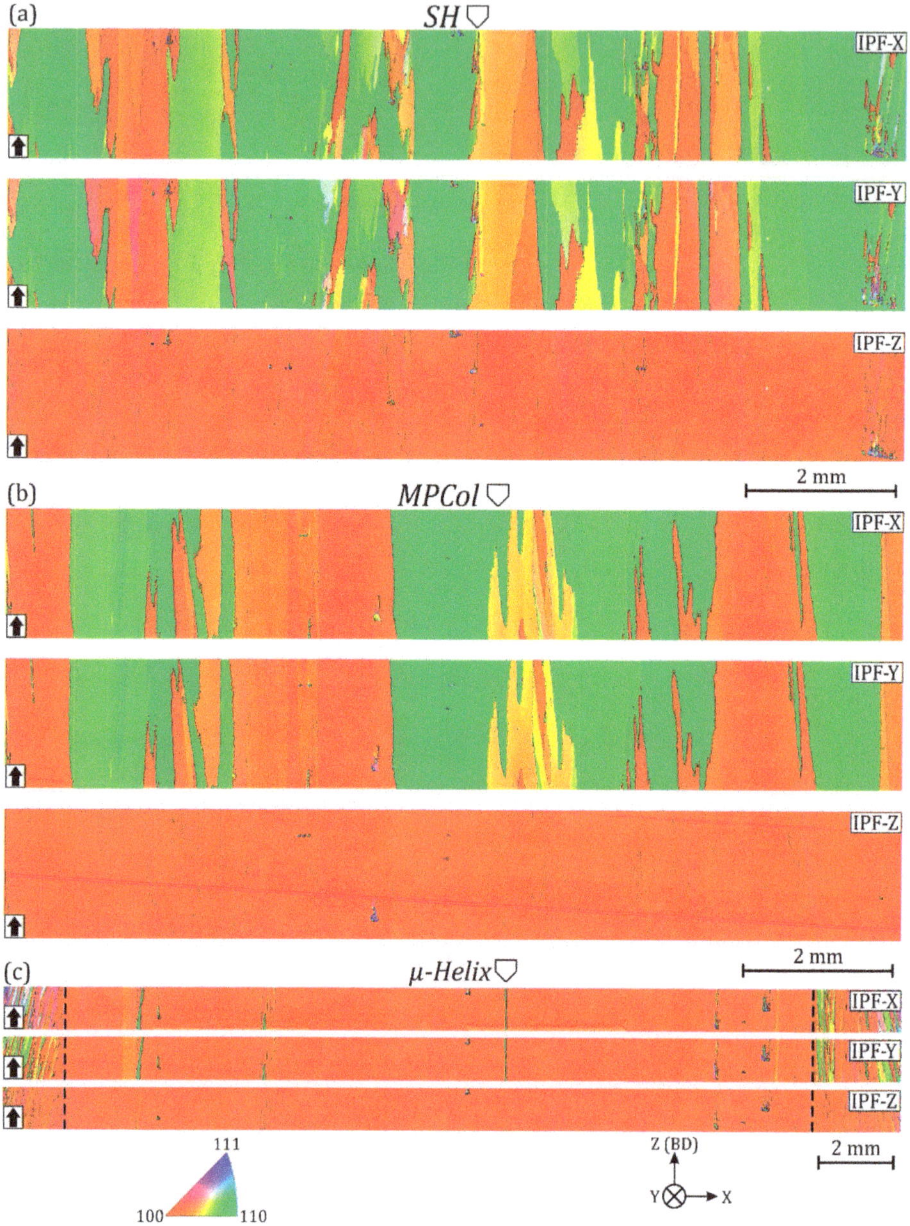

Figure 6. Large area EBSD-IPF maps of trapezoidal prisms manufactured with different SEBM processing parameters: (**a**) SH, (**b**) MPCol, (**c**) μ-Helix. The build direction is indicated by the black arrows. Grain boundaries with a misorientation angle of > 10° are highlighted with black lines. The dashed lines indicate the region of interest for the pole figure of the μ-Helix sample.

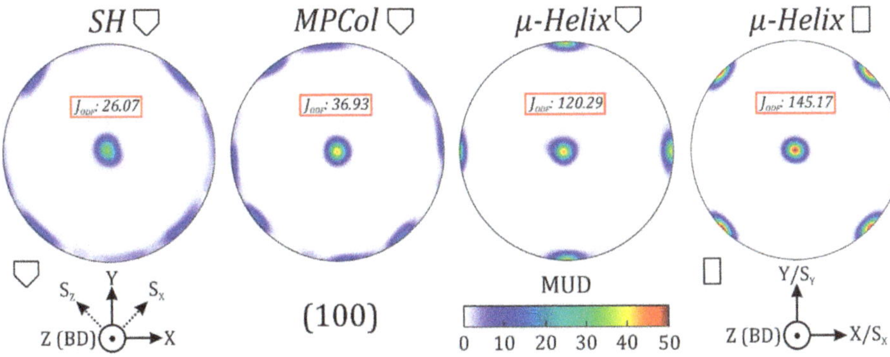

Figure 7. (100) Orientation distribution function (ODF) of pole figures from the EBSD measurements of Figures 6 and 8. The respective texture indices J_{ODF} and hatching directions are given.

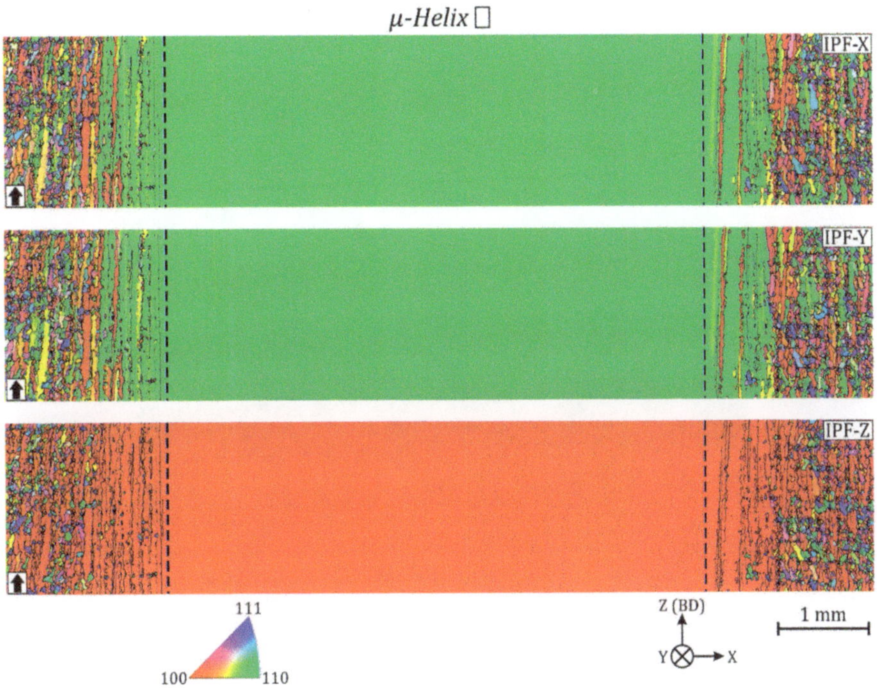

Figure 8. Large area EBSD-IPF maps of a cuboid sample manufactured with the μ-Helix parameter setting. The build direction is indicated by the black arrows. Grain boundaries with a misorientation angle of > 10° are highlighted with black lines. The dashed lines indicate the region of interest for the pole figures.

3.2. Numerical Simulation

Numerical simulations of the SEBM process using our in-house software SAMPLE^{2D} [14] were performed to reveal the solidification conditions for all hatching strategies. The parameters listed in Table 1 were used. The first aim was to assess the resulting melt pools. This is of particular interest, as the melt pool shape mainly determines the orientation of G during solidification, which is perpendicular to the melt pool walls, as depicted in Figure 2b. It has been proven to play a key role in grain structure formation in SEBM [19]. Figure 9 shows results from the SAMPLE^{2D} simulations for the three investigated parameter settings. When comparing the results it is evident, that all melt pools (dark grey) fundamentally exhibit a similar shape, while the size varies. To understand how these rather similarly shaped melt pools can lead to such drastically different grain structures, one has to bare in mind that only a rather small region of interest (ROI) determines the final grain structure, as highlighted in Figure 9. This is due to the fact that the majority of the meltpool will be remelted in subsequent beam passes and layers and, thus, does not contribute to the grain structure of the final part. The height of this ROI is equal to the layer thickness (typically 50 μm) starting from the melt pool bottom. The lateral dimension of the ROI is mainly determined by the line offset, and can be assessed by stringing together a series of melt pools with a distance equal to the line offset. It is evident from these visualizations, that G, on average, diverges from the BD within each layer, as indicated by the solidification paths. One can also derive that the deviation will always start at 0° at the melt pool bottom and increase up to its maximum, φ_{int}, at the very top of the ROI as the melt pool curvature increases continuously with the height. In the following, the solidification conditions are analyzed in detail.

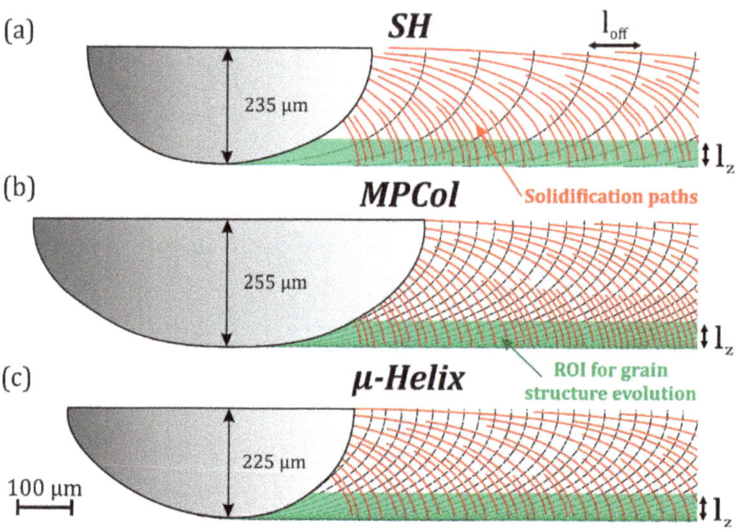

Figure 9. Numerical results from SAMPLE^{2D} (Simulation of Additive Manufacturing on the Powder scale using a Laser or Electron beam). The dark grey areas represent the melt pools for (**a**) SH, (**b**) MPCol, (**c**) μ-Helix. The dashed lines indicate previous pools. Due to remelting in subsequent beam passes and layers, only a small region of interest (ROI) remains for grain structure development (highlighted in green). The red streamlines resemble the orientation of G, i.e., the solidification paths.

We investigated the maximum deviation of G from the BD at the interface between the two layers, φ_{int}, for each parameter setting as a function of the lateral position in the part (see Figure 10). The respective moving averages of φ_{int} are shown. It is evident, that there is a strong influence of the position in the part φ_{int} for all settings. These boundary effects can be understood by considering Figure 10a, where the three simulated layers are arranged schematically, and the respective interface for the evaluation of φ_{int} is highlighted. Here, the aforementioned melt pool initialization phase becomes evident: The melt pool depth increases steadily within the first 2–4 mm (depending on the processing parameters), until it reaches equilibrium. This leads to an overmelting effect of preceding layers in the boundary regions, as highlighted in Figure 10c. On the very left side, where the initialization phase of Layer 1 took place, only Layer 2 effectively remains. It overmelts this portion of Layer 1 entirely. As the interface for the evaluation of φ_{int} was defined on the basis of Layer 1 and the next higher Layer (see Section 2), φ_{int} can not be evaluated here. Consequently, there are no data points available for plotting in Figure 10b. As soon as Layer 1 reaches a sufficient melt depth, φ_{int} increases at the melt depth. Once a stable melt pool with a constant melt depth is established, φ_{int} stays constant and the ROI governs the solidification conditions. Towards the far end of the layer φ_{int} increases again. This is due to the influence of the initialization phase of Layer 2. As a result of the decreasing melt depth, the interface is shifted higher within Layer 1, where higher deviations of G from the BD occur. Subsequently, a second plateau region emerges, as Layer 2 is overmelted completely by Layer 3, which exhibits a stable melt depth at this position. Finally, φ_{int} decreases rapidly and reaches values of up to −90°, i.e., G is tilted towards −X. As the whole melt pool solidifies at the very end of the simulation domain, orientations from across the whole melt pool width show up here. This second boundary region is once again 2–4 mm wide. The described effects explain the phenomena that could be seen on the sidewalls of the trapezoid samples.

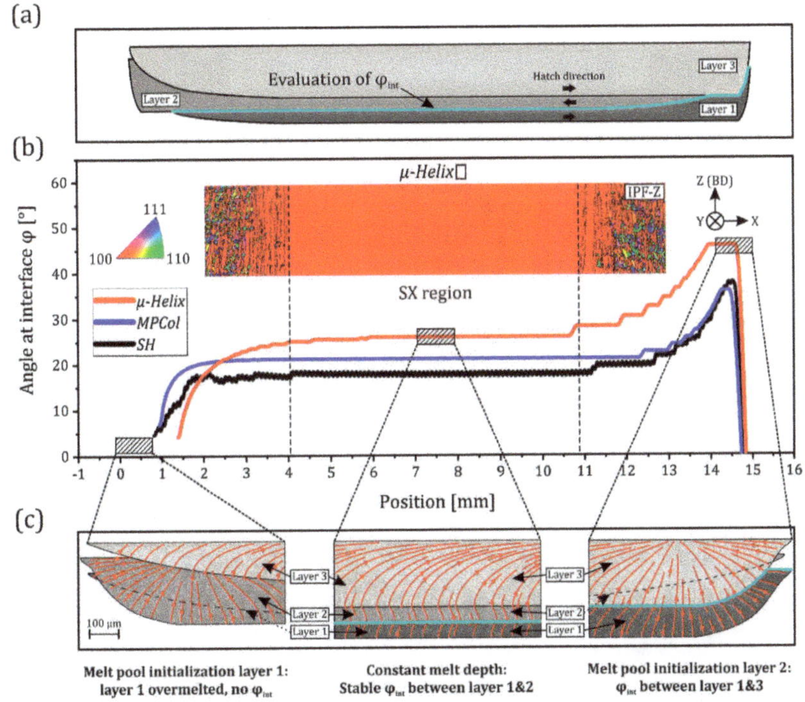

Figure 10. Analysis of the solidification conditions via SAMPLE^{2D}. (**a**) Schematic showing an overlay of three simulated layers. The resulting location for the evaluation of φ_{int} is highlighted. (**b**) Deviation of the thermal gradient G from the BD at the interface between two layers φ_{int}, as a function of the position within the simulation domain. The plotted curves resemble moving averages of the original data for enhanced visibility. A section of the EBSD IPF-Z map of the cuboid μ-Helix sample is fit in for comparison. (**c**) Detailed results of the simulation of μ-Helix. The red streamlines resemble the orientation of G, i.e., the solidification paths.

Comparing the results in the stable mid-section, one can see that SH yields the lowest mean interface angle (18°). The mean φ_{int} of 21° for MPCol is only slightly larger compared to SH. However, there still is a noticeable difference in the resulting grain structure, with MPCol yielding a quantifiably more pronounced grain selection when comparing the texture indices of 26.07 and 36.93, respectively. Employing μ-Helix leads to the highest mean φ_{int} of 26°, and the emergence of an SX. As indicated by the attached IPF-Z maps of the cuboid μ-Helix sample, the intialization phases coincide perfectly with the SX region.

4. Discussion

In order to understand the grain selection process to the single crystal during SEBM it is useful to recall the standard casting process to generate single crystals, see Figure 11a. This two-stage selection process starts with heterogeneous nucleation and growth of randomly oriented grains at a water-cooled chill plate. During the solidification process, the temperature gradient G is oriented parallel to the withdrawal direction. Thus, competitive grain growth based on the primary dendritic orientation prevails within the starter block. This primary selection process results in a columnar grain structure with a pronounced <100> fiber texture. During the following secondary spiral section process, ideally, only one grain survives. Grain selection in this second stage is based on a purely geometric blocking mechanism [22]. As a consequence, the secondary orientation of the selected grain is at random. Thus, it is not possible to select a distinct grain orientation with a spiral grain selector [23].

The experimental results in conjunction with the numerical analysis allow for a deeper understanding of the grain selection process during SEBM. Instead of using directional solidification within a physical mold, a tailored thermal gradient controls the solidification direction in each layer. Combining the progression of the thermal gradient G within one layer with the scan pattern rotation after each layer, G forms a helix on the microscale (μ-Helix) (see Figure 11b). Although G deviates significantly from the building direction in every layer, the resulting grain structure shows a pronounced <100> -type texture. The rotation of the scan pattern after each layer leads to an averaging effect over the course of multiple layers. On average, G is oriented parallel to the BD. In addition, G is oriented in <100> direction at the bottom of the melt pool. Thus, solidification always starts with the <100> direction leading to the pronounced <100> fiber texture commonly observed in parts manufactured by SEBM. As the <100> directions are the fastest growing directions in cubic crystal systems, grains with their primary dendritic orientation close to <100>, with respect to the BD, are systematically favored. This initial selection process is based on the well-known principles of competitive grain growth [18] and is typically finished within the first millimeters of build height, as shown in [21].

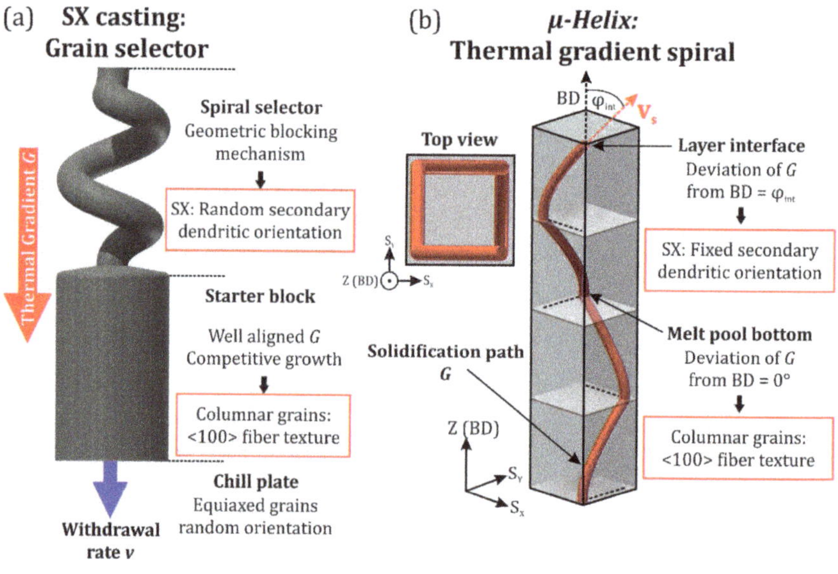

Figure 11. Schematic illustration of different approaches to realize single crystals. (**a**) Physical spiral grain selector used within an investment casting mold, (**b**) μ-Helix selector during SEBM formed by the direction of the local thermal gradient G.

While the mean direction of G explains primary grain selection, the deviation of G from the BD is responsible for the secondary selection process (see Figure 9). Secondary grain selection takes place in the plane perpendicular to the building direction. It occurs by competitive growth of the secondary dendrite arms. At the beginning of solidification, i.e., at the melt pool bottom, G only marginally diverges from the BD. This indicates that only grain selection based on the primary dendritic orientation will take place here. As the deviation of G increases steadily within the layer, grain selection based on the secondary dendritic orientation will attain its maximum at the interface to the subsequent layer. The deviation of G from the BD depends on the processing parameters, see Figure 10. Obviously, the velocity of secondary grain selection increases with increasing deviation of G from the BD. However, a too high deviation of G from the BD will ultimately end in new grain formation [24]. Following the same arguments as for primary grain selection, secondary grain selection in the plane perpendicular to the BD should prefer grains with the secondary dendrite arms oriented parallel to the scan vectors. However, the experimental results do not confirm this expectation. On the contrary, the <110> direction is determined for the secondary orientation, see Figure 8. That is, the secondary dendrite arms are rotated by 45° with respect to the scan vectors. The origin of this phenomenon is not yet understood and is still under investigation.

5. Conclusions

The grain structure of SEBM parts can be strongly influenced by the processing parameters. For this purpose, trapezoidal prism samples from IN718 were produced by SEBM using different processing strategies.

1. Standard hatching and multi passing yield columnar grains with a pronounced <100> fiber texture along the build direction. The novel μ-Helix parameter setting, a special variant of multi passing, allowed for the fabrication of technical single crystals (SXs), which were surrounded by a polycrystalline shell. Within the SX section, minor agglomerations of stray grains were still present. This was overcome in a second build job with improved temperature control, which resulted in high quality cuboid SX samples.
2. In contrast to investment casting, where the secondary orientation of single crystals produced by a spiral grain selector is random, SEBM leads to a defined and reproducible secondary orientation. All SXs show a primary orientation of <100> parallel to the build direction, and a secondary orientation of <110> along the scan vectors.
3. The aperture angle of the trapezoidal prisms determines the height where the SX region reaches its maximum width as new grain orientations emerge as long as the sample cross section still increases.
4. Numerical simulations using our in-house software suite $\mathbb{SAMPLE}^{2D}$ revealed that the deviation of the thermal gradient G from the build direction, right at the interface between two layers, φ_{int}, correlates with the effectiveness of grain selection. The higher φ_{int}, the stronger the grain selection. The *μ-Helix* parameter setting forces the solidification paths along a μm-sized helix over the course of multiple layers.
5. The grain selection mechanisms leading to an SX are two-fold. As G deviates only slightly at the bottom of each melt pool, competitive growth between the grains based on their primary dendritic orientation is triggered, yielding a columnar grain structure. The increasing deviation of G from the BD closer to the layer interface leads to grain selection based on the secondary dendritic orientation. This mechanism is not yet fully understood and still under investigation.

Author Contributions: Conceptualization, M.R.G. and C.K.; methodology, M.R.G., A.M.R. and C.K.; software, A.M.R.; validation, M.R.G., A.M.R. and C.K.; formal analysis, M.R.G., A.M.R. and C.K.; investigation, M.R.G.; resources, C.K.; data curation, A.M.R.; writing—original draft preparation, M.R.G.; writing—review and editing, M.R.G. and C.K.; visualization, M.R.G., A.M.R. and C.K.; supervision, C.K.; project administration, C.K.; funding acquisition, C.K. All authors have read and agreed to the published version of the manuscript.

Funding: This research was funded by the Deutsche Forschungsgemeinschaft (DFG) through the Collaborative Research Center SFB/TR 103, Project B2.

Conflicts of Interest: The authors declare no conflicts of interest. The funders had no role in the design of the study; in the collection, analyses, or interpretation of data; in the writing of the manuscript, or in the decision to publish the results.

References

1. Körner, C. Additive manufacturing of metallic components by selective electron beam melting—A review. *Int. Mater. Rev.* **2016**, *61*, 361–377. [CrossRef]
2. Strondl, A.; Palm, M.; Gnauk, J.; Frommeyer, G. Microstructure and mechanical properties of nickel based superalloy IN718 produced by rapid prototyping with electron beam melting (EBM). *Mater. Sci. Technol.* **2011**, *27*, 876–883. [CrossRef]
3. Helmer, H.E.; Hartmann, N.; Körner, C.; Singer, R.F. Relation Between Processing Strategy, Grain Structure and Mechanical Properties in Superalloy Inconel 718 Processed by Selective Electron Beam Melting. In Proceedings of the DDMC 2014, Fraunhofer Direct Digital Manufacturing Conference, Berlin, Germany, 12–13 March 2014.
4. Sun, S.-H.; Koizumi, Y.; Saito, T.; Yamanaka, K.; Li, Y.-P.; Cui, Y.; Chiba, A. Electron beam additive manufacturing of Inconel 718 alloy rods: Impact of build direction on microstructure and high-temperature tensile properties. *Addit. Manuf.* **2018**, *23*, 457–470. [CrossRef]
5. Strondl, A.; Fischer, R.; Frommeyer, G.; Schneider, A. Investigations of MX and γ'/γ'' precipitates in the nickel-based superalloy 718 produced by electron beam melting. *Mater. Sci. Eng. A* **2008**, *480*, 138–147. [CrossRef]
6. Al-Bermani, S.S.; Blackmore, M.L.; Zhang, W.; Todd, I. The Origin of Microstructural Diversity, Texture, and Mechanical Properties in Electron Beam Melted Ti-6Al-4V. *MMTA* **2010**, *41*, 3422–3434. [CrossRef]
7. Antonysamy, A.A.; Meyer, J.; Prangnell, P.B. Effect of build geometry on the β-grain structure and texture in additive manufacture of Ti6Al4V by selective electron beam melting. *Mater. Charact.* **2013**, *84*, 153–168. [CrossRef]
8. Helmer, H.; Bauereiß, A.; Singer, R.F.; Körner, C. Grain structure evolution in Inconel 718 during selective electron beam melting. *Mater. Sci. Eng. A* **2016**, *668*, 180–187. [CrossRef]
9. Raghavan, N.; Dehoff, R.; Pannala, S.; Simunovic, S.; Kirka, M.; Turner, J.; Carlson, N.; Babu, S.S. Numerical modeling of heat-transfer and the influence of process parameters on tailoring the grain morphology of IN718 in electron beam additive manufacturing. *Acta Mater.* **2016**, *112*, 303–314. [CrossRef]
10. Kirka, M.M.; Lee, Y.; Greeley, D.A.; Okello, A.; Goin, M.J.; Pearce, M.T.; Dehoff, R.R. Strategy for Texture Management in Metals Additive Manufacturing. *JOM* **2017**, *69*, 523–531. [CrossRef]
11. Ramsperger, M.; Carolin, K. Selective Electron Beam Melting Of The Single Crystalline Nickel-Base Superalloy Cmsx-4®: From Columnar Grains To A Single Crystal. In *Superalloys 2016*; John Wiley & Sons: Hoboken, NJ, USA, 2016.
12. Chauvet, E.; Tassin, C.; Blandin, J.-J.; Dendievel, R.; Martin, G. Producing Ni-base superalloys single crystal by selective electron beam melting. *Scr. Mater.* **2018**, *152*, 15–19. [CrossRef]
13. Mainprice, D.; Bachmann, F.; Hielscher, R.; Schaeben, H. Descriptive tools for the analysis of texture projects with large datasets using MTEX: Strength, symmetry and components. *Geol. Soc. Lond. Spec. Publ.* **2014**, *409*, SP409. [CrossRef]
14. Markl, M.; Rausch, A.M.; Küng, V.E.; Körner, C. SAMPLE: A Software Suite to Predict Consolidation and Microstructure for Powder Bed Fusion Additive Manufacturing. *Adv. Eng. Mater.* **2019**, 1901270. [CrossRef]
15. Pottlacher, G.; Hosaeus, H.; Kaschnitz, E.; Seifter, A. Thermophysical properties of solid and liquid Inconel 718 Alloy *. *Scand. J. Metall.* **2002**, *31*, 161–168. [CrossRef]
16. Mills, K.C. *Recommended Values of Thermophysical Properties for Selected Commercial Alloys*; Woodhead: Cambridge, UK, 2002.
17. Mills, K.C.; Youssef, Y.M.; Li, Z.; Su, Y. Calculation of Thermophysical Properties of Ni-based Superalloys. *ISIJ Int.* **2006**, *46*, 623–632. [CrossRef]
18. Walton, D.; Chalmers, B. The origin of the preferred orientation in the columnar zone of ingots. *Trans. Metall. Soc. AIME* **1959**, *215*, 447–457.

19. Harald, E.H. Additive Fertigung durch Selektives Elektronenstrahlschmelzen der Nickelbasis Superlegierung IN718: Prozessfenster, Mikrostruktur und Mechanische Eigenschaften. Ph.D. Thesis, Friedrich-Alexander-Universität Erlangen-Nürnberg, Erlangen, Germany, 2017.
20. Dehoff, R.R.; Kirka, M.M.; List, F.A.; Unocic, K.A.; Sames, W.J. Crystallographic texture engineering through novel melt strategies via electron beam melting: Inconel 718. *Mater. Sci. Technol.* **2015**, *31*, 939–944. [CrossRef]
21. Koepf, J.A.; Gotterbarm, M.R.; Markl, M.; Körner, C. 3D multi-layer grain structure simulation of powder bed fusion additive manufacturing. *Acta Materialia* **2018**, *152*, 119–126. [CrossRef]
22. Dai, H.J.; D'Souza, N.; Dong, H.B. Grain Selection in Spiral Selectors During Investment Casting of Single-Crystal Turbine Blades: Part I. Experimental Investigation. *Metall. Mater. Trans. A* **2011**, *42*, 3430–3438. [CrossRef]
23. Dai, H.J.; Dong, H.B.; D'Souza, N.; Gebelin, J.-C.; Reed, R.C. Grain Selection in Spiral Selectors During Investment Casting of Single-Crystal Components: Part II. Numerical Modeling. *Metall. Mater. Trans. A* **2011**, *42*, 3439–3446. [CrossRef]
24. Rausch, A.M.; Gotterbarm, M.R.; Markl, M.; Körner, C. *Nucleation Mechanism in Powder Bed Fusion Additive Manufacturing*; 2020; Manuscript in preparation.

metals MDPI

Article

Animal Origin Bioactive Hydroxyapatite Thin Films Synthesized by RF-Magnetron Sputtering on 3D Printed Cranial Implants

Diana Chioibasu [1,2], Liviu Duta [1], Gianina Popescu-Pelin [1], Nicoleta Popa [3], Nichita Milodin [3], Stefana Iosub [1,4], Liliana Marinela Balescu [5], Aurelian Catalin Galca [5], Adrian Claudiu Popa [5,6], Faik N. Oktar [7,8], George E. Stan [5,*] and Andrei C. Popescu [1,*]

1 National Institute for Lasers, Plasma and Radiation Physics, RO-077125 Magurele-Ilfov, Romania; diana.chioibasu@inflpr.ro (D.C.); liviu.duta@inflpr.ro (L.D.); gianina.popescu@inflpr.ro (G.P.-P.); stefana.iosub@inflpr.ro (S.I.)
2 Department of Physics, Faculty of Applied Sciences, University Politehnica of Bucharest, RO-060042 Bucharest, Romania
3 National Institute for Mechatronics & Measurement Technique, RO-021631 Bucharest, Romania; nicoleta.popa@incdmtm.ro (N.P.); nichita.milodin@incdmtm.ro (N.M.)
4 Department of Molecular Cell Biology, Institute of Biochemistry of the Romanian Academy, RO-060031 Bucharest, Romania
5 National Institute for Materials Physics, RO-077125 Magurele-Ilfov, Romania; liliana.trinca@infim.ro (L.M.B.); ac_galca@infim.ro (A.C.G.); adrian.popa@gmail.com (A.C.P.)
6 Army Centre for Medical Research, RO-010195 Bucharest, Romania
7 Center for Nanotechnology & Biomaterials Research, University of Marmara, Goztepe Campus, 34722 Istanbul, Turkey; foktar@marmara.edu.tr
8 Department of Bioengineering, Faculty of Engineering, University of Marmara, Goztepe Campus, 34722 Istanbul, Turkey
* Correspondence: george_stan@infim.ro (G.E.S.); andrei.popescu@inflpr.ro (A.C.P.); Tel.: +40-21-241-8128 (G.E.S.); +40-21-457-4550 (A.C.P.)

Received: 18 November 2019; Accepted: 9 December 2019; Published: 10 December 2019

Abstract: Ti6Al4V cranial prostheses in the form of patterned meshes were 3D printed by selective laser melting in an argon environment; using a CO_2 laser source and micron-sized Ti6Al4V powder as the starting material. The size and shape of prostheses were chosen based on actual computer tomography images of patient skull fractures supplied in the framework of a collaboration with a neurosurgery clinic. After optimizations of scanning speed and laser parameters, the printed material was defect-free (as shown by metallographic analyses) and chemically homogeneous, without elemental segregation or depletion. The prostheses were coated by radio-frequency magnetron sputtering (RF-MS) with a bioactive thin layer of hydroxyapatite using a bioceramic powder derived from biogenic resources (Bio-HA). Initially amorphous, the films were converted to fully-crystalline form by applying a post-deposition thermal-treatment at 500 °C/1 h in air. The X-ray diffraction structural investigations indicated the phase purity of the deposited films composed solely of a hexagonal hydroxyapatite-like compound. On the other hand, the Fourier transform infrared spectroscopic investigations revealed that the biological carbonatation of the bone mineral phase was well-replicated in the case of crystallized Bio-HA RF-MS implant coatings. The in vitro acellular assays, performed in both the fully inorganic Kokubo's simulated body fluid and the biomimetic organic–inorganic McCoy's 5A cell culture medium up to 21 days, emphasized both the good resistance to degradation and the biomineralization capacity of the films. Further in vitro tests conducted in SaOs-2 osteoblast-like cells showed a positive proliferation rate on the Bio-HA RF-MS coating along with a good adhesion developed on the biomaterial surface by elongated membrane protrusions.

Keywords: biological origin hydroxyapatite; bioactive layers; cranial mesh implants; selective laser melting; 3D printing; radio-frequency magnetron sputtering

---

1. Introduction

Ti6Al4V is the most popular titanium alloy used in biomedical applications due to its high strength, high fracture toughness, superior corrosion resistance, low density, excellent biocompatibility, and favorable osseointegration [1–3]. Even if Ti6Al4V is considered as a highly biocompatible metallic material, there have been reports of inflammatory reactions occurring around the prostheses upon implantation [4–8]. In order to improve the biocompatibility of the metallic alloys, a bioactive material can be deposited on its surface as a thin film, in order to increase the osseointegration and to ensure the biomimetism of the implant [9–15]. Hydroxyapatite (HA), with the theoretical chemical formula $Ca_{10}(PO_4)_6(OH)_2$, in its non-stoichiometric multi-cation and anion substituted form, represents the main constituent of the mineral part of the bone. By the application of such a biomimetic coating, one can stimulate the osteoblasts to proliferate and to generate osseous matrix, which will culminate with the new bone formation. As stated, biological apatite (further denoted Bio-HA) is similar, but not chemically and structurally identical with the pure synthetic HA. Bio-HA can enclose a series of trace elements (e.g., Na, Mg, Sr, Zn, F), as well as substituting functional groups such as carbonate (CO_3^{2-}), silicate (SiO_4^{3-}), and/or hydrogen phosphate (HPO_4^{-}) [16–19]. Thereby, an apatite compound similar to the one produced in the bones of living mammals is very difficult, if not impossible, to replicate by chemical synthesis. This is why the extraction of Bio-HA from animal bones could represent a viable solution, not only due to the intrinsic natural doping with ions which can promote specific biofunctional traits, but also due to their high availability and reduced costs [17,20]. For biomedical research purposes, bovine [21,22], ovine [23], or swine bones [24] are the most commonly used resources, but also fish [25] and sea-shells [26], as well as more exotic sources such as camel [27] or shark [28] have been explored.

For depositing calcium phosphate-based coatings onto metallic implants, the industrial technique of choice is plasma spray (PS), due to the deposition speed, large area coverage, and inexpensive working atmosphere (i.e., ambient or argon) [29,30]. However, HA coatings produced using this technique could contain residual decomposition phases because of high-processing temperatures, having different dissolution speeds, and thereby an unpredictable biological behavior [31]. Pulsed laser deposition (PLD) is another coating technique which proved reliable for the fabrication of bioactive layers [6,17,32]. The PLD strength comes from its capability to congruently transfer from target-to-substrate even materials with complex stoichiometry, but the biggest shortcoming is the limitation to a very small deposition area of the order of few cm^2 [32]. Radio-frequency magnetron sputtering (RF-MS) was long-time considered a solid alternative to PS since it can produce (by the control of deposition parameters) dense, pure, highly-adherent films on large area substrates (even on ones with complex geometry when coupled with rotation-translation facilities) [33–35]. There are reports in scientific literature on deposition of synthetic pure [7,12,34] or substituted HA [11,13,36] coatings using RF-MS. However, to the best of our knowledge, no information has been reported yet on the successful magnetron sputtering of Bio-HA films and consequently on their structural and biofunctional properties.

Casting and forging are the conventional processing technologies used for implants mass manufacturing. When customized prostheses with dimensions and complex geometries are in demand, such technologies are not reliable because of the time, energy, material consumption, and costs [37]. As an alternative, laser-based additive manufacturing (AM) is a method that allows customized, cost-effective, and rapid production of complex 3D medical parts [38,39].

Selective laser melting (SLM) is an AM process capable of producing metallic, ceramic, plastic, or composite parts using layer-by-layer manufacturing [40–43]. 3D structures are generated from a geometric CAD (computer aided design) file. The model is virtually sliced into 2D contour lines and imported into the AM machine that builds the new part [44–47]. The energy source is a laser beam which scans a selected area of a powder bed. The area scanned by the laser beam is molten and it quickly solidifies. The new solid structure is covered by a powder layer with a leveling system, and subsequently the laser scanner irradiates the next contour and thus, layer-by-layer, the part is completed [48,49].

SLM has been used to manufacture hip, ankle, knee, tibia, maxillo-facial, or spinal prostheses with intricate geometrical parts, based on medical imaging. The correlation between micro-structural features and mechanical properties has been reported in the literature [37,50–53]. Fatigue strength of AM components is the primary mechanism of failure, and for this reason many investigations have been conducted in this area of research [54,55]. Lieverani et al. [56] reported the fabrication of Co-Cr-Mo endoprosthetic ankle devices using SLM. They identified the optimal process parameters for obtaining samples with maximum density and high mechanical strength.

We report in this study the manufacturing in two steps of a cranial prosthesis with bioactive Bio-HA surface with dimensions corresponding to a cranial large defect of a real patient, obtained by computer tomography. First, the prosthesis is generated by 3D printing using the SLM method, while in a second step, a thin film is deposited by RF-MS from a target of biological apatite derived from bovine bones. The focus of this study was to assess for the first time the feasibility of depositing by RF-MS bioceramic coatings of biogenic origin onto SLM custom-fabricated prosthesis. In this respect, morphological, compositional, structural, and biofunctional interrogations were conducted for both the metallic part and the bioceramic coating.

2. Materials and Methods

2.1. Cranial Meshes

2.1.1. 3D Printing of Cranial Meshes

For the manufacturing of meshes, a Ti6Al4V powder with the particle size in the range of 10–30 μm (Eos GmbH, Krailling, Germany) was used. The SLM machine was a model M270 (Eos GmbH, Krailling, Germany).

The laser beam used for melting the metallic powder was generated by a CO_2 laser source (λ = 10.6 μm, continuous wave, Gaussian beam) connected to a scanner. The optimal laser parameters recommended by the producer for SLM of titanium alloys are: 90 W laser power and 450 mm/s scanning speed, and an under 4 bar argon atmosphere. The laser spot on the powder bed was of ~100 μm. For the manufacturing of a 3 mm thick cranial mesh prosthesis a number of 100 successive layers were needed, corresponding to an individual layer thickness of 30 μm.

The cranial meshes were designed in the SolidWorks (Dassault Systems, Vélizy-Villacoublay, France) graphical engineering software, and imported in the EOSPRINT proprietary software of the SLM machine, which generated the movement of the scanner mirrors.

The prostheses were extracted from the Ti plate by electro-erosion cutting using a Robofil 6030 machine (Charmilles Technologies, Schaffhausen, Switzerland) with 0.25 mm thickness of the wire, 13 m/min wire consumption, and a cutting speed of 1 mm/min. Prostheses of 3 mm thickness were produced by SLM, so that after cutting, the thickness of the mesh to be of ~2 mm.

2.1.2. Characterization of the 3D Printed Material

First, bulk prisms of 10 × 10 × 20 mm^3 were 3D printed in view of metallographic characterization, in the same experimental conditions as the meshes. The obtained samples were cut in slices using a disk cutting machine, model Brillant 200 (ATM GmbH, Mammelzen, Germany). The slices were fixed in Bakelite using a machine Opal 410 (ATM GmbH, Mammelzen, Germany) and polished with a Saphir 520 (ATM GmbH, Mammelzen, Germany) semiautomatic machine. The resulting mirror-polished samples were etched with Kroll solution (water 92%, nitric acid 6%, hydrofluoric acid 2%) in order to reveal the metallographic microstructure of the alloy after SLM.

The metallographic analysis was conducted with a BX51M (Olympus, Tokyo, Japan) optical microscope and a scanning electron microscope (SEM) Quanta 450 (FEI Company, Hillsboro, OR, USA). The compositional homogeneity of the prosthesis has been investigated by energy dispersive X-ray spectroscopy (EDXS) measurements.

The micro-hardness of the printed material was evaluated by the Falcon 500 mechanical testing machine (Innovatest, Maastricht, Netherlands) equipped with a Vickers indenter. Ten Vickers indentations have been made on the surface of the specimens using a load of 0.5 kgf/mm^2. The reported micro-hardness value represents the average of the ten independent values ± standard deviation.

2.2. *Coating of the Prosthesis With a Bio-HA Thin Film*

The final step of the manufacturing process was to coat the 3D printed prosthesis with a thin ceramic layer of biological hydroxyapatite (Bio-HA) extracted from bovine bone for increasing the bioactivity of the surface.

2.2.1. Bio-HA Extraction

The extraction of Bio-HA was performed by applying a previously established work protocol [17,23,57], conducted in accordance with the European Regulation 722/2012 and ISO 22442-1:2015—Medical devices using animal tissues and their derivatives—Part 1: Application of risk management. In brief, the bone marrow and the soft tissue residues were removed mechanically from bovine femoral bodies. The femoral shafts were cut into small pieces and ultrasonicated with distilled water. The cleaned parts were deproteinized by immersion for 14 days in an alkali water solution (1% sodium hypochlorite). Next, they were washed with water and dried, and then submitted to a calcination at 850 °C for 4 h in air. In the end, the calcined (white-colored) bone specimens were ball-milled into fine Bio-HA powders.

2.2.2. Deposition of Bio-HA Thin Films by RF-MS

Ti6Al4V cranial meshes manufactured by SLM were used as substrates together with monocrystalline infrared transparent Si <111> wafers (Medapteh, Magurele, Romania). The later ones were required to enable the structural evaluation of deposited films by Fourier transform infrared (FTIR) spectroscopy in transmission mode.

The substrates were first degreased in acetone and further on cleaned in iso-propanol using an ultrasonic bath and a dwell time of 15 min per step. In the end, they were fast-dried by argon gas purging and mounted on a rotating stage of the deposition chamber of the UVN-75R1 (Vacma, Kazan, Republic of Tatarstan, Russian Federation) RF-MS equipment.

The target (with the diameter of 110 mm) has been manufactured by mild-pressing the Bio-HA powder at room-temperature (RT) in a specially designed Ti plate (crucible). Such type of targets are preferred to the compacted ceramic disks (prone to cracking during prolonged sputtering processes), as they provide not only consistency, but also reducing significantly the fabrication costs [58,59].

First, a base pressure of ~3×10^{-3} Pa was achieved in the working chamber, such as to ensure a low contamination level and thus enable the fabrication of good purity films.

Prior to deposition, the substrate and the target were subjected individually to plasma cleaning treatments. The substrates were cleaned in situ for 15 min by argon ion etching following the protocol described in [60]. The target was pre-sputtered in the future deposition conditions for 30 min. This step is essential since it can ensure the removal of contaminants adsorbed on the target surface, and at the same time, it can allow the stabilization of the sputtering processes before the actual deposition [59,61].

The Bio-HA thin film deposition was performed at an argon pressure of 0.3 Pa, without intentional substrate heating (the substrate reached a temperature of ~120 °C owned to radiative plasma bombardment processes only). A 1.78 MHz RF generator and an electrical power of ~100 W were used. The target-to-substrate distance was set at 35 mm. Based on the deposition rate of ~3.7 nm/min (previously estimated by spectroscopic ellipsometry (SE) measurements), Bio-HA films of ~600 nm thickness were deposited. In order to crystallize the initially amorphous as-deposited Bio-HA films, without inducing a strong oxidation of the Ti6Al4V implant substrate, a post-deposition thermal-treatment was applied at 500 °C/1 h in air.

2.3. Characterization of Biological HA Thin Films

2.3.1. Physico–Chemical Characterizations

Since the RF-MS technology produces smooth layers [33,62] and the SEM method does not have enough contrast to distinguish the fine topological details, the morphology of Bio-HA films deposited onto mirror-polished substrates could be better revealed by atomic force microscopy (AFM) analysis. An NTEGRA Probe NanoLaboratory System (NT-MDT, Moscow, Russian Federation) equipped with an NSG01 cantilever with 10 nm tip was used. The AFM images were recorded in non-contact mode on areas of 25 × 25, 10 × 10, 5 × 5, and 2 × 2 μm^2.

The comparative composition of source material and Bio-HA RF-MS deposited film was evaluated by EDXS using an EDAX (Ametek, Berwyn, PA, USA) equipment with a Si-Li detector, operated at 20 kV acceleration voltage. The EDXS measurements were performed in at least four regions with an area of ~240 × 240 μm^2, randomly chosen on the sample surface. For the calibration of Ca/P atomic ratio, a National Institute of Standards and Technology standard reference sample of HA (i.e., NIST-SRM 2910b) was measured as well. Arithmetic means and standard deviations (SD) have been inferred, and statistical analyses were carried out using the unpaired two-tailed Student's t-test, with the differences being considered significant when $p < 0.01$.

The evaluation of the crystalline status of source powder and Bio-HA films was performed by X-ray diffraction (XRD), in the angular range 2θ = 9–54°, using a step of 0.04°. The source material and the heat-treated Bio-HA film were investigated comparatively using symmetrical geometry (Bragg–Brentano) XRD measurements, using a D8 advance diffractometer (Bruker, Karlsruhe, Germany) with $CuK_{\alpha 1}$ radiation. In addition, the films were measured, before and after the post-deposition thermal-treatment, also by grazing incidence XRD (GIXRD) at an incidence angle of 0.5°, using a SmartLab 3 kW diffractometer (Rigaku, Tokyo, Japan), in order to maximize their diffracted signal.

The chemical structure of the source powder and deposited Bio-HA films, with a focus on the short-range order and the transfer of functional groups specific to the bone mineral component, was investigated by Fourier transform infrared (FTIR) spectroscopy, in transmission mode. A Spectrum BX II (Perkin Elmer, Waltham, MA, USA) apparatus was used. The spectra were recorded in the wave numbers range of 400–4000 cm^{-1}, at a resolution of 4 cm^{-1}, and represent the average of 32 individual scans per sample.

2.3.2. In Vitro Tests in Simulated Physiological Solutions

An implant coating has to play a double role: (i) act as a buffer against the release of metallic ions of the metallic substrate in the internal environment and (ii) promote osteoconductivity. In vitro tests in synthetic physiological fluids were performed in order to evaluate the resistance to degradation of the deposited coatings and to explore their biomineralization capability. The tests were carried out following the indications contained in the ISO 23317:2014 protocol: "Implants for surgery—In vitro evaluation for apatite-forming ability of implant material". However, besides the fully inorganic Kokubo's simulated body fluid (SBF), recommended by the ISO standard, we have also used for these tests the complex biomimetic inorganic–organic McCoy's 5A culture medium supplemented with 15% fetal bovine serum (FBS)—the same medium formulation used for the in vitro tests on cell cultures. The immersed samples were kept at 37 °C in a stabilized biology incubator in humid atmosphere (to avoid the medium evaporation) under the accurate homeostatic conditions (5% CO_2) up to 21 days (we note that the ISO 23317:2014 does not provide recommendations for the testing atmosphere, and the implied common use of atmospheric pressure seems faulty). After 7 and 21 days, the Bio-HA films were extracted, gently washed with distilled water, and left to dry in a desiccator. Subsequently, they were investigated by SE, AFM, and FTIR spectroscopy.

The SE measurements were performed using a spectroscopic ellipsometer (J.A. Woollam Co., Lincoln, NE, USA) equipped with a HS-190 monochromator and a goniometer that allows for automatic acquisition of spectra at different angles of incidence. For Bio-HA films before and after immersion in biological medium, the measurements were performed at 65°, 70°, and 75° incidence angles in the spectral range of 1.3–3.5 eV. The Bio-HA material optical properties (refractive index dispersion) were inferred using a Cauchy formula [63].

2.3.3. In Vitro Tests on Cell Cultures

SLM printed prism samples were cut into 10 × 10 mm^2 coupons to serve as substrates for in vitro cell biocompatibility tests. The SLM coupons were polished down to mirror level to exclude any possible morphological incongruences between specimens, and their influence on cell behavior. The Bio-HA coated samples were subjected to sterilization in water vapors at 121 °C for 30 min using an AES-8 autoclave (Raypa, Barcelona, Spain).

Experiments were performed with human osteosarcoma (SaOs-2) cells which were grown in McCoy's 5A medium (Gibco Thermo Fisher Scientific, Waltham, MA, USA) supplemented with 15% FBS (Gibco Thermo Fisher Scientific, Waltham, MA, USA), 50 U/mL penicillin and 50 mg/mL streptomycin (Gibco Thermo Fisher Scientific, Waltham, MA, USA). Cells were seeded on biomaterials at a density of 2000 cells/cm^2. All substrates were then kept in a humid atmosphere with 5% CO_2 at 37 °C for 1, 3, and 7 days, respectively. Borosilicate glass coverslips were used as experimental standard control material.

Cell proliferation was assessed by using a CellTiter 96® AQueous one solution cell proliferation assay (MTS) kit (Promega, Madison, WI, USA). For this, samples were cultured in triplicate for 1, 3, and 7 days. At each time interval, samples were transferred to new wells and incubated at 37 °C for 1 h 30 min with fresh medium containing the MTS reagent. The absorbance values were recorded at 450 nm using a LB 913 Apollo 11 spectrophotometer (Berthold Technologies, Bad Wildbad, Germany). The final data was represented as the average of the measured optical density upon background subtraction. Statistical analysis was performed by using GraphPad Prism 6 (GraphPad, San Diego, CA, USA) in which an unpaired, two-tailed, Student's t-test was calculated. Differences were considered statistically significant for $p < 0.05$.

Cell adhesion was examined by immunofluorescence microscopy. After 24 and 72 h of culture, SaOs-2 cells were fixed for 15 min with 4% paraformaldehyde at RT and stored in phosphate buffered saline solution (PBS) at 4 °C prior to labelling. Fixed cells were then permeabilized for 3 min with 0.2% TritonX-100 and blocked for 1 h in 0.5% bovine serum albumin (BSA). To visualize the actin filaments, cells were labelled with 1:100 diluted Alexa Fluor 488 conjugated Phalloidin (Cell Signaling Technology, Danvers, MA, USA). Cells were subsequently treated with 1 µg/mL Hoechst (Cell Signaling Technology, Danvers, MA, USA) to stain cell nuclei. Upon each incubation, samples were washed three times with PBS. Finally, representative images were acquired on each substrate by using a Leica DM 4000 B LED fluorescence microscope equipped with a DFC 450 C camera (Leica Microsystems GmbH, Wetzlar, Germany) and the appropriate filters.

3. Results

3.1. *Printing of Cranial Mesh Prostheses and Their Structural and Chemical Characterization*

A photograph of a cranial mesh grown by SLM is presented in Figure 1. The pattern design is original, being devised after studying the morphology of several commercial cranial prostheses. Structures of 50 × 50 × 3 mm^3 with the pattern depicted in Figure 1b were built layer-by-layer on a

50 mm thick Ti plate by laser scanning. Every single contour was ~30 μm tall and 100 layers were needed to obtain a final product with the desired height.

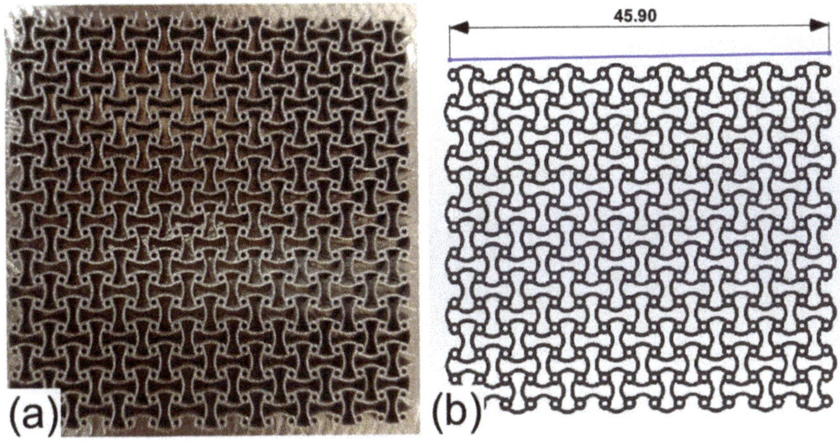

Figure 1. (**a**) Photograph of a Ti6Al4V cranial mesh manufactured by SLM (Selective laser melting) based on a (**b**) three-dimensional (3D) CAD (computer aided design) model generated in the SolidWorks Professional software, version 2013 (Dassault Systems, Vélizy-Villacoublay, France).

After polishing and attacking with a metallographic reagent, the grain structure was exposed and it was studied by optical and SEM microscopies (Figure S1 of Supplemental Information). Optical and SEM micrographs, characteristic to the printed SLM printed Ti6Al4V sample are presented in Figure S1a,b, respectively. Figure S1a displays a martensitic acicular structure that developed in islands in the α phase. More specifically, it is an α′-type martensite that crystallizes in the hexagonal system upon the fast cooling of biphasic titanium alloys (α + β) such as Ti6Al4V. The acicular plates displayed a "clustered" morphology, consisting of structural units with needles exhibiting a certain parallelism. This distribution includes the microstructure of the printed material in the cryptocrystalline martensite category. In general, α′ martensite nucleates at the boundaries of the prior β grains and spreads within their parent β grains, which explains the different orientation of the α′ needles. SEM investigations allowed for a better visualization of the grains, the parallel distribution of lamellar martensite was more obvious, with varying lengths of the grains dependent on the micro-regions from which they were part of (Figure S1b). There are structural units with grains whose average lengths are greater than 100 μm, which correspond to primary α′ martensite, but also units with finer needles, with lengths of about 15–20 μm, corresponding to secondary α′ phase. These later grains were perpendicular to the primary α′ ones.

The hardness tests performed in 10 randomly chosen points of the SLM test sample revealed similar hardness values, which represent an indirect proof of its structural homogeneity. The average hardness of the SLM prints was 391 ± 5 HV, similar to the casted Ti6Al4V alloy [64].

Subsequently, the same sample was analyzed by EDXS. An EDXS characteristic spectrum of a Ti6Al4V 3D printed sample is shown in Figure S2a (of Supplemental Information). The elemental composition was found to be ~90.2 wt% Ti, ~5.6 wt% Al, and ~4.2 wt% V, thus, congruent to the one of the Ti6Al4V starting powder.

Further, the EDXS mapping analysis (Figure S2b–e) attested the presence of Ti, Al, and V, the components of the studied alloy (Figure S2b), and offered qualitative information on the uniform distribution of elements (Figure S2c–e). The brightness is directly proportional to the presence of the element being tracked. Taking into account this principle, it was confirmed that all the components were uniformly distributed over the studied area, with no trace of elemental segregation or depletion.

3.2. *Functionalization of the Cranial Mesh With a Bioceramic Layer*

Illustrative images of a Ti6Al4V cranial mesh, before and after biofunctionalization with a Bio-HA thin film by RF-MS, are presented in Figure 2.

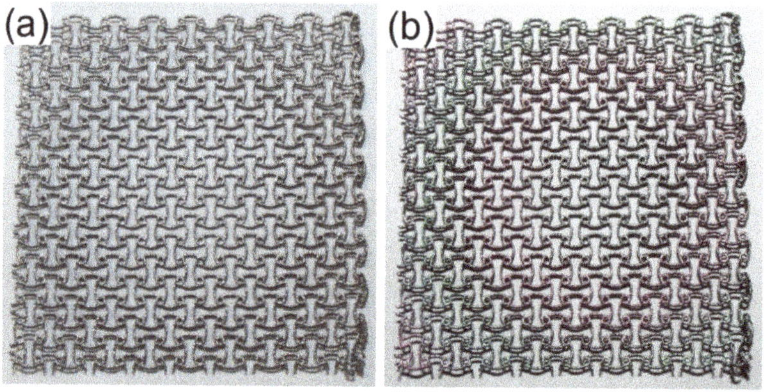

Figure 2. (**a**) Uncoated and (**b**) Bio-HA (derived from biological hydroxyapatite) sputtered cranial mesh of Ti6Al4V fabricated by SLM.

3.2.1. Physico–Chemical Characterization of the Bio-HA Functional Layer

AFM micrographs, collected at different magnifications, are depicted in Figure 3. They highlighted that the crystallized Bio-HA RF-MS film consisted of a compact matrix of polyhedral grains with diameters in the range ~110–230 nm. An average root mean square roughness (R_{rms}) of ~15 nm was inferred, irrespective of the scanned microscopic field area, which advocates for the good uniformity of the deposited film.

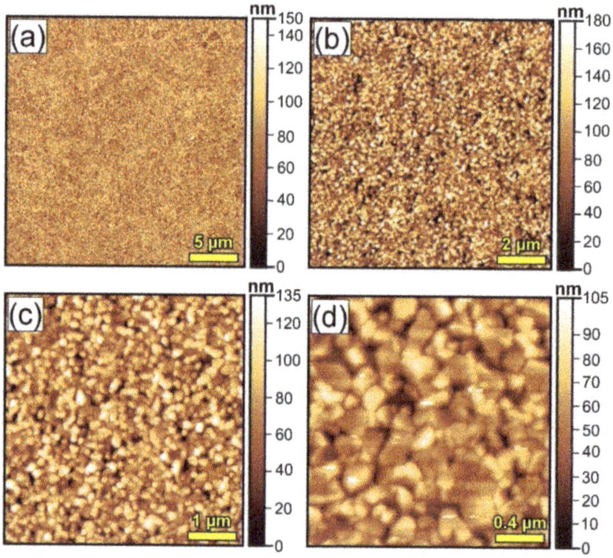

Figure 3. Characteristic AFM (atomic force microscopy) images of the of Bio-HA films surface acquired on scanning areas of: (**a**) 25 × 25; (**b**) 10 × 10; (**c**) 5 × 5; and (**d**) 2 × 2 µm^2, respectively.

Next, the composition of the Bio-HA sputtered film and parent target were determined by EDXS and compared. In addition, the Ca, P, and O, also small contents of Na (~3 at%) and Mg (~2 at%) were detected. Such specific trace elements present in the composition of bone apatite, are known to play important roles in bone metabolism and health [65,66], and thereby, their successful transfer by RF-MS into the deposited layers can be emphasized. The mean elemental concentration ± standard deviations obtained for each of these constituent elements (except oxygen, light element for which the EDXS quantification yields characteristically large errors) of the Bio-HA target and film are shown in Figure 4. Significant compositional differences ($p < 0.01$) were found only in the case of P and Mg. The Ca/P molar ratios for the source material and the Bio-HA film were of 1.63 ± 0.02 and 1.70 ± 0.01, respectively. However, in the case of cation substituted HAs, such as those derived from biogenic resources, the contribution of natural dopants (e.g., Na, Mg) should not be dismissed. In this case, a (Ca + Na + Mg)/P molar ratio could provide further information on the features of the ceramic compounds. The (Ca + Na + Mg)/P molar ratios of the Bio-HA target and film were of 1.76 ± 0.01 and 1.85 ± 0.01, respectively (thus, situated close to the Ca/P ratio of 1.67 of stoichiometric HA, $Ca_{10}(PO_4)_6OH_2$). The higher (Ca + Na + Mg)/P ratios (than 1.67), obtained in the case of both target and sputtered film, are suggesting a deficit of orthophosphate groups, most probably due to their substitution with carbonate functional groups (specific to Bio-HAs). A slight increase of the (Ca + Na + Mg)/P was recorded in the case of the RF-MS film with respect to the parent target. This is connected with the P evaporation phenomena taking place during sputtering deposition [34,59].

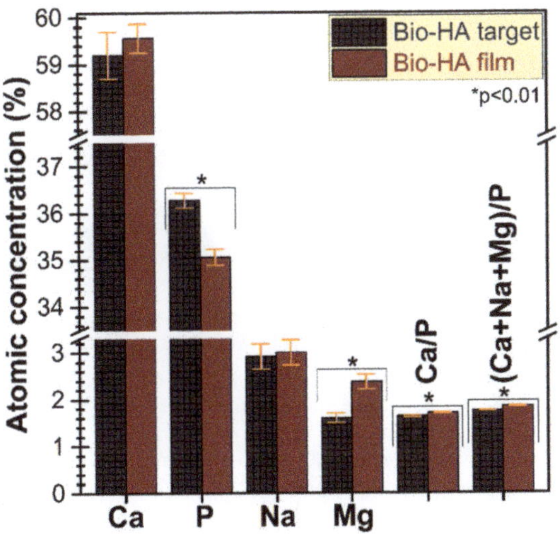

Figure 4. Elemental composition of the Bio-HA film produced by RF-MS (radio-frequency magnetron sputtering) with respect to the source target material. The Ca/P and (Ca + Na + Mg)/P molar ratios of the Bio-HA target and film.

XRD patterns were recorded in symmetric geometry (Figure 5a,b), for both Bio-HA target and crystallized Bio-HA film, to accommodate an equivalent structural comparison and to survey the presence of preferred crystallite orientations (if any). It has been revealed that the HA powder obtained from the bovine cortical bone was monophasic, consisting of a randomly oriented hexagonal hydroxyapatite compound (ICDD-PDF4: 00-009-0432) only. In the case of the thermal-treated (TT) Bio-HA film, besides the intense diffraction peaks of the Ti6Al4V substrate (crystallized in the hexagonal system, ICCD-PDF4: 04-020-7055), only peaks of a well-defined phase of hexagonal HA were evidenced. The increased intensity of the 002 peak of Bio-HA film with respect to the ICDD file indicated that the Bio-HA crystallites elicited a preferred orientation along the *c*-axis, normal to the substrate. The *c*-axis texturing is characteristic to sputtered films synthesized from materials of the hexagonal system [34,36,60,62,67]. The average crystallite size determined based on the full width at the half-maximum (FWHM) of (002) planes reflections was estimated by applying the Scherrer equation [68]. The FWHM was corrected for instrumental breadth using a corundum NIST-SRM 1976a sample. Similar average crystallite sizes of ~167 and ~139 nm were obtained in the case of the source target material and crystalline Bio-HA film, respectively. This finding demonstrated the capability of the chosen thermal treatment to yield excellent crystalline quality. Furthermore, the efficiency of post-deposition thermal treatment performed in air at 500 °C/1 h was highlighted by GIXRD measurements (Figure 5c,d) which showed the complete conversion of the initial amorphous as-sputtered film (see the halo centered at $2\theta \approx 30°$) into a completely crystallized Bio-HA structure.

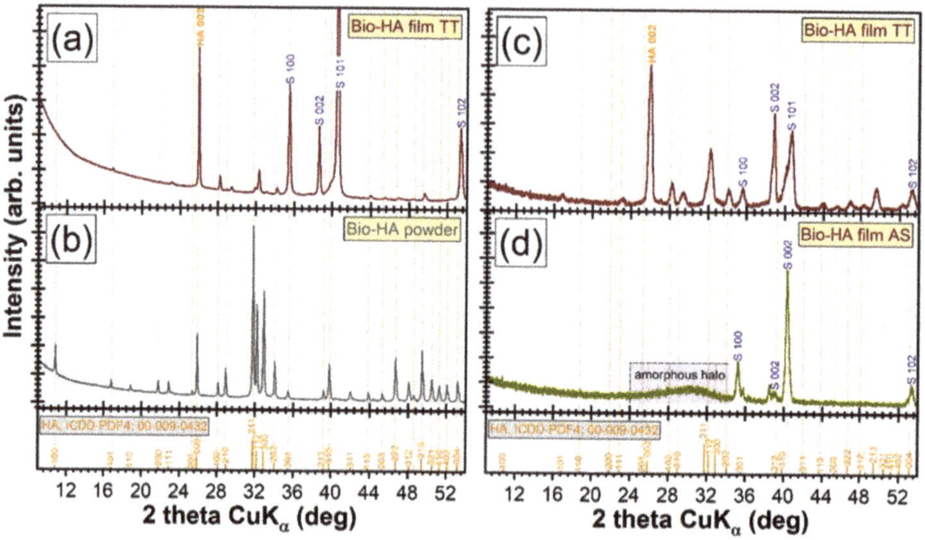

Figure 5. XRD (X-ray diffraction) patterns recorded in (**a**,**b**) symmetric (Bragg–Brentano) and (**b**,**d**) grazing incidence (α = 0.5°) geometries in the case of (**b**) Bio-HA source material and (**d**) as-sputtered (AS) and (**a**,**c**) thermal-treated (TT) Bio-HA film. For comparison, the ICDD-PDF4: 00-009-0432 reference file of pure hexagonal hydroxyapatite is inserted on the bottom of the figures.

The short-range order of the Bio-HA magnetron sputtered film and the presence and type of functional groups incorporated within its structure was surveyed by FTIR spectroscopy measurements (Figure 6). The sharp and well-split IR absorption bands of orthophosphate groups recorded in the case of the Bio-HA film (Figure 6a) testified for it its high crystallinity [69], in full agreement with the XRD data. All the characteristic IR bands of HA were evidenced for the Bio-HA sputtered structure: ν_4 bending (~565 and ~600 cm^{-1}), ν_1 symmetrical stretching (~962 cm^{-1}), and ν_3 asymmetric stretching (~1032 and ~1088 cm^{-1}) vibration modes of orthophosphate groups, and ν_L libration (~632 cm^{-1}) and ν_S stretching (~3753 cm^{-1}) of structural hydroxyl units [26,36,69]. Only slight shifts and intensity modification of the Bio-HA film bands (Figure 6a) with respect to the pure HA reference and Bio-HA source powder (Figure 6b) were noticed. These are denoting minor structural alterations (in terms of bond length and/or angle and short-range order), and are to be expected for a material deposited by magnetron sputtering, thus, under non-equilibrium thermodynamic conditions. Furthermore, it can be emphasized the successful incorporation of carbonate groups in the orthophosphate sites (B-type carbonatation) of the Bio-HA film crystalline lattice, similarly to the source powder: ν_2 symmetrical (~875 cm^{-1}) and ν_3 asymmetrical (~1417 and ~1472 cm^{-1}) stretching vibrations of $(CO_3)^{2-}$ units [69]. Such type of substitution is also characteristic to the mineral phase of the human bone [36,70].

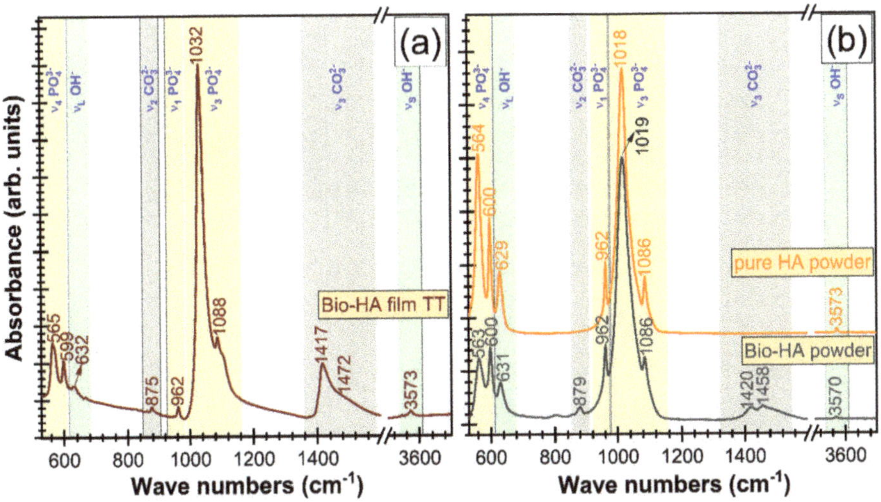

Figure 6. Comparative FTIR (Fourier transform infrared) spectra of (**a**) Bio-HA crystallized film and (**b**) pure commercial HA powder (Acros Organics, Geel, Belgium) and Bio-HA source material.

3.2.2. In Vitro Tests in Simulated Physiological Solutions

The first step in the evaluation of in vitro behavior of Bio-HA coatings was to immerse them in both a fully inorganic SBF solution (as the ISO 23317/2014 standard recommends), as well as in the McCoy's 5A cell culture medium supplemented with 15% FBS (to be used in the in vitro tests on cell cultures). It is our belief that the increased degree of biomimicry of the latter could foster a more truthful interrogation of the material degradation and bioactivity performance. After 7 days of immersion, the films thickness decreased by ~6.5% when immersed in SBF and by ~4% in the case of the tests carried out in the FBS supplemented McCoy's 5A culture medium. After 21 days of immersion a decrease of Bio-HA film thickness with ~14% and ~5% was recorded in SBF and FBS supplemented McCoy's 5A, respectively.

The AFM investigations revealed that radical morphological changes occurred on the surface of Bio-HA sputtered film immersed in SBF and FBS supplemented McCoy's 5A simulated physiological media (Figure 7). After 7 days of immersion in SBF on the surface of Bio-HA films, initially composed of a uniform bed of nano-grains (Figure 3c), was signaled the formation of spheroid conglomerates with a mean diameter of ~1–1.5 μm (Figure 7a). Such type of topography is specific to biomimetic HA layers formed in SBF solution [71,72] and demonstrates (according to ISO 23317:2014) the bioactivity of the Bio-HA sputtered film. For comparison, in Figure 7b a SEM image of a fully-developed biomimetic nano-crystalline HA layer is given, formed in 30 days in SBF on the surface of a bioactive silicate-based glass film [72]. The higher magnification AFM image (Figure 7c) evidenced that the micron-sized spheroid aggregates are constituted of randomly-dispersed small round particles (with sizes in the range of ~35–50 nm). A R_{rms} value of ~4 nm was inferred. After 21 days of immersion in SBF, the AFM analyses (Figure 7e,g) indicated that the formation of the biomimetic calcium phosphate layer progressed, having at this time point a denser morphology (Figure 7g) and an overall lower $R_{rms} \approx 1$ nm.

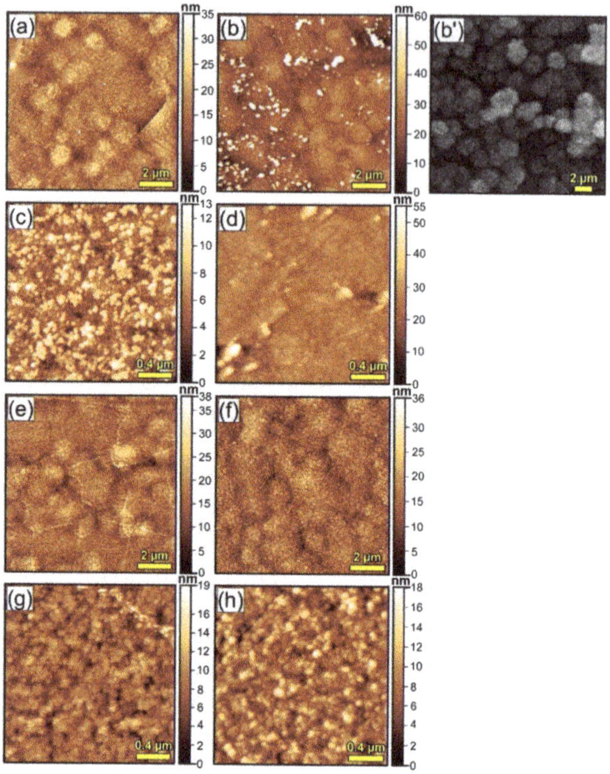

Figure 7. Characteristic AFM images of Bio-HA sputtered films tested in vitro in (**a**,**b**,**e**,**f**) SBF and (**c**,**d**,**g**,**h**) FBS supplemented McCoy's 5A simulated physiological solutions, for (**a**–**d**) 7 and (**e**–**f**) 21 days, acquired on scanning areas of: (**a**,**b**,**e**,**f**) 10 × 10 and (**c**,**d**,**g**,**h**) 2 × 2 μm^2, respectively. (**b'**) SEM images of a fully developed biomimetic nano-crystalline HA layer formed in 30 days in the SBF medium. Figure 7b' is adapted after Figure 2a from Reference 72, with permission from Springer, 2019.

In the case of tests performed in FBS supplemented McCoy's 5A solution, the biomineralization processes seem to take place at a slower pace. After 7 days of immersion the general appearance of the surface was found rather uneven. The presence of the spheroid aggregates is only hinted (Figure 7b), having a smoother morphology with respect to the SBF experiments (Figure 7a). On top of this covering blanket, smaller grains (~0.2–0.5 μm), randomly distributed, were revealed. Consequently, a larger R_{rms} value of ~13 nm was inferred. After 21 days of immersion in the FBS supplemented McCoy's 5A solution, the surface of the samples (Figure 7f,h) closely resembled the one observed in the case of films tested for 7 days in SBF (Figure 7a,c), further suggesting that biomineralization processes are slower in the FBS supplemented McCoy's 5A solution with respect to the ones occurring in SBF. Kindred observations were recently reported by Popa et al. [59], Vladescu et al. [11] and Vranceanu et al. [73] which showed that the (i) biomineralization processes (i.e., formation and structuring of biomimetic HA layers) takes place at a lower rate [59] and (ii) the degradation is significantly reduced [11,73] in media with a complex organic–inorganic composition (such as in the FBS-supplemented McCoy's 5A solution) with respect to those encountered in the inorganic SBF solution.

The FTIR comparative spectra of the Bio-HA films after immersion for 7 and 21 days in SBF and FBS supplemented McCoy's 5A media are presented in Figure 8. A strong diminution of carbonate bands intensity is noticed even after seven days, with the allure and intensity of the orthophosphate and hydroxyl bands basically unaltered. The decrease of the carbonate content of the Bio-HA films could be attributed to the weaker Ca–CO_3 bonds as compared to the Ca–PO_4 ones [74], which makes them more susceptible to be cleaved and released during the ion exchange processes with the surrounding medium [71]. No other remarkable structural changes have been revealed by FTIR measurements (Figure 8) up to 21 days of immersion, which suggests that the modification of the film surface revealed by AFM (Figure 7) is only superficial. However, one should note that the phase discrimination between the top biomimetic HA layer and the Bio-HA sputtered film underneath is impeded by their structural–chemical similarity (i.e., both being nano-structured and having akin compositions).

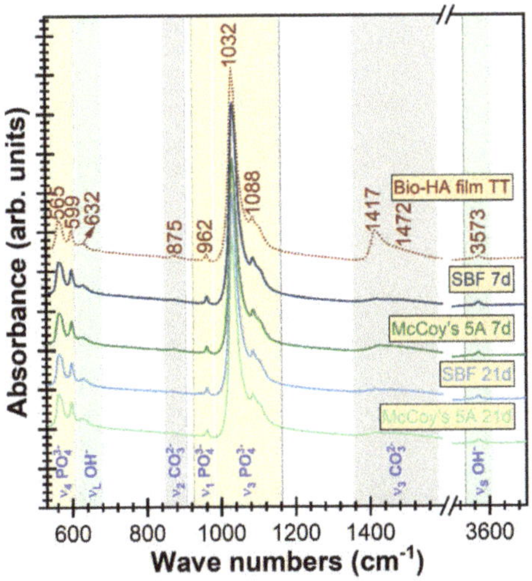

Figure 8. Comparative FTIR spectra of the Bio-HA sputtered film before and after in vitro testing for 7 and 21 days in SBF (simulated body fluid) and FBS (fetal bovine serum) supplemented McCoy's 5A simulated physiological solutions.

Overall, the results suggest a reduced rate of degradation accompanied by a good biomineralization capacity of the Bio-HA films, with the loss of mass being compensated to a certain extent by growth of new calcium phosphate layers in contact with the simulated physiological solutions.

3.2.3. In Vitro Cytocompatibility Studies

To determine the effect of Bio-HA coating on cell viability and proliferation, we have carried out an MTS test at 1, 3, and 7 days after cell seeding. As a control, a borosilicate glass coverslip was used, which is a cell culture standard [75]. The results are represented in Figure 9 and they show an enhanced osteoblasts proliferation on the control up to 72 h. However, after 168 h, the Bio-HA coated substrate was found to encourage cells expansion, as indicated by a significant increase in MTS absorbance value. This result suggests a slower proliferation dynamic on Bio-HA at early time points, which is compensated after cells reach semi-confluency.

The adhesion of SaOs-2 cells on the Bio-HA film and the control surface has been investigated by fluorescence microscopy (Figure 10). The actin filaments were labelled in green to study the cytoskeleton organization, while the cell nuclei were represented in blue. As it can be seen from Figure 10a,b, after one day of culture, the cells have successfully attached on both substrates. However, in the case of the Bio-HA coating (Figure 10a), cells have extended their membrane protrusions and changed their morphology to a more elongated phenotype. This difference in the cytoskeleton structure is more pronounced after three days of culture (Figure 10c,d). Cells grown on the control material (Figure 10d) elicit a more spread cell body compared to those on the Bio-HA substrate where they become rather elongated with multiple developed filipodia (Figure 10c). However, taking into consideration the increased cell number detected after 7 days by MTS (Figure 9) on the Bio-HA sputtered surface, we can conclude that this biomaterial is compatible with bone cell adhesion and expansion, which are key early steps important for osseointegration.

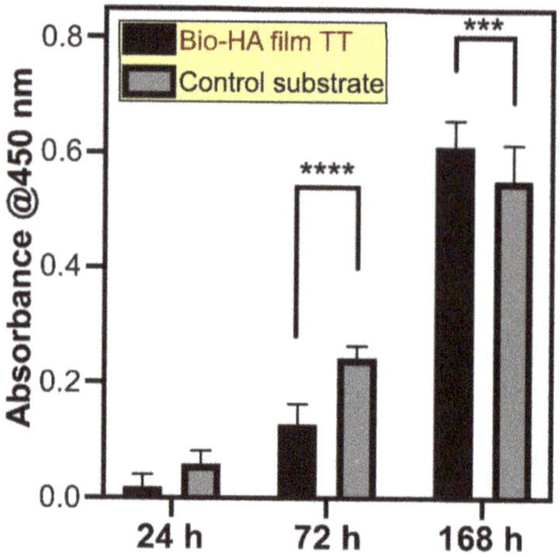

Figure 9. SaOs-2 cells proliferation on Bio-HA sputtered film and control substrate, after 1, 3, and 7 days of culture (*** $p < 0.001$; **** $p < 0.0001$).

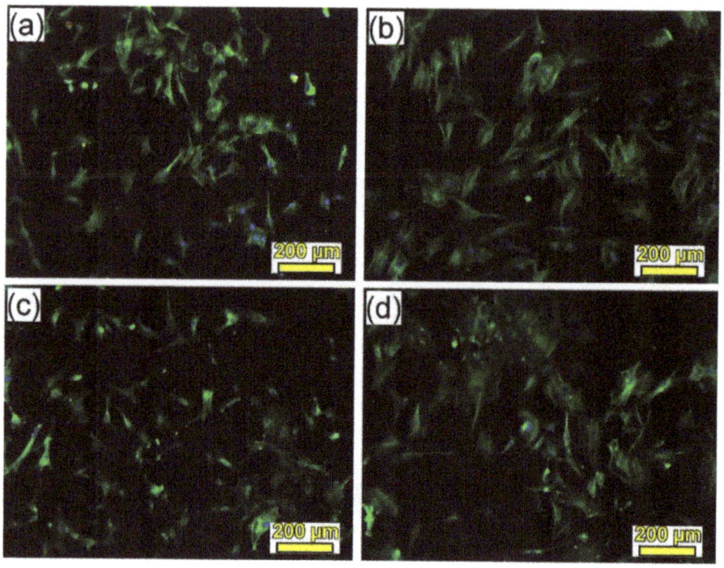

Figure 10. Immunofluorescence microscopy images of SaOs-2 cells grown on the surface of the Bio-HA coated 3D printed Ti6Al4V substrates (**a**,**c**) and control substrate (**b**,**d**), after one day (**a**,**b**) and three days (**c**,**d**) of seeding. The actin filaments are stained in green and the cell nuclei are labeled in blue. Scale bar = 200 µm.

4. Discussion

3D printing is a technique that resonates with medicine. The prospect of custom metal implants that can be produced rapidly for specific bone fractures is highly appealing. In general, metal printing techniques are not cheap and neither is prosthesis manufacturing. While 3D printing is not envisaged for mass production due to its high costs, it is an excellent option for manufacturing prototypes and complex shape items that cannot be obtained by conventional casting. 3D printed metal implants fall into this category, as they can be patient specific, fabricated in a low number for targeted needs and can support the high manufacturing cost. Among laser metal printing techniques, SLM and laser melting deposition (LMD) are the main options. In the case of cranial meshes, LMD is out of discussion, as it lacks the resolution for producing lines with submillimeter width. With the LMD experimental setup in our laboratory, we have not succeeded to produce lines thinner than 2.5 mm. This width increases in the case of multiple layer patterns, as material flows during deposition. In the case of SLM, this problem was non-existent, the minimum width of obtained lines being of ~200 μm, thus, allowing to obtain even cranial prostheses with intricate patterns.

However, one pending drawback of titanium-based implants, in spite of their excellent biocompatibility and mechanical performance, is their slow osseointegration. One solution to tackle this downside is the application of coatings of bioactive ceramics [34,76], an ever-growing and dynamic field of research if taking into account the recent statistics on yearly distribution of kindred scientific articles published in Reference 76. For this purpose, radio-frequency magnetron sputtering was selected as the deposition method of choice, due to its ability to generate high purity, dense, and adherent coatings on large area substrates (dependent only on the cathode gun size), making it scale-up and mass production ready. This study focused on sputtered hydroxyapatite films derived from biogenic sources (i.e., cortical bovine bones), which have both cost-efficient and biofunctionality valances. On the one hand, repurposing highly-abundant waste materials (such as animal bones, resulting in large quantities from the food industry) represents an inexpensive, sustainable, and environmentally-preferred alternative. On the other hand, biological apatite is substituted with low contents of cations (e.g., Na^+, Mg^{2+}) and anions (e.g., CO_3^{2-}, F^-, SiO_4^{4-}), each with specific biofunctional roles [17,18,65,66,74,76–78]. Such trace elements can be directly responsible for the modifications in crystallinity, degradation speed, biomineralization capacity, or mechanical performance of apatite [18,74,76].

Thereby, the current study is congruent with the current efforts taking place worldwide to delineate patient-customized implants capable to warrant a rapid osseointegration process. Furthermore, the importance of this study stems also from being the first attempt to assess RF-MS technique ability (to the best of our knowledge) to transfer biological apatite compounds in the form of thin films. The physical–chemical analysis results showed great promise, with the composition (presenting only slight deviations, to be expected in the case of a physical vapor deposition technique such as RF-MS) and structure (similar crystallite size, conservation of the B-type carbonatation) of the source biogenic apatite material being well reproduced into the crystallized sputtered films. The *c*-axis preferential orientation of the Bio-HA sputtered films could constitute a supplemental attribute to be further explored, since several studies have stressed the superiority of such textured structured with respect to the randomly oriented ones in terms of chemical stability in physiological environments [79], wear resistance [80], hardness and Young modulus [81], and bioactivity [81].

The in vitro degradation/bioactivity tests performed in the complex organic–inorganic McCoy's 5A FBS-supplemented medium (for a more rigorous evaluation), indicated on the one part the stability of Bio-HA film (a thickness decrease of only ~5% was recorded after three weeks of immersion) and the good biomineralization capacity highlighted by the formation of the characteristic spherulitic deposits of biomimetic calcium phosphates. The bone bonding mechanism of dense hydroxyapatite constructs (here including also magnetron sputtered films) is considered to consist of five consecutive reaction stages: (i) slight decrease of homeostatic pH at the interface between HA—biological environment; (ii) dissolution/precipitation processes culminating with the formation of the biomimetic carbonated HA;

(iii) production of extracellular matrix by collagen and adhesive proteins; (iv) concurrent mineralization of collagen fibrils and incorporation of carbonated HA within the new bone remodeling process; and (v) interlocking and strengthening of host bone— HA interface by mineralized collagen [82,83]. The first two stages can be evaluated in vitro in media with ion concentrations equivalent to those of human blood plasma [59,84], and currently, it is generally viewed that the in vitro capacity of a material to generate the formation of a biomimetic layer on its surface is a predictor of in vivo bone-bonding ability [84]. The partial dissolution process (in vivo mediated by extracellular fluids and/or phagocytosis by osteoclasts) of HA leads to a local concentration increase of Ca^{2+} and PO_4^{3-} ions which will upsurge even more the saturation of the micro-environment (already saturated with respect to calcium and phosphate) and trigger the nucleation and precipitation of apatitic crystals, which will incorporate in time carbonate, hydroxyl, and various cations in its further crystallization and structuring process [85,86]. Thereby, a bioactive material will always need to dissolve slightly in order to induce the formation of biological apatite which will facilitate the chemical bond with the bone [87]. A too fast or extreme degradation could lead to the loss of the osteogenic coating before fulfilling its role. However, in our case, after 21 days of immersion in the simulated body media solutions Bio-HA film experienced only a minor thickness reduction (i.e., of ~14% and 5%, in SBF and SBF and FBS-supplemented McCoy's 5A, respectively). The formation of biomimetic calcium phosphates layers (whose nucleation already started at 7 days, and fully developed at 21 days), shown for the Bio-HA RF-MS film, in both SBF and FBS-supplemented McCoy's 5A, already started to compensate for any further implant coating mass loss, and will hinder its further degradation. Thereby, it is expected for the Bio-HA films to preserve one of its designed roles (i.e., buffer against the release of metallic ions of the metallic substrate in the internal environment) until the bone regeneration process (typically taking 4–6 weeks) is complete [88].

The biomineralization processes were suggested to be slower in the medium with increased biomimicry with respect to purely inorganic SBF solution. This could be related to the unnatural high pH (i.e., ~8) attained typically by the SBF solution with respect to the FBS-supplemented McCoy's 5A (which had a constant homeostatic pH of 7.3–7.4 throughout the testing period) and/or to the adsorption of proteins from the medium [59] on the surface of the Bio-HA film, forming a capping layer which slightly retards the biomineralization processes. The development at a slower pace of HA in complex organic–inorganic testing media could be also connected with (i) the inhibitory effect of amino-acids which are known to chelate Ca^{2+} and PO_4^{3-} ions or bind to nuclei of calcium phosphate [89] and (ii) the proficiency of testing medium proteins to increase the free energy of nucleation and thereby, to reduce the rate of calcium phosphate nucleation by decreasing the supersaturation level [90]. However, since there is no interaction of an implant with purely inorganic medium such as SBF, it is our belief that only tests performed in solutions with an increased biomimicry degree could provide accurate gauges of their real in situ performances.

Furthermore, the cytocompatibility tests indicated that besides their bioactive behavior, the films were also non-toxic and stimulated the proliferation of osteoblast-like cells. Roughness was long advertised to play an important role on the biological response of implants [91,92]. Since the Bio-HA sputtered films were rather smooth ($R_{rms} \approx 15$ nm) and uniform, the in vitro biological response (i.e., bioactivity, cell shape modifications, and their accelerated proliferation rate) could be associated prominently to the chemical and structural nature of the biological HA coatings.

Overall the physico–chemical properties and preliminary biological behavior of the sputtered Bio-HA films hold promise, and thus encourage us to further and more insightfully assess their potential from mechanical and biofunctional points of view.

5. Conclusions

3D cranial meshes of Ti6Al4V were successfully manufactured by SLM. The microstructure of Ti6Al4V deposited by SLM was martensitic α'. The irradiation parameters were optimized in order to obtain shapes without defects, such as pores or cracks and with homogeneous distribution of alloy elements in their volume.

In a second technological step the implants produced by SLM were coated by radio-frequency magnetron sputtering with a ~600 nm layer of hydroxyapatite derived from biogenic sustainable resources (i.e., calcined cortical bovine bones), which could constitute a cost-efficient, environmental-friendly, and highly promising biofunctional solution. Remarkably, the composition and structure of the biological apatite material were well-replicated into the smooth ($R_{rms} \approx 15$ nm) and uniform sputtered post-deposition crystallized films. Consequently, the Bio-HA films consisted of a single B-type carbonated hydroxyapatite phase with traces of Na^+ and Mg^{2+}, having a *c*-axis preferential texturing, similar to the bone mineral phase.

The Bio-HA sputtered films elicited promising biological performances:

1. Chemical stability in simulated physiological solutions (after 21 days the film thickness was reduced by ~14% in SBF and ~5% in FBS-supplemented McCoy's 5A medium);
2. Biomineralization capacity (Bio-HA generated the formation of spherulitic deposits of biomimetic calcium phosphates in contact with both types of simulated physiological solutions);
3. Absent cytotoxicity;
4. Optimal osteoblast cell proliferation and adhesion.

3D printing is now viewed as a highly appealing approach for the manufacturing of personalized implants. The most important functional request of such devices should be their capacity to stimulate fast bone regeneration. Thereby, the materials of choice would be bioceramics that highly resemble the composition and structure of the mineral bone phase. Unfortunately, such stand-alone bioceramics lack suitable mechanical properties risking implant failure. Furthermore, they are difficult to be 3D printed. Therefore, the compromise solution is the fabrication of customized implants by 3D printing from metallic materials, followed by their coating with bioactive ceramic thin films, closely mimicking the fine and intricate features of the metallic biomedical device. In this study we aimed to show that RF-MS could be a candidate deposition technique for the coating of such medical devices with thin films of biological apatite, on the one hand due to its now demonstrated ability to preserve the physico–chemical properties of the source material, and on the other hand due to its recognized capability to synthesize, in a controlled manner, dense and adherent thin films on large area substrates.

Supplementary Materials: The following are available online at http://www.mdpi.com/2075-4701/9/12/1332/s1, Figure S1: (a) Optical microscopy image and (b) SEM micrograph of the Ti6Al4V sample produced by the SLM method. (c) A typical EDXS spectrum recorded for the Ti6Al4V sample 3D printed by SLM. Figure S2: (a) A typical EDXS spectrum recorded for the Ti6Al4V sample 3D printed by SLM. Compositional EDXS maps for (b) all constitutive elements (overlayed), (c) Ti, (d) Al and (e) V, collected for the area highlighted in the microscopic field presented in (Figure S1b).

Author Contributions: Conceptualization, A.C.P., G.E.S., F.N.O. and L.D.; SLM manufacturing, N.P. and N.M.; Metallographic investigations and writing—original draft preparation, D.C. and A.C.P.; SE, SEM, XRD, FTIR, AFM, EDXS, and in vitro degradation/bioactivity testing, G.P.-P., L.M.B., A.C.G., A.C.P., and G.E.S.; In vitro cytocompatibility testing, S.I.; writing—review and editing, L.D., G.E.S., and A.C.P.

Funding: INFLPR authors acknowledge the funding of the Core Program contract no. 16N/2019 and of the Romanian Ministry of Research and Innovation in the framework of the projects PN-III-P2-2.1-PED-2016-1309 (PED241/2017), PN-III-P1-1.1-TE-2016-2015 (TE136/2018), and PN-III-P1-1.1-PD-2016-1568 (PD6/2018). NIMP authors are thankful for the support of PCCDI—UEFISCDI, in the framework of project PN-III-P1-1.2-PCCDI-2017-0062 (contract no. 58PCCDI/2018)/component project no. 2 and of Core Program (contract no. 21N/2019). This work was also supported by a grant of the Romanian Ministry of Research and Innovation, PCCDI— UEFISCDI, project number PN-III-P1-1.2-PCCDI-2017-0224 (contract no. 77PCCDI/2018)—DigiTech, within PNCDI III.

Acknowledgments: In vitro studies were conducted at the Biology Lab of CETAL Department.

Conflicts of Interest: The authors declare no conflict of interest.

References

1. Yan, L.; Yuan, Y.; Ouyang, L.; Li, H.; Mirzasadeghi, A.; Li, L. Improved mechanical properties of the new Ti-15Ta-xZr alloys fabricated by selective laser melting for biomedical application. *J. Alloys Compd.* **2016**, *688*, 156–162. [CrossRef]
2. Banerjee, D.; Williams, J.C. Perspectives on titanium science and technology. *Acta Mater.* **2013**, *61*, 844–879. [CrossRef]
3. Agius, D.; Kourousis, K.I.; Wallbrink, C. A review of the as-built SLM Ti-6Al-4V mechanical properties towards achieving fatigue resistant designs. *Metals* **2018**, *8*, 75. [CrossRef]
4. Sargeant, A.; Goswami, T. Hip implants-Paper VI-Ion concentrations. *Mater. Des.* **2007**, *28*, 155–171. [CrossRef]
5. Cui, C.; Shen, Y.; Li, Y.; Sun, J.-B. Fabrication, microstructure, and mechanical properties of tip/Al composite. *Adv. Eng. Mater.* **2003**, *5*, 725–729. [CrossRef]
6. Koch, C.; Johnson, S.; Kumar, D.; Jelinek, M.; Chrisey, D.; Doraiswamy, A.; Jin, C.; Narayan, R.J.; Mihailescu, I.N. Pulsed laser deposition of hydroxyapatite thin films. *Mater. Sci. Eng. C* **2007**, *27*, 484–494. [CrossRef]
7. Nelea, V.; Morosanu, C.; Iliescu, M.; Mihailescu, I. Microstructure and mechanical properties of hydroxyapatite thin films grown by RF magnetron sputtering. *Surf. Coat. Technol.* **2003**, *173*, 315–322. [CrossRef]
8. Mahammadi, S.; Wictorin, L.; Ericsonetal, L.E. Cast titanium as implant material. *J. Mater. Sci. Mater. Med.* **1995**, *6*, 435–444. [CrossRef]
9. Corpe, R.S.; Steflik, D.E.; Whitehead, R.Y.; Wilson, M.D.; Young, T.R.; Jaramillo, C. Correlative experimental animal and human clinical retrieval evaluations of hydroxyapatite (HA)-coated and non-coated implants in orthopaedics and dentistry. *Crit. Rev. Biomed. Eng.* **2000**, *28*, 395–398. [CrossRef]
10. Sakkers, R.; Dalmeyer, R.; Brand, R.; Rozing, P.; Van Blitterswijk, C. Assessment of bioactivity for orthopedic coatings in a gap-healing model. *J. Biomed. Mater. Res.* **1998**, *36*, 265–273. [CrossRef]
11. Vladescu, A.; Parau, A.; Pana, I.; Cotrut, C.M.; Constantin, L.R.; Braic, V.; Vranceanu, D.M. In vitro activity assays of sputtered HAp coatings with SiC addition in various simulated biological fluids. *Coatings* **2019**, *9*, 389. [CrossRef]
12. Surmeneva, M.A.; Ivanova, A.A.; Tian, Q.; Pittman, R.; Jiang, W.; Lin, J.; Liu, H.H.; Surmenev, R.A. Bone marrow derived mesenchymal stem cell response to the RF magnetron sputter deposited hydroxyapatite coating on AZ91 magnesium alloy. *Mater. Chem. Phys.* **2019**, *221*, 89–98. [CrossRef]
13. Prolosov, K.A.; Belyavskaya, O.A.; Linders, J.; Loza, K.; Prymak, O.; Mayer, C.; Rau, J.V.; Epple, M.; Sharkeev, Y.P. Glancing angle deposition of Zn-doped calcium phosphate coatings by RF magnetron sputtering. *Coatings* **2019**, *9*, 220. [CrossRef]
14. Aydin, I.; Bahcepinar, A.I.; Kirman, M.; Cipiloglu, M.A. HA coating on Ti6Al7Nb alloy using as electrophoretic deposition method and surface properties examination of the resulting coatings. *Coatings* **2019**, *9*, 402. [CrossRef]
15. Popa, A.C.; Stan, G.E.; Enculescu, M.; Tanase, C.; Tulyaganov, D.U.; Ferreira, J.M.F. Superior biofunctionality of dental implant fixtures uniformly coated with durable bioglass films by magnetron sputtering. *J. Mech. Behav. Biomed. Mater.* **2015**, *51*, 313–327. [CrossRef] [PubMed]
16. Rey, C.; Combes, C.; Drouet, C.; Glimcher, M.J. Bone mineral: Update on chemical composition and structure. *Osteoporos. Int.* **2009**, *20*, 1013–1021. [CrossRef] [PubMed]
17. Duta, L.; Popescu, A.C. Current status on pulsed laser deposition of coatings from animal origin calcium phosphate sources. *Coatings* **2019**, *9*, 335. [CrossRef]
18. Graziani, G.; Boi, M.; Bianchi, M. A review on ionic substitutions in hydroxyapatite thin films: Towards complete biomimetism. *Coatings* **2018**, *8*, 269. [CrossRef]
19. Miculescu, F.; Miculescu, M.; Ciocan, L.T.; Ernuteanu, A.; Antoniac, I.; Pencea, I.; Matei, E. Comparative studies regarding heavy elements concentration in human cortical bone. *Dig. J. Nanomater. Biostruct.* **2011**, *6*, 1117–1127.
20. Akram, M.; Ahmed, R.; Shakir, I.; Ibrahim, W.A.W.; Hussain, R. Extracting hydroxyapatite and its precursors from natural resources. *J. Mater. Sci.* **2013**, *49*, 1461–1475. [CrossRef]
21. Popescu, A.C.; Florian, P.E.; Stan, G.E.; Popescu-Pelin, G.; Zgura, I.; Enculescu, M.; Oktar, F.N.; Trusca, R.; Sima, L.E.; Roseanu, A.; et al. Physical-chemical characterization and biological assessment of simple and lithium-doped biological-derived hydroxyapatite thin films for a new generation of metallic implants. *Appl. Surf. Sci.* **2018**, *439*, 724–735. [CrossRef]

22. Bianchi, M.; Gambardella, A.; Graziani, G.; Liscio, F.; Maltarello, M.C.; Boi, M.; Berni, M.; Bellucci, D.; Marchiori, G.; Valle, F.; et al. Plasma-assisted deposition of bone apatite-like thin films from natural apatite. *Mater. Lett.* **2017**, *199*, 32–36. [CrossRef]
23. Duta, L.; Mihailescu, N.; Popescu, A.C.; Luculescu, C.R.; Mihailescu, I.N.; Cetin, G.; Gunduz, O.; Oktar, F.N.; Popa, A.C.; Kuncser, A.; et al. Comparative physical, chemical and biological assessment of simple and titanium-doped ovine dentine-derived hydroxyapatite coatings fabricated by pulsed laser deposition. *Appl. Surf. Sci.* **2017**, *413*, 129–139. [CrossRef]
24. Liu, R.; Qiao, W.; Huang, B.; Chen, Z.; Fang, J.; Li, Z.; Chen, Z. Fluorination enhances the osteogenic capacity of porcine hydroxyapatite. *Tissue Eng. Part A* **2018**, *24*, 1207–1217. [CrossRef] [PubMed]
25. Boutinguiza, M.; Pou, J.; Comesaña, R.; Lusquiños, F.; de Carlos, A.; León, B. Biological hydroxyapatite obtained from fish bones. *Mater. Sci. Eng. C* **2012**, *32*, 478–486. [CrossRef]
26. Mocanu, A.C.; Stan, G.E.; Maidaniuc, A.; Miculescu, M.; Antoniac, I.V.; Ciocoiu, R.C.; Voicu, S.I.; Mitran, V.; Cimpean, A.; Miculescu, F. Naturally-derived biphasic calcium phosphates through increased phosphorous-based reagent amounts for biomedical applications. *Materials* **2019**, *12*, 381. [CrossRef] [PubMed]
27. Jaber, H.L.; Hammood, A.S.; Parvin, N. Synthesis and characterization of hydroxyapatite powder from natural Camelus bone. *J. Aust. Ceram. Soc.* **2018**, *54*, 1–10. [CrossRef]
28. Lopez-Alvarez, M.; Vigo, E.; Rodriguez-Valencia, C.; Outeirino-Iglesias, V.; Gonzalez, P.; Serra, J. In vivo evaluation of shark teeth-derived bioapatites. *Clin. Oral Implant. Res.* **2016**, *28*, 91–100. [CrossRef]
29. Levingstone, T.J.; Khaled Benyounis, M.A.; Looney, L.; Stokes, J.T. Plasma sprayed hydroxyapatite coatings: Understanding process relationships using design of experiment analysis. *Surf. Coat. Technol.* **2015**, *283*, 29–36. [CrossRef]
30. Roy, M.; Bandyopadhyay, A.; Bose, S. Induction plasma sprayed nano hydroxyapatite coatings on titanium for orthopaedic and dental implants. *Surf. Coat. Technol.* **2011**, *205*, 2785–2792. [CrossRef]
31. Heimann, R.B. Plasma-sprayed hydroxylapatite-based coatings: Chemical, mechanical, microstructural, and biomedical properties. *J. Therm. Spray Technol.* **2016**, *25*, 827–850. [CrossRef]
32. Eason, R. (Ed.) *Pulsed Laser Deposition of Thin Films: Applications-Lead Growth of Functional Materials*; Wiley & Sons: New York, NY, USA, 2007.
33. Kelly, P.J.; Arnell, R.D. Magnetron sputtering: A review of recent developments and applications. *Vacuum* **2000**, *56*, 159–172. [CrossRef]
34. Surmenev, R.A. A review of plasma-assisted methods for calcium phosphate-based coatings fabrication. *Surf. Coat. Technol.* **2012**, *206*, 2035–2056. [CrossRef]
35. Fernandes, H.R.; Gaddam, A.; Rebelo, A.; Brazete, D.; Stan, G.E.; Ferreira, J.M.F. Bioactive glass and glass-ceramics for healthcare application in bone regeneration and tissue engineering. *Materials* **2018**, *11*, 2530. [CrossRef]
36. Sima, L.E.; Stan, G.E.; Morosanu, C.O.; Melinescu, A.; Ianculescu, A.; Melinte, R.; Neamtu, J.; Petrescu, S.M. Differentiation of mesenchymal stem cells onto highly adherent radio frequency-sputtered carbonated hydroxyalapatite thin films. *J. Biomed. Mater. Res. A* **2010**, *95*, 1203–1214. [CrossRef]
37. Attar, H.; Calin, M.; Zhang, L.C.; Scudino, S.; Eckert, J. Manufacture by selective laser melting and mechanical behavior of commercially pure titanium. *Mater. Sci. Eng. A* **2014**, *593*, 170–177. [CrossRef]
38. Kruth, P.; Vandenbroucke, B.; Van Vaerenbergh, J.; Naert, I. *Virtual Prototyping and Rapid Manufacturing—Advanced Research in Virtual and Rapid Prototyping*; Taylor and Francis: London, UK, 2005; pp. 139–146.
39. Zhang, L.C.; Klemm, D.; Eckert, J.; Hao, Y.L.; Sercombe, T.B. Manufacture by selective laser melting and mechanical behavior of a biomedical Ti-24Nb-4Zr-8Sn alloy. *Scr. Mater.* **2011**, *65*, 21–24. [CrossRef]
40. Bose, S.; Ke, D.; Sahasrabudhe, H.; Bandyopadhyay, A. Additive manufacturing of biomaterials. *Prog. Mater. Sci.* **2018**, *93*, 45–111. [CrossRef]
41. Singh, S.; Ramakrishna, S.; Singh, R. Material issues in additive manufacturing: A review. *J. Manuf. Process.* **2017**, *25*, 185–200. [CrossRef]
42. Weißmann, V.; Drescher, P.; Bader, R.; Seitz, H.; Hansmann, H.; Laufer, N. Comparison of single Ti_6Al_4V struts made using selective laser melting and electron beam melting subject to part orientation. *Metals* **2017**, *7*, 91. [CrossRef]
43. Zhang, P.; He, A.N.; Liu, F.; Zhang, K.; Jiang, J.; Zhang, D.Z. Evaluation of low cycle fatigue performance of selective laser melted titanium alloy Ti–6Al–4V. *Metals* **2019**, *9*, 1041. [CrossRef]

44. Jardini, A.L.; Larosa, M.A.; Kaasi, A.; Kharmandayan, P. Additive manufacturing in medicine. In *Reference Module in Materials Science and Materials Engineering*; Hasmi, S., Ed.; Elsevier: Amsterdam, The Netherlands, 2017; pp. 1–21. [CrossRef]
45. Jardini, A.L.; Lasora, M.A.; Macedo, M.F.; Bernardes, L.F.; Lambert, C.S.; Zavaglia, C.A.C.; Maciel, R.; Calderoni, D.R.; Ghizoni, E.; Kharmandayan, P. Improvement in cranioplasty: Advanced prosthesis biomanufacturing. *Procedia CIRP* **2016**, *49*, 203–208. [CrossRef]
46. Ataee, A.; Li, Y.; Brandt, M.; Wen, C. Ultrahigh-strength titanium gyroid scaffolds manufactured by selective laser melting (SLM) for bone implant applications. *Acta Mater.* **2018**, *158*, 354–368. [CrossRef]
47. Sun, J.; Zhang, F.Q. The application of rapid prototyping in prosthodontics. *J. Prosthodont.* **2012**, *21*, 641–644. [CrossRef]
48. Zhang, L.C.; Sercombe, T.B. Selective laser melting of low-modulus biomedical Ti-24Nb-4Zr-8Sn Alloy: Effect of laser point distance. *Key Eng. Mater.* **2012**, *520*, 226–233. [CrossRef]
49. Gu, D.D.; Meiners, W.; Wissenbach, K.; Poprawe, R. Laser additive manufacturing of metallic components: Materials, processes and mechanisms. *Int. Mater. Rev.* **2012**, *57*, 133–164. [CrossRef]
50. Mohammed, M.T. Mechanical properties of SLM-titanium materials for biomedical applications: A review. *Mater. Today Proc.* **2018**, *5*, 17906–17913. [CrossRef]
51. Attar, H.; Bönisch, M.; Calin, M.; Zhang, L.; Scudino, S.; Eckert, J. Selective laser melting of in situ titanium–titanium boride composites: Processing, microstructure and mechanical properties. *Acta Mater.* **2014**, *76*, 13–22. [CrossRef]
52. Attar, H.; Löber, L.; Funk, A.; Calin, M.; Zhang, L.; Prashanth, K.; Scudino, S.; Zhang, Y.; Eckert, J. Mechanical behavior of porous commercially pure Ti and Ti–TiB composite materials manufactured by selective laser melting. *Mater. Sci. Eng. A* **2015**, *625*, 350–356. [CrossRef]
53. Krakhmalev, P.; Fredriksson, G.; Yadroitsava, I.; Kazantseva, N.; du Plessis, A.; Yadroitsev, I. Deformation behavior and microstructure of Ti6Al4V manufactured by SLM. *Phys. Procedia* **2016**, *83*, 778–788. [CrossRef]
54. Kajima, Y.; Takaichi, A.; Nakamoto, T.; Kimura, T.; Yogo, Y.; Ashida, M.; Doi, H.; Nomura, N.; Takahashi, H.; Hanawa, T.; et al. Fatigue strength of Co–Cr–Mo alloy clasps prepared by selective laser melting. *J. Mech. Behav. Biomed.* **2016**, *59*, 446–458. [CrossRef] [PubMed]
55. Leuders, S.; Thöne, M.; Riemer, A.; Niendorf, T.; Tröster, T.; Richard, H.A.; Maier, H.J. On the mechanical behaviour of titanium alloy TiAl6V4 manufactured by selective laser melting: Fatigue resistance and crack growth performance. *Int. J. Fatigue* **2013**, *48*, 300–307. [CrossRef]
56. Liverani, E.; Fortunato, A.; Leardini, A.; Belvedere, C.; Siegler, S.; Ceschini, L.; Ascari, A. Fabrication of Co–Cr–Mo endoprosthetic ankle devices by means of Selective Laser Melting (SLM). *Mater. Des.* **2016**, *106*, 60–68. [CrossRef]
57. Duta, L.; Chifiriuc, M.C.; Popescu-Pelin, G.; Bleotu, C.; (Pircalabioru) Gradisteanu, G.; Anastasescu, M.; Achim, A.; Popescu, A. Pulsed laser deposited biocompatible lithium-doped hydroxyapatite coatings with antimicrobial activity. *Coatings* **2019**, *9*, 54. [CrossRef]
58. Popa, A.C.; Marques, V.M.F.; Stan, G.E.; Husanu, M.A.; Galca, A.C.; Ghica, C.; Tulyaganov, D.U.; Lemos, A.F.; Ferreira, J.M.F. Nanomechanical characterization of bioglass films synthesized by magnetron sputtering. *Thin Solid Films* **2014**, *553*, 166–172. [CrossRef]
59. Popa, A.C.; Stan, G.E.; Husanu, M.A.; Mercioniu, I.; Santos, L.F.; Fernandes, H.R.; Ferreira, J.M.F. Bioglass implant-coating interactions in synthetic physiological fluids with varying degrees of biomimicry. *Int. J. Nanomed.* **2017**, *12*, 683–707. [CrossRef]
60. Galca, A.C.; Stan, G.E.; Trinca, L.M.; Negrila, C.C.; Nistor, L.C. Structural and optical properties of c-axis oriented aluminum nitride thin films prepared at low temperature by reactive radio-frequency magnetron sputtering. *Thin Solid Films* **2012**, *524*, 328–333. [CrossRef]
61. Popa, A.C.; Stan, G.E.; Besleaga, C.; Ion, L.; Maraloiu, V.A.; Tulyaganov, D.U.; Ferreira, J.M.F. Submicrometer hollow bioglass cones deposited by radio frequency magnetron sputtering: Formation mechanism, properties, and prospective biomedical applications. *ACS Appl. Mater. Interfaces* **2016**, *8*, 4357–4367. [CrossRef]
62. Besleaga, C.; Dumitru, V.; Trinca, L.M.; Popa, A.C.; Negrila, C.C.; Kołodziejczyk, Ł.; Luculescu, C.R.; Ionescu, G.C.; Ripeanu, R.G.; Vladescu, A.; et al. Mechanical, corrosion and biological properties of room-temperature sputtered aluminum nitride films with dissimilar nanostructure. *Nanomaterials* **2017**, *7*, 394. [CrossRef]
63. Cauchy, A.L. Sur la refraction et la réflexion de la lumière. *Bull. Sci. Math.* **1830**, *14*, 6–10.

64. Ti6Al4V Properties Datasheet. Available online: http://asm.matweb.com/search/SpecificMaterial.asp?bassnum=MTP642 (accessed on 1 November 2019).
65. Zofkova, I.; Nemcikova, P.; Matucha, P. Trace elements and bone health. *Clin. Chem. Lab. Med.* **2013**, *51*, 1555–1561. [CrossRef] [PubMed]
66. Gaffney-Stomberg, E. The impact of trace minerals on bone metabolism. *Biol. Trace Elem. Res.* **2019**, *188*, 26–34. [CrossRef] [PubMed]
67. Besleaga, C.; Stan, G.E.; Galca, A.C.; Ion, L.; Antohe, S. Double layer structure of ZnO thin films deposited by RF-magnetron sputtering. *Appl. Surf. Sci.* **2012**, *258*, 8819–8824. [CrossRef]
68. Patterson, A. The Scherrer formula for X-ray particle size determination. *Phys. Rev.* **1939**, *56*, 978–982. [CrossRef]
69. Markovic, M.; Fowler, B.O.; Tung, M.S. Preparation and comprehensive characterization of a calcium hydroxyapatite reference material. *J. Res. Natl. Inst. Stand. Technol.* **2004**, *109*, 553–568. [CrossRef]
70. Rey, C.; Collins, B.; Goehl, T.; Dickson, I.R.; Glimcher, M.J. The carbonate environment in bone mineral: A resolution-enhanced Fourier transform infrared spectroscopy study. *Calcif. Tissue Int.* **1989**, *45*, 157–164. [CrossRef]
71. Hench, L.L. The story of Bioglass®. *J. Mater. Sci. Mater. Med.* **2006**, *17*, 967–978. [CrossRef]
72. Stan, G.E.; Pina, S.; Tulyaganov, D.U.; Ferreira, J.M.F.; Pasuk, I.; Morosanu, C.O. Biomineralization capability of adherent bio-glass films prepared by magnetron sputtering. *J. Mater. Sci. Mater. Med.* **2010**, *21*, 1047–1055. [CrossRef]
73. Vranceanu, D.M.; Parau, A.C.; Cotrut, C.M.; Kiss, A.E.; Constantin, L.R.; Braic, V.; Vladescu, A. In vitro evaluation of Ag doped hydroxyapatite coatings in acellular media. *Ceram. Int.* **2019**, *45*, 11050–11061. [CrossRef]
74. Šupová, M. Substituted hydroxyapatites for biomedical applications: A review. *Ceram. Int.* **2015**, *41*, 9203–9231. [CrossRef]
75. Popescu, A.C.; Sima, F.; Duta, L.; Popescu, C.; Mihailescu, I.N.; Capitanu, D.; Mustata, R.; Sima, L.E.; Petrescu, S.M.; Janackovic, D. Biocompatible and bioactive nanostructured glass coatings synthesized by pulsed laser deposition: In vitro biological tests. *Appl. Surf. Sci.* **2009**, *255*, 5486–5490. [CrossRef]
76. Tite, T.; Popa, A.C.; Balescu, L.M.; Bogdan, I.M.; Pasuk, I.; Ferreira, J.M.F.; Stan, G.E. Cationic substitutions in hydroxyapatite: Current status of the derived biofunctional effects and their in vitro interrogation methods. *Materials* **2018**, *11*, 2081. [CrossRef] [PubMed]
77. Wopenka, B.; Pasteris, J.D. A mineralogical perspective on the apatite in bone. *Mater. Sci. Eng. C* **2005**, *25*, 131–143. [CrossRef]
78. Mihailescu, N.; Stan, G.E.; Duta, L.; Chifiriuc, M.C.; Bleotu, C.; Sopronyi, M.; Luculescu, C.; Oktar, F.N.; Mihailescu, I.N. Structural, compositional, mechanical characterization and biological assessment of bovine-derived hydroxyapatite coatings reinforced with MgF_2 or MgO for implants functionalization. *Mater. Sci. Eng. C* **2016**, *59*, 863–874. [CrossRef] [PubMed]
79. Wang, Y.; Liu, X.; Fan, T.; Tan, Z.; Zhou, Z.; He, D. In vitro evaluation of hydroxyapatite coatings with (002) crystallographic texture deposited by micro-plasma spraying. *Mater. Sci. Eng. C.* **2017**, *75*, 596–601. [CrossRef] [PubMed]
80. Fu, J.; He, C.; Xia, B.; Li, Y.; Feng, Q.; Yin, Q.; Shi, X.; Feng, X.; Wang, H.; Yao, H. c-axis preferential orientation of hydroxyapatite accounts for the high wear resistance of the teeth of black carp (*Mylopharyngodon piceus*). *Sci. Rep.* **2019**, *6*, 23509. [CrossRef]
81. Kim, H.; Camata, R.P.; Chowdhury, S.; Vohra, Y.K. In vitro dissolution and mechanical behavior of c-axis preferentially oriented hydroxyapatite thin films fabricated by pulsed laser deposition. *Acta Biomater.* **2010**, *6*, 3234–3241. [CrossRef]
82. Nordstrom, E.G.; Sanchez Munoz, O.L. Physics of bone bonding mechanism of different surface bioactive ceramic materials in vitro and in vivo. *Bio Med. Mater. Eng.* **2011**, *11*, 221–231.
83. LeGeros, R.Z.; Daculsi, G. In vitro transformation of biphasic calcium phosphate ceramics: Ultrastructural and physicochemical characterizations. In *Handbook of Bioactive Ceramics, Vol. II. Calcium Phosphate Ceramics*; Yamamuro, T., Hench, L.L., Wilson, J., Eds.; CRC Press: Boca Raton, FL, USA, 1990; pp. 17–28.
84. Kokubo, T.; Takadama, H. How useful is SBF in predicting in vivo bone bioactivity? *Biomaterials* **2006**, *27*, 2907–2915. [CrossRef]

85. Daculsi, G.; LeGeros, R.Z.; Nery, E.; Lynch, K.; Kerebel, B. Transformation of biphasic calcium phosphate ceramics in vivo: Ultrastructural and physicochemical characterization. *J. Biomed. Mater. Res.* **1989**, *23*, 883–894. [CrossRef]
86. Porter, A.E.; Patel, N.; Skepper, J.N.; Best, S.M.; Bonfield, W. Comparison of in vivo dissolution processes in hydroxyapatite and silicon-substituted hydroxyapatite bioceramics. *Biomaterials* **2003**, *24*, 4609–4620. [CrossRef]
87. Kattimani, V.S.; Kondaka, S.; Lingamaneni, K.P. Hydroxyapatite—Past, present, and future in bone regeneration. *Bone Tissue Regen. Insights* **2016**, *7*, 9–19. [CrossRef]
88. Lim, J.; Lee, J.; Yun, H.-S.; Shin, H.-I.; Park, E.K. Comparison of bone regeneration rate in flat in long bone defects: Calvarial and tibial bone. *Tissue Eng. Regen. Med.* **2013**, *10*, 336–340. [CrossRef]
89. Tavafoghi, M.; Cerruti, M. The role of amino acids in hydroxyapatite mineralization. *J. R. Soc. Interface* **2016**, *13*, 20160462. [CrossRef]
90. Wang, K.; Leng, Y.; Lu, X.; Ren, F.; Ge, X.; Ding, Y. Theoretical analysis of protein effects on calcium phosphate precipitation in simulated body fluid. *CrystEngComm* **2012**, *14*, 5870–5878. [CrossRef]
91. Zareidoost, A.; Yousefpour, M.; Ghaseme, B.; Amanzadeh, A. The relationship of surface roughness and cell response of chemical surface modification of titanium. *J. Mater. Sci. Mater. Med.* **2012**, *23*, 1479–1488. [CrossRef]
92. Hatamleh, M.M.; Wu, X.; Alnazzawi, A.; Watson, J.; Watts, W. Surface characteristics and biocompatibility of cranioplasty titanium implants following different surface treatments. *Dent. Mater.* **2018**, *34*, 676–683. [CrossRef]

metals

MDPI

Article

Evaluation of Support Structure Removability for Additively Manufactured Ti6Al4V Overhangs via Electron Beam Melting

Wadea Ameen [1,2,*], Muneer Khan Mohammed [2] and Abdulrahman Al-Ahmari [1,2]

1 Industrial Engineering Department, King Saud University, Riyadh P.O. Box 800-Riyadh 11421, Saudi Arabia; alahmari@ksu.edu.sa

2 Raytheon Chair for Systems Engineering (RCSE Chair), Advanced Manufacturing Institute, King Saud University, Riyadh P.O. Box 800-Riyadh 11421, Saudi Arabia; muneerkm@ksu.edu.sa

* Correspondence: wadeaameen@gmail.com; Tel.: +966114697372

Received: 22 October 2019; Accepted: 9 November 2019; Published: 11 November 2019

Abstract: The addition of support structures is essential for the successful fabrication of overhang structures through additive manufacturing (AM). The support structures protect the overhang portion from distortions. They are fabricated with the functional parts and are removed later after the fabrication of the AM part. While structures bearing insufficient support result in defective overhangs, structures with excessive support result in higher material consumption, time and higher post-processing costs. The objective of this study is to investigate the effects of design and process parameters of support structures on support removability during the electron beam melting (EBM)-based additive manufacturing of the Ti6Al4V overhang part. The support design parameters include tooth parameters, no support offset, fragmentation parameters and perforation parameters. The EBM process parameters consist of beam current, beam scan speed and beam focus offset. The results show that both support design and process parameters have a significant effect on support removability. In addition, with the appropriate selection of design and process parameters, it is possible to significantly reduce the support removal time and protect the surface quality of the part.

Keywords: additive manufacturing; support structures; electron beam melting; support structure removability

1. Introduction and Literature Review

Additive manufacturing (AM) is a manufacturing technique used to fabricate three-dimensional parts in a layer-by-layer manner directly from a computer-aided design (CAD) model. In powder-bed-fusion (PBF) processes, such as electron beam melting (EBM), and in selective laser sintering and melting, a thin powder layer is selectively melted using an electron or laser beam [1,2]. Once the powder layer is molten, a new powder layer is spread and melted based upon the CAD data, the steps are repeated until the full component is fabricated. Although PBF technology can fabricate complex geometries, it has some limitations in the fabrication of structures with overhang geometries. The fabrication of structures with overhang geometries that do not contain support structures results in warpage deformation due to the thermal gradient and residual stress.

Support structures are essential in the manufacturing of defect-free overhang structures through the PBF process [3]. By facilitating the removal of heat from the structure, support structures prevent part curling, distortion, sagging, cracks, shrinkage, and/or other deformations that result from thermal stresses. They also serve to hold the fabricated part in the construction platform during the manufacturing process [4]. Current support generation methods usually involve the use of certain types of structures to cover the overhang space. These methods may result in overestimation of the support volume or the placement of a large number of supports, which may be unnecessary and increase the post-processing time [5]. Support structures have many design rules that are to be incorporated by the designers. For example, the support structures should be designed to engage in minimal contact with the overhang surface parts [6]. The design of the support structure must also consider the post-processing costs. For instance, minimising the number of supports, which leads to a minimal number of marks on the surface and reduces the amount of part clean-up and post-processing [7]. Support structures are the non-functional part of the structure and are removed after the construction process. Optimisation of the support structure is an important consideration for the successful fabrication of parts. Accurate selection of support structure design is critical, as it affects the production cost, time and quality of construction. For example, applying a large contact area support to the overhang part would result in increased difficulty in support removal, and it might damage the part surface during removal [8,9]. The support structures are used to build the overhanging geometries and are later removed by breaking them away. Often, commercial software like Magics (Materialise, Leuven, Belgium) is used to automatically generate support structures. After the parts are fabricated using AM technology (EBM), the excess powder and support structures are removed. In PBF, the support structures are mechanically removed by the application of force or cutting (often manually) [10]. In some cases, the inability to remove the support structures leads to the inability to manufacture the desired structures. Various methods are used to remove metal support structures, depending on the material and geometry. Wire electrical discharge machining, saws [10,11], Dremel handheld power tools [12] and pliers [13] are popular methods and tools for removing the support structures from PBF parts [14]. Sandpaper is a quick tool that is used to remove the witness marks leftover by the supports on the bottom surface.

For metal supports, a wire EDM machine, milling equipment, and bandsaw may be required to remove the part support from the baseplate and then remove the supports from the part [7]. Most of the AM cost models consider support structures as main parts and as a factor in the estimate of the total cost. Furthermore, some cost models consider the density of the support structure to estimate the cost of AM [15,16]. Calignano et al. investigated the effect of block support structures fabricated by selective laser melting of aluminium and titanium alloys. The results show that the removal of the support structures from the titanium samples is an easier task as compared to the aluminium samples. In addition, the support design has an effect on the removability of the applied support structures [17]. Kuo et al. used topology optimisation to design an easily removable support structure that minimises support material and fabrication time, and results in the fewest artefacts on the surface of the specimen. A cost-based formulation, comprising a simple and straightforward method, is employed to find a compromise between the cost and surface profile error induced by specimen weight [18]. To select the type of support structure design for optimal support removability and the surface quality of the parts, Järvinen et al. compared the properties of two types of support structures (web and tube) during the evaluation of removability. The results showed that the web supports provided superior removability as compared to the tube supports [19]. All block support structure parameters (tooth, no support offset parameters, fragmentation, and perforation parameters) are shown in Figure 1.

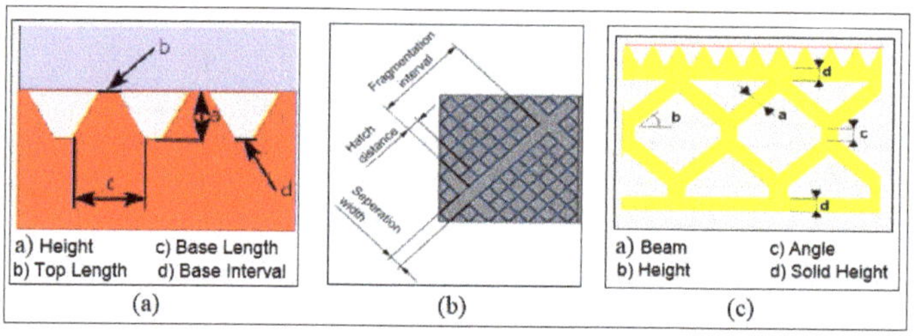

Figure 1. Design parameters: (**a**) tooth parameters, (**b**) fragmentation parameters, (**c**) perforation parameters.

In the EBM process, parameters such as beam current (I), beam scanning speed (V), line offset (LO) and beam focus offset (FO) are the most relevant process parameters [20]. Line offset refers to the distance between two hatch lines, and focus offset is the offset of the focal plane from its zero position. Increased line offset results in lower energy densities, which forms voids in Ti-6Al-4V specimens. An increase in the focus offset increases the beam diameter and reduces the energy density, which in turn, results in an increase in the porosity [21]. From the above parameters, the energy density (J/mm^2) can be calculated as follows [22]:

$$Q = \frac{\Pi \times V_{acc} \times I_{scan} \times \gamma_{abs}}{4 \times D_{beam} \times V_{scan}} \tag{1}$$

where Q is energy input/mm^2 (J/mm^2), $\Pi = 3.14$, V_{acc} is accelerating voltage (V) = 60,000 V (standard accelerating voltage in Arcam EBM), I_{scan} is beam current (mA), V_{scan} is beam scan speed (mm/s), D_{beam} is beam spot diameter (mm) and P1 is focus offset (mA). D_{beam} is calculated from the focus offset (P1) by using quadratic nonlinear regression approximation (Source: Arcam AB):

$$D_{beam} = 0.000102\,(P1)^2 + 0.00151\,(P1) + 0.131, \tag{2}$$

where γ is percentage of energy absorbed by the system, $1 - \varepsilon back$. Mahale [22] by Monte Carlo simulation in CASINO software, obtaining a value of $\varepsilon back = 0.0025$; thus, $\gamma abs = 0.9975$.

EBM is a metal additive manufacturing process that utilises stiffness support structures for the production of overhangs. In this study, the effect of support structure parameters (design and process) is envisaged to develop the design rules for the support structure of Ti6Al4V parts fabricated by EBM.

2. Experimental Procedures

In the current study, ledge overhangs with an overhang portion length of 15 mm and a thickness of 5 mm were designed using block support structures in Magics software (Materialise, Leuven, Belgium). ARCAM A2 (Arcam EBM, Mölnlycke, Gottenburg, Sweden) was used to construct the overhang specimens. The support structure design and process parameters were varied to evaluate their effect on support structure removability during the fabrication of the overhang structures. Table 1 lists the design and process parameters, and their corresponding settings.

In the EBM process, after the completion of the build, the specimens were surrounded by sintered powder within the build envelope, as shown in Figure 2a. The build envelope was subjected to a powder recovery system (powder blasting process) to obtain parts free from sintered powder, as shown in Figure 2b.

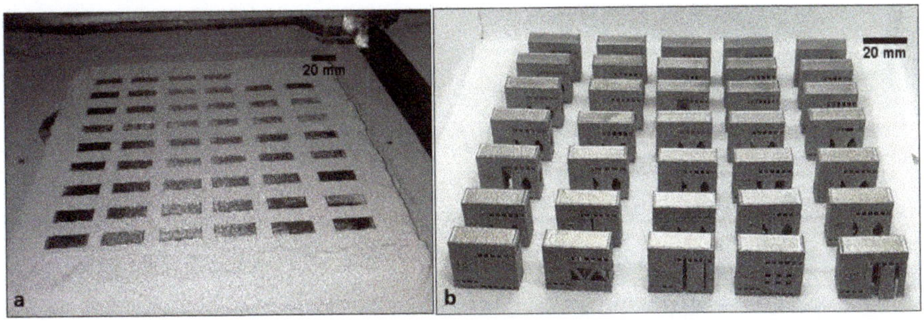

Figure 2. Fabricated specimens: (**a**) specimens on sintered powder; (**b**) specimens after sand blasting.

The fabricated specimens were then subjected to the support structure removal process. To evaluate the effect of support structure design and process parameters on the removability of the support structures, the support structures were removed manually using simple pliers, and the time consumed in removing the support structures for each specimen was recorded. To maintain uniformity, the support structure removal pattern was kept the same for all the specimens and rest time was given after each specimen to avoid the fatigue. The support removal process was started by removing the support tooth and then the support body to make the pliers accessible to the interior tooth. Finally, the supports adhered to the body of the overhang part were removed.

Post processing such as powder and support removal as performed as a last step of the fabrication stage. However, small support removal burrs remain on the supported part surface, which in turn effect the surface quality of the produced part. To evaluate the effect of the addition of a support structure and its parameters on surface quality, surface roughness analysis was carried out by utilizing the Dektak XT surface profiler (Bruker, Germany), as shown in Figure 3. The support teeth, fragmentation and process parameters were considered in this investigation. The surface profiles were measured at three different locations on the supported surface with evaluation length of 10 mm, as shown in Figure 3b, and then the averages were calculated.

Table 1. Support structure design and process parameters.

Support Structure Parameters			**Level 1**	**Level 2**	**Level 3**	**Level 4**
Design parameters	Tooth and No Support Offset Parameters	Tooth height (mm)	1.2	1.80	2.4	3
		Tooth top length (mm)	0.05	0.1	0.15	0.2
		Tooth base length (mm)	0.5	0.7	0.90	1.1
		Tooth base interval	0.5	1.50	2.5	3.5
		Z offset (mm)	0	0.5	0.1	0.25
		No support offset (mm)	0	2	4	6
	Hatching and Fragmentation Parameters	X and Y hatching (mm)	3	6	12	15
		Fragmentation	0	1	-	-
		Fragmentation interval (mm)	2.5	5	10	-
		Separation width (mm)	0.2	0.4	0.8	1.6
	Perforation Parameters	Perforation style	0	1 (Diam.)	1 (Rect.)	-
		Perforation beam (mm)	1.75	3.5	7	-
		Perforation angle	30	45	60	-
		Perforation height (mm)	1	2	4	-
		Perforation solid height (mm)	1	2	4	-
Process parameters	Processes Parameters	Current (mA)	1.5	2.5	5	10
		Scan speed (mm/s)	1000	1400	1800	2200
		Focus offset (mA)	1	2	4	4

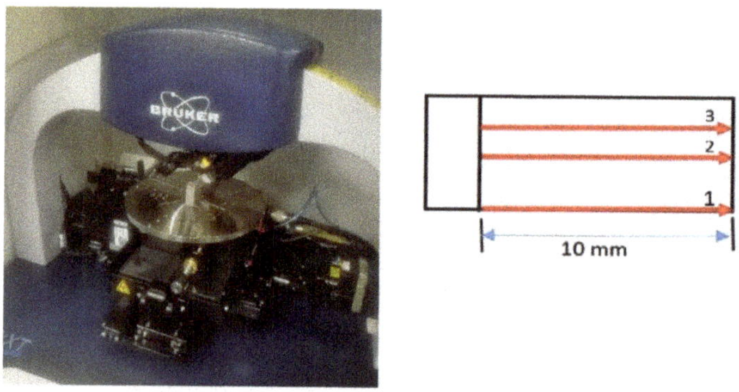

(a) (b)

Figure 3. Surface roughness measuring (**a**) Dektak XT surface profiler; (**b**) specimen measured surface.

3. Results Analysis

3.1. *Effect of Design Parameters*

During the support removal process, it was found that the time taken to remove the support structure tooth is very small as compared to the time consumed in removing the support structure body, particularly when some of the support body is adhered to the part surface. In addition, the supports, which are very close to the part surface, affect the surface quality of the part, as shown in Figure 4. Therefore, it is recommended to offset the side support body from the part surface, even by a small distance (no support offset).

Figure 4. Overhang side surface quality: (**a**) No support offset; (**b–d**) layers of support adhesion on the overhang wall.

It was observed that the support structures with a hatching distance of 6 mm consumed more time to remove than the supports with a hatching distance of 12 mm. This is because of the increase in the number of the support teeth for the 6 mm hatching, which decreases the accessibility of the cutting tool during the removal process. Therefore, the support structures with 6 mm hatching are removed in smaller segments, which result in increased removal time. It was also observed that the sintered powder behind the support walls act as an obstacle to support removability, as shown in Figure 5. To remove the effect of the sintered powder on support structure removability, the powder-blasting process is carried out as and when it is required.

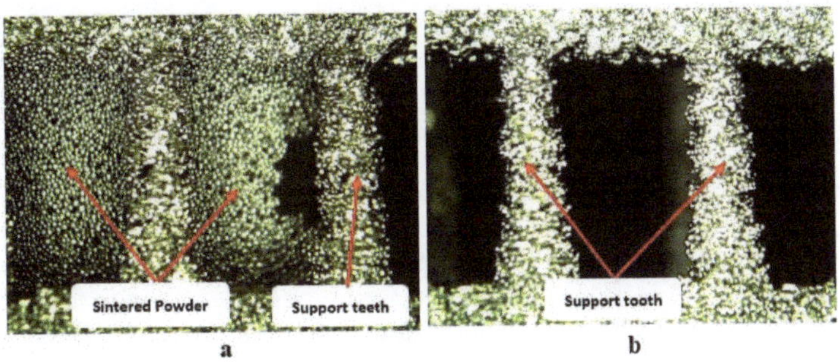

Figure 5. Support tooth structures: (**a**) before powder removal; (**b**) after powder removal.

The influence of support structure teeth parameters, hatching, fragmentation parameters and perforation parameters were evaluated in this section. To study the effect of the support tooth parameters on support structure removability, overhangs with different support tooth parameters were constructed and analysed. Teeth parameters after the removal of the sintered powder are presented in Figure 6.

Figure 6. Support structure teeth parameters.

The effect of support design parameters on the support removal time for 6 and 12 mm hatching distances supports is shown in Figure 7. Tooth height, tooth top length, tooth base interval, no support offset and separation width were found to have significant effect on the support removal time. Whereas the other factors had little or no effect on the removability. The support removal time decreases with increase in the tooth height, this is because the increase in the tooth height increases the accessibility of the cutting tool (pliers) and reduces the need to remove the support body to reach all teeth. In addition, with increased tooth height, the non-melted powder around the tooth is easily removed during powder blasting, thereby easing the support removal. Regarding the effect of the tooth top and base length, the support removal time increases with increase in tooth top length. This is because the tooth strength increases with increase in tooth top length. However, the tooth base length did not have a significant effect on support removal time. The results also found that the support tooth interval has a significant effect on the support removal time. As the tooth interval increases, the support removal time decreases. This is because as the support tooth interval increases, the total number of teeth decrease, and the removal tool accessibility increases, which results in easier support removal.

Reducing the support structure volume through the increase of no support offset has a significant effect on the support removal time. As the no support offset increases, the support removal time decreases significantly. This is due to the decrease in the supported area and increase in the accessibility of the cutting tool during the support removal. In addition, the elimination of the contact between the support structure wall and side surface of the part facilitates the support removal process, which results in reduced support removal time.

The effect of the hatching distance in the x- and y-directions on support structure removability is shown in Figure 8. It can be seen that at lower hatching distance, the support removal time is significantly high. This is because the lower hatching distance results in greater support structure, and the support-part contact area decreases the removability of the non-melted powder and accessibility of the support removal tool (pliers).

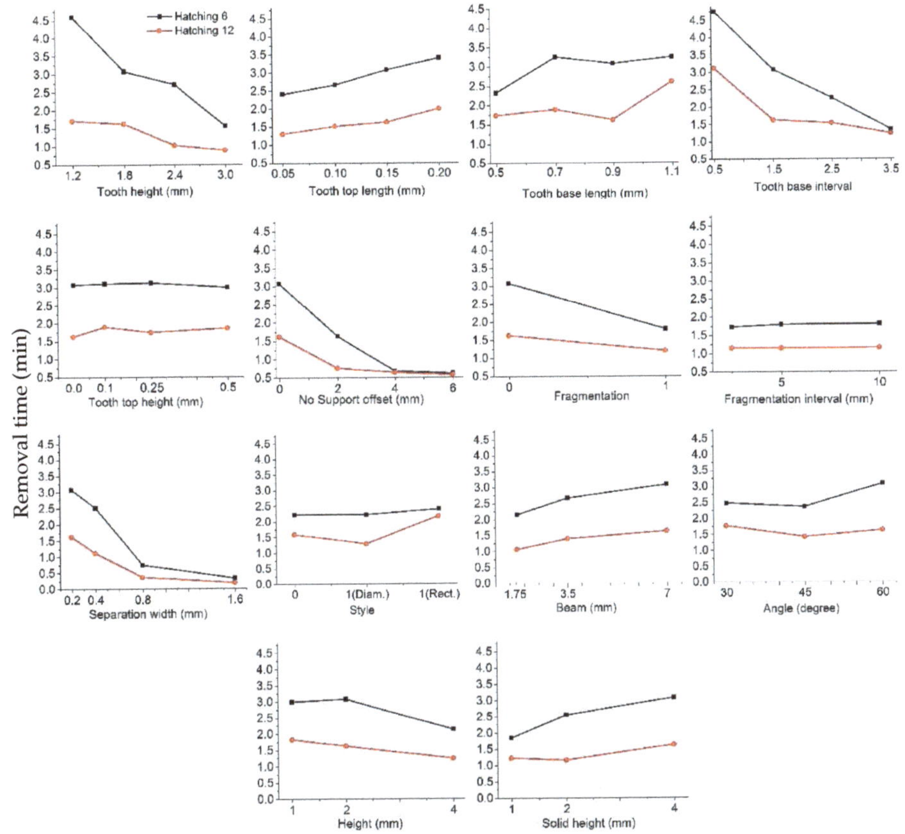

Figure 7. Effect of support design parameters on support removability.

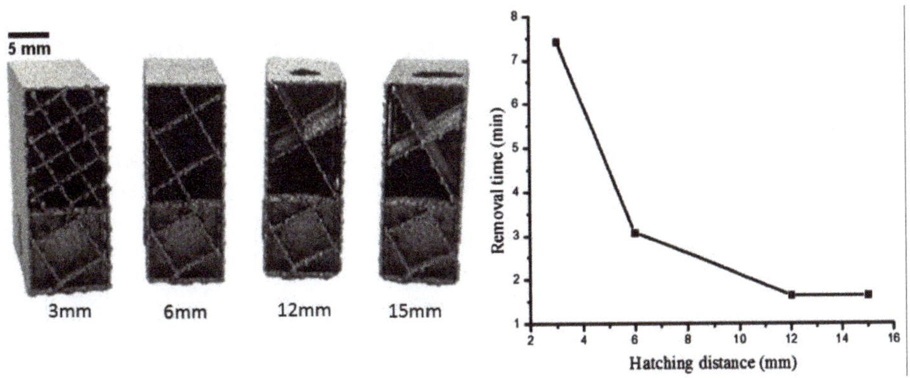

Figure 8. Effect of hatching distance on support removability.

Implementing the fragmentation design strategy results in subdivision of support structures and slightly improves its removability. It can be seen in Figure 7 that the improvement is higher in the case of the 6 mm hatching distance support, because without the fragmentation strategy, the support body is solid; therefore, the addition of fragmentation strategy results in subdivided support blocks which ease the support removal. Whereas in the case of the 12 mm hatching distance, the perforation strategy is present by default even without the fragmentation strategy. The application of the fragmentation strategy results in the elimination of the perforation strategy. Therefore, there is very little effect on support removal time.

In addition, it was found that the fragmentation separation width has a significant effect on the removability of support structures. As the fragmentation separation width increases, the support removal time decreases. This is due to the decrease in the supported surface, increasing the accessibility of the removal tool and separating the supports into contiguous blocks that are easy to remove, as shown in Figure 9.

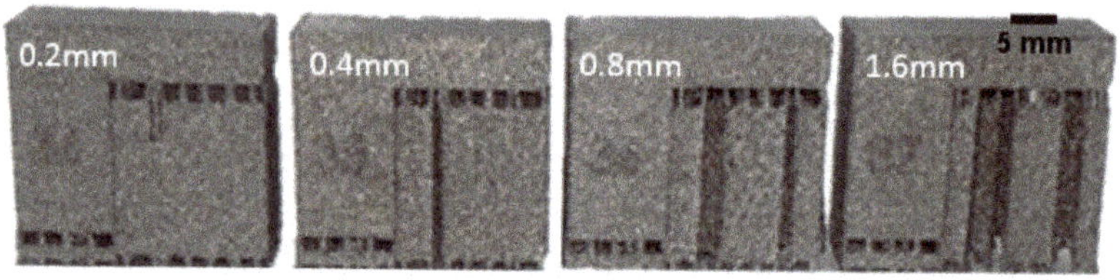

Figure 9. Support structures with various fragmentation separation widths.

Application of perforation strategy results in new support structure configurations, wherein the supports have holes, which make them weaker as compared to the solid supports and they consume less material. The effect of perforation style, beam, angle, height, and solid height on support removability in the 6 and 12 mm hatching distance specimens is shown in Figure 7. The diamond-style perforation had no effect on support removal time in the 6 mm hatching case, because there is no change in the geometry of the support. However, in the case of the 12 mm hatching distance support, the diamond-style perforation resulted in a decrease in the support removal time. Rectangular-style perforation was found to have a negative effect on support removal time, owing to the disintegration of the body of the built support into small pieces, which consumes more time.

Figure 7 shows the influence of the perforation beam on support removability. As the perforation beam increases, the time required to remove the support increases. This is due to the reduction in the perforation holes, which in turn results in the increase in support body strength. It was also found that the default perforation angle of 45° performed better than other angles in terms of support removal time. Regarding the effect of perforation height and perforation solid height, it is found that an increase in the perforation height resulted in the slight decrease in support removal time, causing a decrease in solid-wall support and an increase in hole size, and vice versa, for perforation solid height, as shown in Figure 7.

An ANOVA test with 95% confidence was performed to determine the significance of the selected parameters. Results of 6 and 12 mm hatching are presented in Tables 2 and 3, respectively.

In case of 6 mm hatching distance supports, most of the selected factors were found to have significant effects ($p < 0.05$) on the support removal time. The factors such as tooth top length, tooth base length, tooth top height and fragmentation interval were found to have no significant effect on the support removal time, as shown in Table 2. However, in case of 12 mm hatching distance supports, all the factors except the fragmentation interval were found to have significant effect on the support removal time.

Table 2. ANOVA for support removability with respect to the support design parameter (hatching 6).

Source	*df*	Sum of Squares	Mean Square	F-Value	*p*-Value Prob > F
Model	34	36.411	1.071	13.88	<0.0001
Linear	34	36.411	1.071	13.88	<0.0001
Tooth height	3	4.698	1.566	20.29	<0.0001
Tooth top length	3	0.613	0.204	2.65	0.093
Tooth base length	3	0.575	0.192	2.48	0.107
Tooth base interval	3	6.341	2.114	27.39	<0.0001
Tooth top height	3	0.022	0.007	0.09	0.962
No support offset	3	10.558	3.519	45.60	<0.0001
Fragmentation	1	2.255	2.255	29.23	<0.0001
Fragmentation interval	2	0.089	0.045	0.58	0.575
separation width	3	5.021	1.674	21.69	<0.0001
Style	2	0.837	0.419	5.42	0.019
Beam	2	0.753	0.377	4.88	0.026
Angle	2	0.592	0.296	3.84	0.049
Height	2	0.712	0.356	4.61	0.031
Solid height	2	1.368	0.684	8.86	0.004
Error	13	1.003	0.077	-	-
Total	47	37.415	-	-	-

Table 3. ANOVA for support removability with respect to the support design parameter (hatching 12).

Source	*df*	Sum of Squares	Mean Square	F-Value	*p*-Value Prob > F
Model	34	12.275	0.361	45.08	<0.0001
Linear	34	12.275	0.361	45.08	<0.0001
Tooth height	3	0.746	0.249	31.06	<0.0001
Tooth top length	3	0.250	0.083	10.42	0.001
Tooth base length	3	0.989	0.330	41.17	<0.0001
Tooth base interval	3	2.371	0.790	98.70	<0.0001
Tooth top height	3	0.141	0.047	5.85	0.009
No support offset	3	2.133	0.711	88.77	<0.0001
Fragmentation	1	0.469	0.469	58.52	<0.0001
Fragmentation interval	2	0.000	0.000	0.00	0.997
separation width	3	1.490	0.497	62.03	<0.0001
Style	2	0.310	0.155	19.38	<0.0001
Beam	2	0.293	0.147	18.31	<0.0001
Angle	2	0.061	0.031	3.82	0.049
Height	2	0.176	0.088	10.96	0.002
Solid height	2	0.284	0.142	17.71	<0.0001
Error	13	0.104	0.008	-	-
Total	47	12.379	-	-	-

3.2. The Influence of Process Parameters

The effect of the EBM process parameters on support structure removability is shown in Figure 10. It was found that as the beam current increases, the support removal time increases. This is due to the fact that as the beam current increase, the input energy into the supports increase, which in turn, increases the support strength, thereby increasing the support removal time. In case of scan speed, it was found that as the scan speed increases the support removal time decreases. Focus offset was found to have a little or no effect on the removal time, as shown in Figure 10.

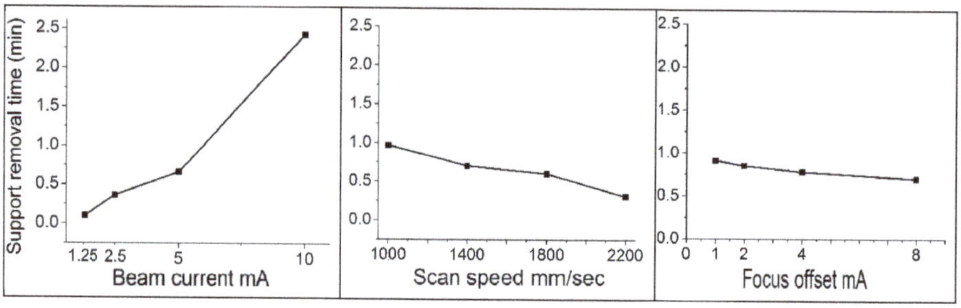

Figure 10. Effect of EBM process parameters on support removability.

A beam current of 1.25 mA in the specimen labelled C1 resulted in supports wherein the supports were broken during the powder-blasting process, as shown in Figure 11. The support remains were easily removed by hand owing to the low strength of supports due to the low input energy. It was observed that the increase in current from 1.25 to 2.5 mA resulted in supports that could withstand powder blasting but were very easy to remove. A further increase in the beam current resulted in supports with considerable strength and significantly higher support removal times.

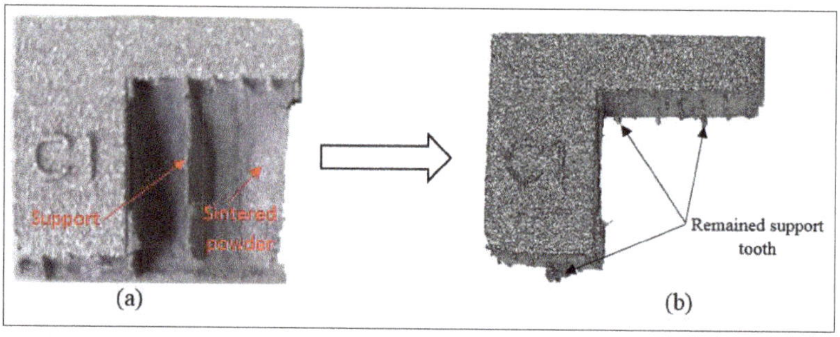

Figure 11. Specimen C1 (current = 1.25 mA) (**a**) before powder-blasting process, and (**b**) after powder-blasting process.

Regarding the effect of the scan speed as the scan speed increases, the energy flux decreases, which reduces the support strength. The support structure of specimen S4 (speed = 2200 mm/s) had the lowest strength and was most easy to remove. Some of its support was broken during the powder-blasting process.

The results of ANOVA tests for the effect of the EBM process parameters on support removability are shown in Table 4. It is clear that the beam current and scan speed have a significant effect on the removal time. Whereas, focus offset was found to have no significant effect on support removal time.

Table 4. ANOVA for the effect of EBM process parameters on support removability.

Source	*df*	Sum of Squares	Mean Square	F-Value	*p*-Value Prob > F
Model	9	3.691	0.410	109.84	0.009
Linear	9	3.691	0.410	109.84	0.009
Current	3	3.354	1.118	299.48	0.003
Scan speed	3	0.228	0.076	20.33	0.047
Focus offset	3	0.040	0.013	3.61	0.224
Error	2	0.007	0.004	-	-
Total	11	3.698	-	-	-

3.3. *Surface Roughness*

Support structure parameters were found to have an effect on the surface roughness of the overhangs. In general, 6 mm hatching support structures resulted in higher surface roughness, as compared to 12 mm hatching support structures, as shown in Figure 12. This is because of the presence of a greater number of support teeth with the 6 mm hatching supports, than with the 12 mm hatching supports. Tooth top length, tooth base length and separation width were found to have significant effect on the surface roughness. The surface roughness decreases with increases in tooth top length and separation width. Whereas it increases with increases in tooth base length, as shown in Figure 12.

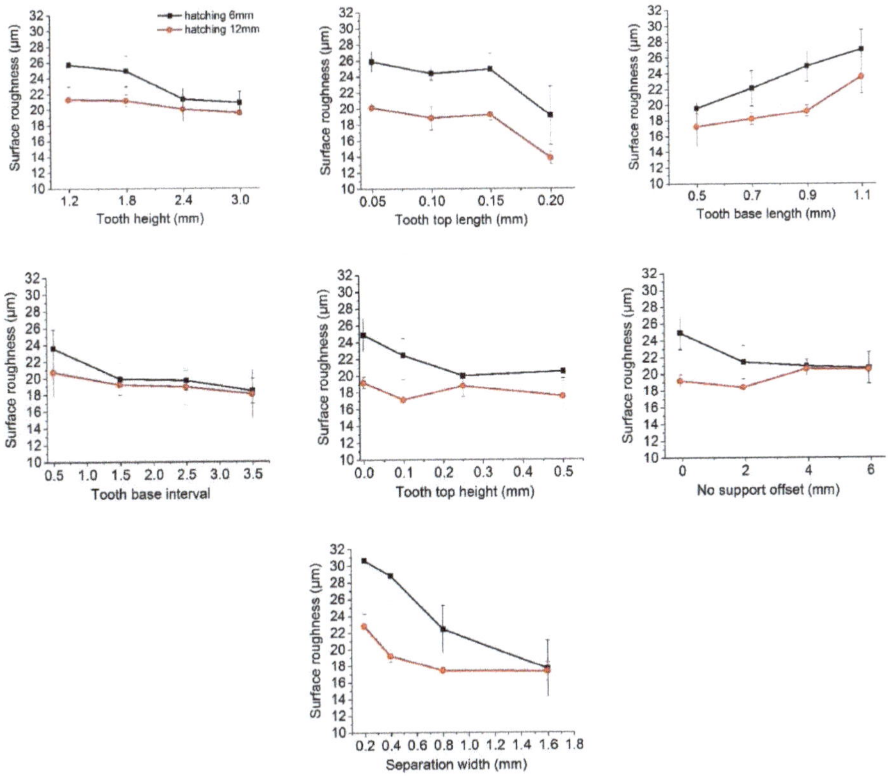

Figure 12. Effect of support parameters on surface roughness.

The surface profiles of the samples with minimum support removal time (0.2 min) and maximum support removal time (4.77 min) are shown in Figure 13a,b, respectively. The surface roughness results showed the Ra of 21.9 µm for the sample with minimum support removal time, whereas the sample with maximum support removal time resulted in Ra of 20.68 µm. It was found that the support structure parameters that resulted in minimum support removal time did not have any adverse effect on the surface quality of the overhangs and that the surface roughness was similar to the samples built with other parameters.

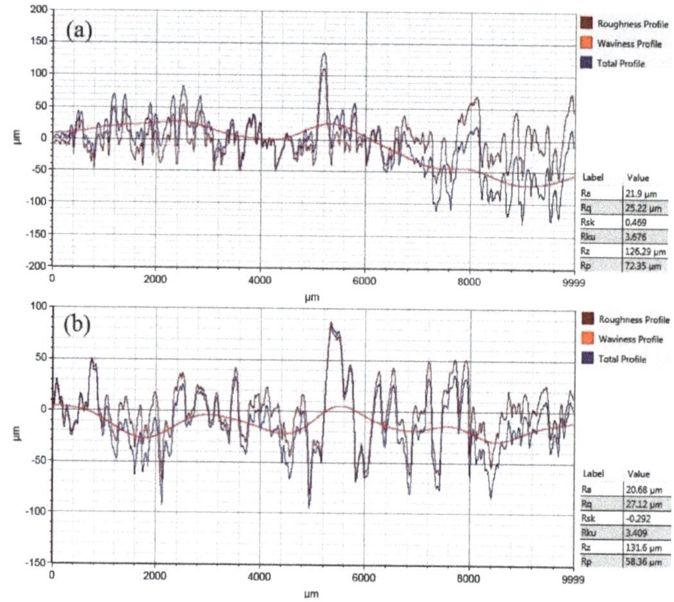

Figure 13. Surface roughness profiles: (**a**) minimum support removal time; (**b**) maximum support removal time.

4. Conclusions

In the current study, the effect of support design and process parameters on support structure removability was evaluated during the additive manufacturing of Ti6Al4V overhangs via electron beam melting. In general, the results show that the design and process parameters of the support structures have a significant effect on support removability. It was found that with the appropriate selection of design and process parameters for the support structures, it is possible to reduce the support removal time and protect the part surface quality. The results show that support removability can be improved significantly with help of proper support design parameters by reducing the supported area, decreasing the strength of the part/support contact area, and decreasing the strength of the support structure body. Process parameters, such as beam current and scan speed, were found to have a more significant effect on support structure removability as compared with beam focus offset.

Based on the current study the following conclusions can be drawn:

1. As compared to the default support tooth parameters, 3 mm tooth height, 0.05 mm tooth top length and 3.5 mm tooth base interval were found to significantly reduce the support removal time.
2. The fragmented support structure with separation width of 1.6 mm was found to decrease the support removal time by about 80%.
3. The perforation parameters such as 1.75 mm perforation beam, 60° angle, 4 mm height, and 1 mm solid height resulted in around 20% reduction in support removal time, as compared to the default parameters.
4. In case of EBM process parameters, as compared to the default parameters, the beam current of 1.25 mA and scan speed of 2200 mm/sec resulted in more than 80% reduction of support removal time.

Author Contributions: W.A. and A.A.-A. conceived and designed the study. W.A. and M.K.M. conducted the EBM experiments. W.A. conducted the measurements and analysed the results. W.A. and M.K.M. wrote the manuscript and reviewed it. A.A.-A. supervised the work.

Funding: This study received funding from the Raytheon Chair for Systems Engineering. The authors are grateful to the Raytheon Chair for Systems Engineering for funding.

Acknowledgments: The authors thank the Deanship of Scientific Research and RSSU at King Saud University for their technical support.

Conflicts of Interest: The authors declare no conflict of interest.

References

1. Gu, D.; Meiners, W.; Wissenbach, K.; Poprawe, R. Laser additive manufacturing of metallic components: Materials, processes and mechanisms. *Int. Mater. Rev.* **2012**, *57*, 133–164. [CrossRef]
2. Bobbio, L.D.; Qin, S.; Dunbar, A.; Michaleris, P.; Beese, A.M. Characterization of the strength of support structures used in powder bed fusion additive manufacturing of Ti-6Al-4V. *Addit. Manuf.* **2017**, *14*, 60–68. [CrossRef]
3. Umer, U.; Ameen, W.; Abidi, M.H.; Moiduddin, K.; Alkhalefah, H.; Alkahtani, M.; Al-Ahmari, A. Modeling the Effect of Different Support Structures in Electron Beam Melting of Titanium Alloy Using Finite Element Models. *Metals* **2019**, *9*, 806. [CrossRef]
4. Poyraz, Ö.; Yasa, E.; Akbulut, G.; Orhangul, A.; Pilatin, S. *Investigation of Support. Structures for Direct Metal. Laser Sintering (DMLS) of IN625 Parts*; International Solid Freeform Fabrication Symposium: Austin, TX, USA, 2015.
5. Cheng, B.; Chou, Y.K. Overhang Support Structure Design for Electron Beam Additive Manufacturing. In Proceedings of the ASME 2017 12th International Manufacturing Science and Engineering Conference Collocated with the JSME/ASME 2017 6th International Conference on Materials and Processing, Los Angeles, CA, USA, 4–8 June 2017.
6. Jhabvala, J.; Boillat, E.; André, C.; Glardon, R. An innovative method to build support structures with a pulsed laser in the selective laser melting process. *Int. J. Adv. Manuf. Technol.* **2012**, *59*, 137–142. [CrossRef]
7. Gibson, I.; Rosen, D.W.; Stucker, B. *Additive Manufacturing Technologies*; Springer: Berlin, Germany, 2010.
8. Thomas, D. *The Development of Design Rules for Selective Laser Melting*; University of Wales: Wales, UK, 2009.
9. Pullin, J.; Offen, A. Back to the Drawing Board-Addressing the design issues of RM. In Proceedings of the Time Compression Technologies Rapid Manufacturing Conference 2008, Coventry, West Midlands, October 2008.
10. Sames, W.J.; List, F.; Pannala, S.; Dehoff, R.R.; Babu, S.S. The metallurgy and processing science of metal additive manufacturing. *Int. Mater. Rev.* **2016**, *61*, 315–360. [CrossRef]
11. Das, P.; Mhapsekar, K.; Chowdhury, S.; Samant, R.; Anand, S. Selection of build orientation for optimal support structures and minimum part errors in additive manufacturing. *Comput. Aided Des. Appl.* **2017**, *14*, 1–13. [CrossRef]
12. Bibb, R.; Eggbeer, D.; Williams, R. Rapid manufacture of removable partial denture frameworks. *Rapid Prototyp. J.* **2006**, *12*, 95–99. [CrossRef]
13. Ameen, W.; Al-Ahmari, A.; Mohammed, M.K.; Mian, S.H. Manufacturability of Overhanging Holes Using Electron Beam Melting. *Metals* **2018**, *8*, 397. [CrossRef]
14. Williams, R.E.; Melton, V.L. Abrasive flow finishing of stereolithography prototypes. *Rapid Prototyp. J.* **1998**, *4*, 56–67. [CrossRef]
15. Byun, H.-S.; Lee, K.H. Determination of the optimal part orientation in layered manufacturing using a genetic algorithm. *Int. J. Prod. Res.* **2005**, *43*, 2709–2724. [CrossRef]
16. Piili, H.; Happonen, A.; Väistö, T.; Venkataramanan, V.; Partanen, J.; Salminen, A. Cost Estimation of Laser Additive Manufacturing of Stainless Steel. *Phys. Procedia* **2015**, *78*, 388–396. [CrossRef]
17. Calignano, F. Design optimization of supports for overhanging structures in aluminum and titanium alloys by selective laser melting. *Mater. Des.* **2014**, *64*, 203–213. [CrossRef]
18. Kuo, Y.-H.; Cheng, C.-C.; Lin, Y.-S.; San, C.-H. Support structure design in additive manufacturing based on topology optimization. *Struct. Multidiscip. Optim.* **2017**, *57*, 1–13. [CrossRef]
19. Järvinen, J.P.; Matilainen, V.; Li, X.; Piili, H.; Salminen, A.; Mäkelä, I.; Nyrhilä, O. Characterization of effect of support structures in laser additive manufacturing of stainless steel. *Phys. Procedia* **2014**, *56*, 72–81. [CrossRef]
20. Safdar, A.; He, H.; Wei, L.-Y.; Snis, A.; Chavez de Paz, L.E. Effect of process parameters settings and thickness on surface roughness of EBM produced Ti-6Al-4V. *Rapid Prototyp. J.* **2012**, *18*, 401–408. [CrossRef]

21. Schwerdtfeger, J.; Singer, R.F.; Körner, C. In situ flaw detection by IR-imaging during electron beam melting. *Rapid Prototyp. J.* **2012**, *18*, 259–263. [CrossRef]
22. Mahale, T.R. Electron Beam Melting of Advanced Materials and Structures, Mass Customization, Mass Personalization. Ph.D. Thesis, North Carolina State University, Raleigh, CA, USA, 2009.

MDPI
St. Alban-Anlage 66
4052 Basel
Switzerland
Tel. +41 61 683 77 34
Fax +41 61 302 89 18
www.mdpi.com

Metals Editorial Office
E-mail: metals@mdpi.com
www.mdpi.com/journal/metals

www.ingramcontent.com/pod-product-compliance
Lightning Source LLC
Chambersburg PA
CBHW061500210926
51850CB00008B/1777